Matt @ Matt Hyland
January 2016

Y0-CBY-666

Beyond
Horizons

Beyond Horizons

A Half Century of Air Force Space Leadership

Revised Edition

David N. Spires

Senior Editor
George W. Bradley III

Associate Editors
Rick W. Sturdevant
Richard S. Eckert

Air Force Space Command
in association with
Air University Press

1998

Library of Congress Cataloging-in-Publication Data

Spires, David N.
    Beyond Horizons: a half century of Air Force space leadership / David N. Spires; Senior editor,
George W. Bradley III; associate editors, Rick W. Sturdevant, Richard S. Eckert.
    Includes bibliographical references and index.
    1. Astronautics, Military—United States. 2. Space Warfare. 3. Space Weapons.
    98-125908

ISBN 1-58566-060-4

First Printing July1998
Second Printing September 2001
Third Printing July 2002
Fourth Printing April 2003
Fifth Printing September 2004
Sixth Printing February 2007

Air University Press
131 West Shumacher Avenue
Maxwell AFB AL 36112-6615
http://aupress.maxwell.af.mil

To the men and women
who made Air Force space history
and those who recorded their achievements.

Contents

FOREWORD

In the late summer of 1992, Chief of Staff of the Air Force General Merrill A. McPeak asked me to chair a panel to study the role of the Air Force in space into the 21st century. This second Blue Ribbon Panel on space, which came four years after a similar study completed in the late 1980s, had as its primary objectives to conduct a comprehensive review of the Air Force's existing space policy, organization, and infrastructure, to define the service's future role in space, to develop a strategy to carry out that role, and to make appropriate recommendations to the senior leader-ship of the Air Force. The Chief believed recent political, military, and economic developments necessitated a new look at military space operations. These included the collapse of the Soviet Union and the end of the Cold War, the emergence of a "multi-polar" world, a shift in national security strategy, the implications of Operation Desert Storm, and the worldwide proliferation of sophisticated weapons. Changing domestic priorities, declining defense budgets, and Congressional interest in military roles and missions also contributed to the need for an evaluation of the development, acquisition, and operation of space systems. Our panel, which consisted of some thirty Air Force officers and civilians, met at Maxwell Air Force Base from early September to early November 1992. Early the following year the Chief approved and released a report of our findings and recommendations.

Among our recommendations was one that called for making "integrated aerospace employment a fundamental principle...in all training and education programs." We urged the Air Force to examine all of its training, education, and personnel policies to develop a comprehensive approach to teaching space to the aviation community, and conversely, introducing space personnel to the principles and requirements of more traditional air warfare. This book represents a major step toward fulfilling the first of these two goals.

In the aftermath of the panel's report I asked Dr. Richard Hallion, the Air Force Historian, to add a history of the Air Force in space to his program's book-writing plans. Subsequent discussions of the project led to a decision to produce the study through a contract let by Air Force Space Command's Directorate of History. Since contracting out such studies was a familiar practice in the Air Force History Program, it promised to give us an academic-quality book in a reasonable amount of time. Mr. George W. "Skip" Bradley, Director of History at Air Force Space Command, led the team which ultimately selected Dr. Dave Spires to write the study.

Beyond Horizons is by no means the first attempt to tell the story of the Air Force in space, although it may be first to present that story to a wide audience, both within the service and in the general public. Official organizational histories and monographs prepared by the civilian and blue-suit historians of the Air Force History Program have recorded and documented the evolution of the service's space programs since their earliest days in the post-World War II era. Classification issues and the nature of the history program itself, however, limited readers of these works primarily to those already well aware of the Air Force space story or to the actual participants in these efforts. Perhaps of greater importance, the way the Air Force organized and managed its space effort created an environment somewhat detached and insulated from the mainstream flying Air Force. As a result, knowledge of this vital part of the service's history and heritage remained closeted and to a certain extent inaccessible to both service members and scholars of Air Force history. It was my intention in requesting the preparation of this study to open up the story of the Air Force in space to a much wider audience and by doing so to generate a level of interest in the subject area that would result in additional, more focused monographs and papers.

The publication of *Beyond Horizons* comes at a significant point in the history of the Air Force, one with implications well beyond the coincident recognition of the service's 50th anniversary. Recently the service announced its vision for the Air Force of the 21st century. Central to this vision, the leadership of today's Air Force agrees, is a transition from an *air* force to an *air and space* force on an evolutionary path to a *space and air* force. Clearly, as the service moves in this direction over the coming years, awareness of the roots of the Air Force's space heritage must increase and broaden. For service members, the transition necessitates a greater appreciation of this part of our history to foster an understanding of the changes currently taking place or emerging on the horizon. Scholars of Air Force history and others in the public at large similarly will gain insights into issues and events either minimized or omitted from mainstream Air Force history.

Beyond Horizons promises to open the door somewhat wider to a story that to date has, for various reasons, not received the attention it deserves and requires. Unquestionably, the growing availability of official records from the earlier years of the Air Force space program will allow researchers to fill in details missing from this study and offer new interpretations of some issues and events. As the Air Force moves into its second half century, this added knowledge, together with what we already know from the work of Dave Spires and others, can only help us understand better the foundation upon which the Air Force of the 21st century is emerging.

THOMAS S. MOORMAN, JR.
General, USAF
Vice Chief of Staff of the Air Force

PREFACE

Beyond Horizons: A Half Century of Air Force Space Leadership is a study of the United States Air Force in space. Of all the military services, the Air Force has been preeminently involved for the past fifty years in initiating, developing, and applying the technology of space-based systems in support of the nation's national security. Yet there has been no single-volume overview of the Air Force space story to serve as an introduction and guide for interested readers. In November 1992, a high-level Air Force Blue Ribbon Panel on Space, chaired by then Lieutenant General Thomas S. Moorman, Jr., commander of Air Force Space Command, concluded there was a specific need to better educate people, both in the service and among the general populace, about the history of Air Force space activities. *Beyond Horizons* has been written to meet this need.

Beyond Horizons begins with a review of pre-World War II rocketry developments and the forging of the important partnership between the Army Air Forces' Brigadier General Henry H. "Hap" Arnold and noted Cal Tech aerodynamicist Theodore von Kármán. Wartime provided important momentum in establishing the foundation for later Air Force space efforts. At Arnold's initiative, von Kármán, late in the war, formed what became the United States Air Force Scientific Advisory Board and produced *Where We Stand*. This seminal study provided the Air Force a research and development agenda for the future. Equally important, the Air Force-sponsored Rand Corporation, in early 1946, issued a report on the feasibility of artificial satellites that would lead to the important Project Feedback reports of the early 1950s. Although the von Kármán and Rand studies produced no immediate rush to develop space systems, the ground had been prepared.

Chapter 1 focuses on space and missile efforts prior to the launch of the Soviet Sputnik satellites in late 1957. Beginning with analysis of the Rand satellite report, the chapter examines the policy, organizational, and funding constraints, based largely on inter- and intra-service rivalries, that Air Force missile and space advocates had to overcome during the late 1940s and early 1950s in order to establish an effective enterprise. In a sense, the Air Force entered the space age on the coattails of intercontinental ballistic missile (ICBM) development and President Dwight D. Eisenhower's determination to protect the nation from surprise attack. Operational ballistic missiles could also serve as satellite boosters, while a reconnaissance satellite could provide strategic intelligence on Soviet capabilities. Along with the other services, the Air Force pursued missile and satellite development by establishing the Western Development Division and giving its commander, Brigadier General Bernard A. Schriever, wideranging responsibilities to produce an operational ICBM and a military reconnaissance satellite. Eventually, these efforts would lead to the Lockheed Agena booster-satellite, the infrared missile warning satellite, and the reconnaissance satellites of the National Reconnaissance Office.

Chapter 2 focuses on the important policy and organizational steps taken after Sputnik which helped the Air Force achieve leadership of the nation's military space activities. Initial Air Force hopes of leading a national space program ended with the establishment of the National Aeronautics and Space Administration (NASA). At the same time, NASA's absorption of Army and Navy space assets left the Air Force preeminent in military space and the new civilian agency dependent on the service for the immediate future. During the second Eisenhower administration, the Air Force initiated the first of several unsuccessful "campaigns" to receive formal recognition as executive agent for all military space efforts with approval to lead an expanded space program. Forced to share space responsibilities with the other services and agencies, Air Force leaders also chafed under an Eisenhower space policy that downplayed military space activities and prohibited deployment of weapons in space.

Chapter 3 describes Air Force efforts to achieve a dominant role in space through its support of NASA and its attempts to acquire a military manned spaceflight mission and approval for development of space-based weapons. Expectations were high at the outset of the Kennedy administration when Defense Secretary Robert McNamara designated the Air Force the military service responsible for space research and development, and the service established Air Force Systems Command to lead the way. Yet, by the end of the 1960s, NASA basked in the glow of its lunar landing, while cancellation of the Air Force's Manned Orbiting Laboratory ended hopes for a military manned space mission. Moreover, earlier it had become clear that space policy would continue to restrict space-based weapons to the study phase. Despite the seemingly bleak outlook for an Air Force space future by the early 1970s, however, two developments would reinvigorate the Air Force space program—the success of instrumented satellites and the Space Shuttle.

Chapter 4 examines the Air Force's leadership role in the emergence of artificial earth satellites during the 1960s for communications, navigation, meteorology, and surveillance and reconnaissance. These mission functions had been identified in the late 1950s and would remain the bedrock of space activities for the remainder of the century. Booster and infrastructure support paralleled the rise of unmanned satellites. The Air Force developed more powerful launch vehicles and established worldwide networks for ground-based control of satellites, space surveillance, and missile warning. By the end of the decade, unmanned military spacecraft had demonstrated important operational applications including, during the Vietnam conflict, the first use of satellites to support military requirements in wartime.

Chapter 5 discusses the complex interplay of space policy, organizational, and operational issues that culminated in the formation of the Air Force's Space Command. The maturing of unmanned satellites and the advent of the Space Shuttle compelled the service to confront and reassess its fragmented organization for space and the heretofore dominant role of the space research and development community. With the increasing importance of space for operational commanders, the central questions became whether the research and development commands should continue to launch spacecraft and provide on-orbit control, and whether the service should create an operational command for its space activities. The debate led to the establishment of a major command for space operations in September 1982.

Chapter 6 describes the efforts of Air Force Space Command in the 1980s to consolidate its control over space systems and move the Air force from an "operational agenda" for space to the creation of an "operational mindset" for space. Along the way the command had to achieve an effective working relationship with a new unified United States Space Command and deal with the space launch crisis resulting from the *Challenger* disaster. By the end of the decade Air Force leaders increasingly referred to the "operationalization" of space in making space systems critical to the warfighter.

Chapter 7 focuses on the role of space in the Persian Gulf War in early 1991. This conflict represented the coming of age of military space by demonstrating the value of an "operational mindset" for space. During Desert Storm, space systems that traditionally had supported strategic requirements proved sufficiently flexible to provide essential tactical support to the warfighter.

The final chapter serves as both a summary of the Air force space story and a point of departure for assessing Air Force space prospects for the new century. The Gulf War provided the momentum for the Air Force to take advantage of the further technological growth and refinement of military space systems and the emerging trends toward greater military use of civil and commercial space capabilities in order to better institutionalize space within the Air Force. The study concludes with an assessment of the Air Force's leadership position in the ongoing debate over service roles and missions and its vision for the nation's space program as the United States prepared to enter the 21st century.

In preparing this study I received help from many quarters. Above all, I wish to thank the historians at Air Force Space Command—Director of History Mr. George W. "Skip" Bradley, and Dr. Rick W. Sturdevant and Dr. Rick Eckert. All three read the entire manuscript and provided wise counsel and unstinting encouragement. Skip Bradley directed the project with a firm hand and provided full access to the wealth of information in the command's historical archives. Rick Sturdevant tracked down many documents and labored mightily to have classified material downgraded and made available for my use. The knowledge he shared through many hours of discussion contributed substantially to my understanding of key policy and technical issues. Of special note, early in the project we elected to defer more complete coverage of the Air Force-National Reconnaissance Office relationship until a larger portion of the historical record is accessible. Rick Eckert offered important suggestions from his perspective as the primary author of the space chapters in the command's periodic histories. He also performed the final editing of the manuscript as well as completed the design and page layout in preparation for printing. I also wish to acknowledge the outstanding administrative support provided by Ms. Karen Martin of the command's Office of History.

I am especially indebted to three historians who agreed to read and comment on the initial draft for accuracy and clarity. Mr. R. Cargill Hall, the person responsible for contract histories at the Center for Air Force History, offered many insights based on his extensive knowledge and long experience in the civilian and military space communities. NASA historian Dr. Roger Launius provided valuable suggestions on the portions of the study dealing with early rocket developments and issues affecting NASA. I also greatly benefited from the comments of Dr. Donald R. Baucom, Ballistic Missile Defense Organization historian, whose understanding of missile defense and the Strategic Defense Initiative is second to none. They, of course, are not responsible for my interpretation of the Air Force space story.

Individuals at two major military archives also deserve special thanks. Dr. Timothy C. Hanley and Dr. Harry N. Waldron, III of the Space and Missile Systems Center at Los Angeles Air Force Base, California, generously allowed me extensive use of their important document collection that begins with records of the Western Development Division in the early 1950s. Colonel Richard S. Rauschkolb, commander of the Air Force Historical Research Agency at Maxwell Air Force Base, Alabama, also went beyond the call of duty to support my research efforts. As a result, I benefited from the knowledge and helpfulness of the agency's outstanding group of archivists and historians. I also wish to acknowledge Dr. Thomas Fuller, United States Space Command historian who furnished useful documents on contemporary space issues as well as his perspective on issues affecting the unified command. Additionally, I am grateful to Lee D. Saegesser, NASA Headquarters History Office archivist, who provided sound advice and access to his substantial holdings on Air Force-NASA issues.

Special thanks are owed to two individuals central to the Air Force story. General Bernard A. Schriever, the "father" of the Air Force space program, gave me the benefit of his views on the early years, and former Air Force Secretary and Director of the National Reconnaissance Office John L. McLucas helped broaden my understanding of space programs and issues during the 1960s and 1970s.

Finally, it should be recognized that a book of this nature could not have been completed without the benefit of the work done by the Air Force space pioneers and the historians who documented and recorded the Air Force story. We who are their heirs are forever in their debt.

David N. Spires
Spring 1997

EDITOR'S NOTE

In early January 1992, Lieutenant General Thomas S. Moorman, vice commander of Air Force Space Command, called me to his office to discuss a project he had in mind. Specifically, he asked me to look into the possibility of having the Office of Air Force History prepare a history of the Air Force's role in space since its beginnings shortly after World War II. In March 1993, General Moorman and I met with Dr. Richard Hallion, Chief of Air Force History, in his office at Bolling AFB, Washington, D.C. General Moorman outlined the project to Dr. Hallion and several of his staff members. What General Moorman proposed was not only visionary but also hard to do. He wanted a comprehensive academic-quality book that would, for the first time, put into an unclassified text a survey history of the entire range of activities conducted by the Air Force in space. Not only did he request a high quality study but he wanted it written in less than three years and published as soon after completion as possible. After discussing several ways of producing the book in-house at the Office of Air Force history, Dr. Hallion suggested contracting-out the writing of the history to a qualified historian and author. Although the Office of Air Force History had a dedicated historian who managed contract histories, Mr. R. Cargill Hall, Dr. Hallion had a different managerial scheme in mind for this project. He proposed that the Air Force Space Command History Office, of which I was Chief, manage the contract to ensure timeliness and quality as well as ensure that the author selected had access to all the materials necessary to complete the project. Although I had never managed a project of this nature, I felt that retaining control at Headquarters Air Force Space Command would be of considerable benefit since much of the documentation and corporate memory on the subject resided at the headquarters. Moreover, I felt that by keeping the book's management in Colorado Springs, I could ensure that the study would remain faithful to the goals and expectations of the leadership of the command who had generously agreed to fund the project.

I would like to make the first of many acknowledgments at this point. This project simply would not be as it is without the guiding hand of General Moorman. He not only conceived the idea for the book and set its initial direction, but he also spent many hours with me explaining the history of the Air Force in space. General Moorman, an historian himself, if he had the time, could certainly have authored this study. As it was, he patiently worked with me to develop a project outline that eventually become the basis for the content portion of the contract's statement of work. I owe a great debt of gratitude to General Moorman who not only gave me his vision of the Air Force's role in space but inspired me to tackle this project with enthusiasm and excitement.

We began the process of contracting with an author (or authors) in May 1993 and submitted a Request for Proposals (RFP) in September of that year. After releasing the RFPs we received a number of excellent proposals. I'd like to make another acknowledgment here. The contracting process is much more complicated than I ever imagined, and I developed a great deal of admiration and respect for the dedicated contracting officials at Peterson AFB's 21st Space Wing who provided the expertise to complete the contract. Unfortunately, the contracting officials had little or no experience in contracting for the writing of an academic quality history book, and we learned together the fine nuances of this unique process. What amazed me was that despite the fact that this project involved a relatively small amount of money compared to what contracting officials normally managed, they treated my small workload with as much concern and dedication as any of the other large scale and multi-million dollar tasks they normally completed. I am indebted to the 21st Contracting Squadron for the outstanding support they gave the project from the first day of work to the very final day of contract completion. In particular, I am especially indebted to two contracting officials, Ms. Geraldine Humphrey and Ms. Donna Tiernan. Their professional expertise, willingness to understand the requirements and standards I insisted on, and patience were critical to the success of this endeavor. "Gerry" Humphrey worked with me from the beginning to the end of the project, and I am grateful for her constant support and interest.

Selecting a contractor was no easy task as both Ms. Humphrey and Ms. Tiernan warned me. The selection team that assisted me was invaluable. Mr. R. Cargill Hall of the Office of Air Force History and Dr. Rick Sturdevant of the Air Force Space Command History Office spent many hours reviewing and evaluating proposals.

I can not overestimate Mr. Hall's help as his experience in contracting historical studies at the Office of Air Force History was invaluable at all stages of this project. Dr. Sturdevant's knowledge of space history and his wide-ranging publication record ensured that I had an expert's breadth of knowledge in selecting the correct contractor. After many months of work, the contract was finally awarded in December 1993 to Dr. David N. Spires who teaches history at the University of Colorado in Boulder. Dr. Spires was uniquely qualified. As an Air Force officer he taught history at the Air Force Academy, and he has also authored a number of books on Air Force history as a contract author for the Office of Air Force History. He not only proved to be an able writer, but has demonstrated a real personal interest in the successful completion of this project.

Both the Foreword and Preface have given amplification to the nature of this study. I would like to add that this work was completed on schedule and as budgeted. This was accomplished in no small measure because of the dedication of a number of people, many of whom I have already named. I would like to acknowledge several others who may not have been mentioned previously. Lieutenant Colonel William Semmler, an Individual Mobilization Augmentee assigned to Air Force Space Command's Directorate of History, helped select photos, edited copy, and produced the glossary and index. His several readings of the narrative assisted us in eliminating a number of errors and inconsistencies. 2nd Lieutenant Denise Bostick, a reservist working in the Directorate of History for a time, took great pains to find and reproduce a number of photos which appear in this book and assisted in a number of administrative tasks in support of its completion. Colonel Billy G. Meazell, Inspector General at Air Force Space Command, generously contributed his time and talent to create the dust jacket art. Despite an extremely busy schedule, he donated his spare time to create an artistic representation of the history of the Air Force in space. Ms. Freda Norris and Ms. Karen Martin, Editorial Assistants in the Directorate of History, accomplished numerous administrative tasks not only in the production of the book but also in the contracting process as well as the contract management aspect of this task. Freda and Karen made a much more significant contribution than their normal modesty allows them to admit. Dr. Spires has already mentioned the contributions of our three outside reviewers, Mr. R. Cargill Hall of the Office of Air Force History; Dr. Roger Launius, Chief of the NASA History Office, and Dr. Donald R. Baucom, historian for the Ballistic Missile Defense Organization. I would like to add my personal appreciation to them. They spent many hours advising me on the management of this project as well as giving Dr. Spires the benefit of their vast professional expertise in space history. I am indebted to them for their willingness to spend both their professional time and, in many cases, their personal time, to review and comment on the manuscript.

At this point, I need to acknowledge two people who have labored unceasingly to help complete this study: Dr. Rick Sturdevant and Dr. Rick Eckert. Both Dr.

Sturdevant and Dr. Eckert are Staff Historians in the Air Force Space Command Directorate of History and have worked with me since my appointment as Chief of that office in spring 1992. They are both longtime Air Force historians and have worked in Air Force Space Command for many years. Their knowledge of space history and the command has proved invaluable at every step of the way. Dr. Spires has graciously acknowledged their contributions, but I need to thank them even more. They not only spent much professional time providing research material to Dr. Spires, guiding him to other sources, and reviewing and editing the book, but have counseled me numerous times in every phase of the management of this project. They have performed jobs too numerous to name, but I would like to acknowledge specifically their contributions as Associate Editors. As Senior Editor, I chose to adopt a seminaring method for reviewing each chapter. Dr. Spires agreed, and it was during these seminars, held each time Dr. Spires completed a draft chapter, that they made especially significant contributions to this projects. Their insights and comments were not only useful to Dr. Spires as he completed final chapter drafts but served to provide an historical framework that helped mold the context and subtext of the project. Dr. Sturdevant was especially critical in ensuring that many classified documents were downgraded for Dr. Spires' use. He is one of the most dedicated professionals serving in the Air Force History Program, and one can not praise his contributions enough. Dr. Eckert has been unquestionably the driving force in completing this study. He has probably read the manuscript more than any of us as an editor. It is due to his personal dedication that I have been freed from the mundane task of copy-editing. Fortunately for the project, his knowledge of space history also allowed him to make stylistic and content changes of great value to the author. Moreover, Dr. Eckert's expertise in word processing proved invaluable as he developed the layout and completed the page proofs as well as the electronic final "disk" copy of the book. He worked diligently to ensure that all aspects of the printing process were completed with the highest standards. His work with the Defense Printing Service was critical to the timely production of this volume. This book would not have been possible without the dedicated services of my two colleagues, Dr. Sturdevant and Dr. Eckert, and I thank them for their dedication, professionalism, and advice that was given freely throughout this long four-year process.

I also need thank the headquarters staff who supported the Directorate of History's efforts to produce this study. A special thanks goes to the Director of Field Support, Colonel Robert Koenig, who as Chairman of the command's 50th Anniversary Committee, ensured that funds were available to print this book. So many staff members at the headquarters provided support toward the completion of this study that it is impossible to name them all. I would, however, like to especially thank those staff members who read the manuscript as part of the security and policy review process overseen by the command's Directorate of Public Affairs.

Most importantly, a project of this nature can not be successfully concluded without support from the top. From the first day this project began, the leadership at Air Force Space Command has provided unfailing support both in terms of funding and managerial guidance. I would especially like to thank the two Vice Commanders who supported this project in every phase of its accomplishment: Lieutenant General Thomas S. Moorman, Jr. and Lieutenant General Patrick P. Caruana. They provided leadership to ensure the project was appropriately supported, and their personal interest inspired all those participating in the effort. I would also like to thank the commanders of Air Force Space Command, each of whom in turn never wavered in their support and interest for this study: General Charles A. Horner, General Joseph W. Ashy, and General Howell M. Estes III. Their comments and advice each time I briefed the project to them were always positive and frequently insightful.

Finally, while Air Force Space Command has provided funding and support for the completion of this study, the command has exercised only security and policy review of the manuscript. The content and conclusions are solely those of the author and do not necessarily reflect the views of the Directorate of History, Air Force Space Command, or the United States Air Force.

George W. Bradley
Director of History
Air Force Space Command

INTRODUCTION
The Dawn of the Space Age

In making the decision as to whether or not to undertake
construction of such a [space]craft now [1946], it is not
inappropriate to view our present situation as similar to
that in airplanes prior to the flight of the Wright brothers.
We can see no more clearly all the utility and implications
of spaceships than the Wright brothers could see fleets of
B-29s bombing Japan and air transports circling the globe.[1]

In 1946, the authors of the first Air Force-sponsored Project Rand (Research
and Development) study on the feasibility of artificial earth satellites aptly
characterized the challenge and uncertainty surrounding the country's initial
foray into the space age. Postwar skeptics dismissed proposed satellite and missile
projects as excessively costly, technologically unsound, militarily unnecessary, or
simply too "fantastic," while space advocates themselves remained hard pressed to
convince opponents and stifle their own self-doubts. Space represented a "new
ocean," a vast uncharted sea yet to be explored. The dawn of the space age
brought many questions but offered few answers. Could satellites be successfully
produced, launched, and orbited? If technically feasible, what military—or
civilian scientific—functions should they perform? How should space functions
be organized? What space policy would best integrate space into the national
security agenda? What should be the Air Force role in space?

In view of the uncertainties involved, the period from the close of the Second World War to the launching of the first Sputnik in the fall of 1957 proved to be the conceptual phase of the nation's space program. Only by the mid-1950s, a full decade after the 1946 Rand study, could observers identify two sides of a national space policy that would characterize the American space program from the Eisenhower presidency to the present day. One side comprised a civilian satellite effort, termed Project Vanguard, designed to launch a scientific satellite by the end of 1958 as part of the International Geophysical Year. The other, an Air Force-led military initiative, sought to place into earth orbit a strategic reconnaissance satellite capable of providing vital intelligence about Soviet offensive forces.[2]

The Air Force played a central role during the formative era before Sputnik and afterward when the nation's leaders established space policy and organized to confront the Sputnik challenge. The National Space Act of 1958 created the civilian agency, the National Air and Space Administration (NASA), to operate the civilian space effort, while the Air Force and other military services and agencies jockeyed for position within the Defense Department and the overall national space program. Although the Air Force won the contest for military "supremacy" among the services, it seemed to many Air Force leaders that the policy of promoting the "peaceful uses of space" meant a diminished role for Air Force space interests and a threat to the nation's security. Nevertheless, by the end of the Eisenhower administration, the Air Force space program revealed the basic defense support mission characteristics it would retain for the remainder of the century.

Arnold and von Kármán Form a Partnership

The Air Force space saga began with the partnership of General Henry H. "Hap" Arnold, Commanding General of the Army Air Forces (AAF), and the brilliant scientist, Dr. Theodore von Kármán, Director of the Guggenheim Aeronautical Laboratory at the California Institute of Technology (GALCIT). Together they provided the emerging Air Force with a strong research and development focus and championed Air Force interests in the new missile and satellite fields. Their legacy would endure.

Hap Arnold and Dr. von Kármán first met in 1935, when Arnold visited his friend, Dr. Robert Millikan, head of the California Institute of Technology (Cal Tech) in Pasadena, California, while serving as commander of the First Wing, General Headquarters Air Force, at neighboring March Field. The two men could hardly have appeared more different. Arnold radiated physical energy and heartiness from his large frame, while the short, slender intellectual Hungarian émigré exuded a quieter, less forceful presence. Yet the two men took to each other immediately. The Air Corps brigadier general's long-standing interest in aviation technology and association with the National Advisory Committee for Aeronautics (NACA) helped spark an immediate personal and professional friendship. Back in the First World

War Arnold had participated in primitive pilotless aircraft tests, and later served as a military representative to the NACA. For his part, renowned aerodynamicist von Kármán later recalled that while Arnold had no significant technical background or training, he possessed an appreciation for what science could contribute to aviation and the "vision" to persevere against long odds.[3]

After their first meeting, Arnold often visited Cal Tech to observe wind tunnel experiments and discuss with von Kármán various aeronautical and, especially, rocket propulsion initiatives Cal Tech had just undertaken. Von Kármán, who had established his reputation in structures and fluid dynamics as well as aerodynamics, showed the foresight to support a research project first proposed in 1936 by his bright graduate student, Frank Malina. Malina and his four-man team, known as the "suicide squad," had formed the GALCIT Rocket Research Group to develop both high-altitude sounding rockets and rocket-powered airplanes along the lines described by Austrian theorist, Dr. Eugen Saenger. With Cal Tech's move into rocketry, von Kármán's research placed him squarely at the center of the two areas of propulsion that would take the Air Force to "the fringes of space." One was the aerodynamic approach, represented by the NACA, which involved jet-propelled airbreathing "cruise" missiles; the other, astronautical approach, encompassed rocket-powered "ballistic" missiles.[4]

NACA and the Rocketeers Lay the Groundwork

Since its founding in 1915, the National Advisory Committee for Aeronautics had served as the major government agency performing experiments in basic aviation technology and advanced flight research. During the 1920s and 1930s its research engineers worked closely with the Army, Navy, the Bureau of Standards, and the infant aircraft industry to improve aircraft design and performance. Relying primarily on wind tunnels at its Langley research laboratory in Virginia, its research led to use of retractable landing gear, engine cowlings, laminar flow airfoils, and low-winged all-metal monoplanes. It developed an outstanding reputation for its work in aero-dynamics and with aerodynamic loads. Chartered to benefit both civil and military aviation, the NACA generally performed the research and left to the military services and industry the practical development of aircraft design and production. During the 1930s, the country's focus on Depression issues and budget retrenchment convinced the NACA to remain a small agency with interests primarily in aerodynamics. On the eve of World War II, Chairman Vannevar Bush's organization employed only 523 people and operated one research laboratory at Langley Field.[5]

Wartime, however, brought major expansion in the number of personnel, broader responsibilities in the area of structural materials and powerplants, and the addition of two new laboratories, one adjacent to Cleveland's municipal airport, and the other next door to the naval air station at Moffett Field forty miles south-west of San Francisco. During the war the NACA served as the "silent partner of US airpower," and solved a host of aeronautical problems. Alarmed by reports of German turbojet developments in 1940, for example, the NACA established a Special Committee on Jet Propulsion, and followed in 1944 with a Special Committee on Self-Propelled Guided Missiles. Although the NACA intended to work diligently with the Navy and Army Air Forces on these threats, the need to provide "quick fixes" throughout the conflict meant that basic research became secondary. At war's end the NACA proved eager to learn from the war by continuing its cooperative research efforts with the military. In an agreement signed between the NACA and the services in 1946, the parties agreed that "the effects of accelerated enemy research and development in preparation for war helped to create an opportunity for aggression which was promptly exploited. This lesson is the most expensive we ever had to learn. We must make certain that we do not forget it."[6]

The NACA's postwar vision embraced support of American supersonic flight probes by means of small solid-propellant sounding rockets, and the "X" series of high-altitude, rocket-propelled research aircraft. The first rocket-powered aircraft, Bell Laboratory's X-1, broke the sound barrier on 14 October 1947 with Captain Charles "Chuck" Yaeger at the controls. His historic flight became the first of many increasingly higher and faster experimental aircraft flights toward the fringes of space. The last, the single-place X-20A Dyna-Soar (named for dynamic soaring), would be the Air Force's best hope to launch a manned boost-glide rocket aircraft to the border of space. Although it did not become operational after initial development in the late 1950s, the Dyna-Soar served as a precursor of the Space Shuttle of the 1980s. Although the NACA expressed interest in rocket propulsion, its focus remained centered on aerodynamic experiments and manned flight within the earth's atmosphere. Space research seemed wholly outside its experience and interests.[7]

Rocketeers Lead the Way

Spaceflight represented a challenge far more daunting than traditional aviation. Although future Air Force leaders would lay claim to spaceflight as a logical extension of Air Force operations in the atmosphere, aviation technology offered only limited solutions on the road to outer space. Although the technical advances that led from reciprocating to jet turbine engines powered aircraft higher into the upper atmosphere, the oxygen-dependent airplane remained confined to the atmosphere. Rockets, on the other hand, operate independent of the atmosphere by relying on their own internal propellants: fuel and oxidizer. In their flight through increasingly thinner atmosphere on the way to airless space, rockets become progressively more efficient. Although the post-World War II American rocket research airplanes could provide useful information on the guidance and control challenges facing vehicles in the upper atmosphere, their small rockets could never break the bonds of gravity, and they remained primarily aerodynamic vehicles. To operate either manned spacecraft or instrumented satellites in outer space, rockets needed sufficient thrust to boost their payloads into orbit where centrifugal force balanced the earth's gravitational field.[8]

The challenge of manned spaceflight had captivated the imaginations of dreamers for centuries. Yet their ideas remained only idle musings until technological progress in the late 19th century led serious enthusiasts to consider liquid-propellant rockets as "boosters" of spacecraft. Among the pioneers of liquid-propellant rocket research linked to visions of manned spaceflight, three men—Russian Konstantin Tsiolkovsky, German-Romanian Hermann Oberth, and American Robert Goddard—paved the way for the successful military and civilian space programs of the second half of the 20th century. While their research initially led to production of bombardment rockets for use by their respective military forces in the Second World War, they all remained committed to visions of spaceflight.[9]

The earliest of the space triumvirate, mathematics teacher Konstantin Eduardovich Tsiolkovsky, in 1895 published the first technical essays on artificial earth satellites. By the end of the century, he had worked out the theory of a liquid-fueled rocket dependent on kerosene to achieve sufficient exhaust velocity. For the next 20 years he immersed himself in theoretical studies but remained largely unknown to the world outside Russia. Yet, by the time of his death in 1935, his pioneering work had helped the Soviets establish a strong prewar rocket and jet-powered aircraft development program which led to the space program of the postwar era.

Although Hermann Oberth also taught mathematics and produced important theoretical studies on spaceflight, he assumed the role of publicist for rocketry and space exploration to enthusiastic European audiences after World War I. In 1923 he established his reputation in the new field of astronautics with the seminal publication, "The Rocket into Interplanetary Space," in which he described the technical requirements for propelling satellites into earth orbit. In 1927 he helped found the German Society for Space Flight, which became the most influential of the numerous rocket societies in Europe. By 1931, Oberth's work with the Society came to the attention of the German Army, which saw in sponsorship of the young rocket scientists a means of obtaining bombardment rockets for an army sorely constrained by the Versailles Treaty. Among the Society members who joined the Army project in 1932 was a 20-year old engineer named Wernher von Braun. After 1933, the Nazi regime expanded the Wehrmacht program and in 1937 began developing the Peenemuende experimental site on the Baltic coast under supervision of Captain Walter Dornberger. Although von Braun and his colleagues had now to focus on long-range rockets to help fuel Germany's military expansion, they continued to dream of manned spaceflight. During the Second World War, while the Luftwaffe produced the V-1 aerodynamic pulse-jet "cruise" missile, the Wehrmacht's Peenemünde rocketeers developed the far more impressive "big rocket," the V-2. Known as the A-4 to the rocket specialists, the V-2 measured 46 feet in length, weighed 34,000 pounds, and approached a range of 200 miles under 69,100 pounds of thrust produced by its liquid-propellant engine. To the Allies the V-2 presented a frightening weapon that could not be thwarted with any known defense. After the war Americans discovered that German plans had called for an intercontinental ballistic missile to strike New York by 1946. To the German rocketeers, however, the A-4 always represented the first rung on the ladder to space. After the war, the American Army's Operation Paperclip brought Dornberger, von Braun, and a host of other German rocket experts to the United States, where they joined the Army's rocket program—with their visions of spaceflight still alive.[10]

The German rocket specialists freely acknowledged their debt to American rocket pioneer, Robert H. Goddard. Unlike his Russian and German contemporaries, Goddard immediately moved beyond theoretical studies to practical experimentation. He always found applied research more exciting than theoretical studies. From his post as a physics professor at Clark University, Goddard began experimenting with powder rockets, and in 1914 received a patent for his liquid-propellant rocket engine. In 1920 the Smithsonian released his highly technical paper, "A Method of Reaching Extreme Altitudes," which described various rocket-propelled experiments that could be conducted as high as 50 miles in altitude. His paper also included a theoretical argument for rocketing a payload of flash powder to the moon, which drew public censure after a *New York Times* reporter ridiculed the idea in print. The experience left Goddard badly scarred and more than ever inclined to focus on private research. By 1926 he had built and tested the first liquid-propellant rocket, and in 1935 successfully launched a gyroscopic-stabilized rocket to an altitude of 7000 feet. Eventually, the prolific experimenter amassed an amazing 214 patents for his designs and devices. But Goddard preferred working alone and jealously guarded his work from other space enthusiasts like the intrepid members of the fledgling American Rocket Society.

In the 1930s Goddard moved his increasingly complex liquid propellant experiments from Massachusetts to the New Mexico desert, where he worked with his wife and various assistants supported by grants from the Guggenheim Foundation. Guggenheim officials quite naturally sought to bring Goddard and von Kármán's Cal Tech Rocket Research Project team together. Characteristically, Goddard proved reluctant, and von Kármán refused to collaborate without full disclosure of Goddard's research results.[11]

Despite the acknowledged importance of Goddard's work for future rocket development, active collaboration between von Kármán and Goddard might well have placed the postwar American rocket program on better technical footing and created more incentive for the Air Force to promote research in ballistic rather than aerodynamic missiles after the war. Cooperation between the two camps would certainly have helped the neophyte rocket group at Cal Tech, which had developed convincing theories about rocket flight but had no experimental data to work with. Moreover, as Malina recalled, in the 1930s most scientists generally considered rocket experiments a part of science fiction. With so little available practical data, Goddard's assistance would have been welcomed by von Kármán and the young rocketeers, who proceeded largely independently of Goddard.[12]

Wartime Provides the Momentum—Arnold and von Kármán Establish the Foundation

Meanwhile, von Kármán and his Cal Tech rocket team continued their research into high-altitude sounding rockets and jet-assisted takeoff (JATO) devices by examining potential fuel types, rocket nozzle shapes, reaction principles, and thrust measurements. They managed to keep their experiments afloat with very little money until General Arnold came to the rescue in 1938. Late that year, Arnold, now chief of the Army Air Corps, helped convince the National Academy of Sciences to provide initial funding for the Cal Tech project. Shortly thereafter, in January 1939, the Air Corps assumed direction of the program, and in June awarded the researchers a $10,000 contract. Von Kármán explained that the program's label, "Air Corps Jet Propulsion Research, GALCIT #1," included the word "jet" rather than "rocket" because of wide-spread skepticism among his colleagues. As one of them told him, he was welcome to the "Buck Rogers" job.[13]

Malina wisely committed his team to explore both liquid- and solid-propellant rocket engine research. The team made rapid progress once they developed the first relatively long-duration, controlled-explosion solid-propellent engine. In August 1941, the Cal Tech engineers carried out their first flight tests in which Captain Homer Boushey, using four JATO canisters attached to his Ercoupe monoplane, rapidly climbed to an altitude of 20 feet. Malina was ecstatic. Continued test successes brought in a JATO contract from the Navy, and von Kármán and Malina in 1942 decided to capitalize on their growing project by forming a private company, Aerojet Engineering Company, to produce the jet canisters and work on other rocket-related contracts they expected to receive.[14]

In late 1943, after reviewing intelligence reports on German rocket development, von Kármán wrote a brief paper entitled, "Memorandum on the Possibilities of Long-Range Rocket Projectiles," in which he proposed that the AAF support development of a 10,000-pound air-breathing missile with a seventy-five-mile range as an extension of JATO research. When the AAF demurred, the Army Ordnance Department stepped in and offered von Kármán a far more challenging contract. The scientist readily agreed to the Army's project, which called for producing a 20,000-pound liquid-propellant rocket with a burn time of sixty seconds and a range of nearly forty miles. Organized under Frank Malina, the large project became known as ORDCIT (representing Ordnance, California Institute of Technology), until renamed the Jet Propulsion Laboratory (JPL) in November 1944. Their work would lead to the successful launching of the WAC Corporal series of liquid-propellant sounding rockets after the war. Meanwhile, as the Army's Ordnance Department focused primarily on rockets, the AAF's Air Materiel Command preferred to stress air-breathing missiles.[15]

During much of the war, von Kármán served as an aeronautical troubleshooter for Hap Arnold, the Commanding General of the AAF. By 1944 Arnold had become convinced that the next war, unlike the last, would demand far more technical competence. As Chief of the Army Air Forces, he said, his job was to

> project [himself] into the future; to get the best brains available, have them use as a background the latest scientific developments in the air arms...and determine what steps the United States should take to have the best Air Force in the world twenty years hence.[16]

In September of that year he called on von Kármán to lead a study group comprised of civilian and military experts to chart a course for the Air Force future. Arnold outlined his objectives for the group in a 7 November 1944 memorandum, "AAF Long Range Development Program." In order to place Air Force research and development programs on a "sound and continuing basis," he called for a plan whose farsighted thinking would provide a sound prescription for preparing Air Force research and development programs as well as congressional funding requests. Because "our country will not support a large standing Army" and "personnel casualties are distasteful, we will continue to fight mechanical rather than manpower wars." Given these constraints, he said, how can science be used to provide the Air Force with the best means to ensure the nation's security?[17]

With Arnold's strong support to overcome any bureaucratic impediments, von Kármán began work immediately, and by December had brought together a group of twenty-two renowned scientists and engineers. Calling itself the Army Air Forces Scientific Advisory Group, it would remain in place and continue as the Scientific Advisory Board after the Air Force became a separate service in September 1947. Following field trips to Europe and Russia to assess the current state of research, von Kármán's group on 22 August 1945 issued a preliminary report, *Where We Stand*, which explored the "fundamental realities" of future air power. The report argued that technological advances led by Germany during the war set the stage for an air force that must embrace supersonic flight, long-range guided missiles with highly destructive payloads, and jet propulsion to achieve air superiority. Von Kármán viewed government-supported research centers on the German model as a major element in the postwar national defense structure. *Where We Stand* raised crucial questions about the future of air power, and the Scientific Advisory Group intended to provide answers in its final report to General Arnold due at the end of the year.[18]

Meanwhile, while von Kármán and his team in late 1945 gathered additional field data and prepared their final report to the AAF chief, Arnold took additional steps to shape the future Air Force's scientific focus. Two of the most important involved the creation of Project Rand and a new Air Staff office to establish and direct the Army Air Forces' research and development agenda.

In September of 1945, Franklin Collbohm of the Douglas Aircraft Company proposed that the AAF establish a research project to provide it with long-range strategic planning based on ongoing scientific and technological advances. Collbohm's ideas had taken shape during his wartime association with Dr. Edward L. Bowles, who had served as General Arnold's special technical consultant. Late that month, Arnold and Bowles flew to California, where at Hamilton Field, north of San Francisco, they met with Collbohm and Donald Douglas, who strongly supported the proposal. At their meeting, Arnold decided to divert $10 million from the fiscal year 1946 procurement budget for Douglas Aircraft to organize a group of civilian scientists and engineers at Santa Monica, California, which would function independently of the company's existing research and engineering division. It would serve as a technical consultant group charged with operations analysis and long-range planning to examine future warfare and the best way the Air Force could perform its missions. Shortly thereafter, the Air Materiel Command (AMC) and Douglas Aircraft agreed to a three-year, $10 million contract for Project Rand to begin operating in May 1946.[19]

To provide an Air Staff focus for Project Rand and other research activities, General Arnold also created the office of Deputy Chief of Air Staff for Research and Development. The new position, which became effective 5 December 1945, drew criticism from the powerful Air Materiel Command, which heretofore tightly controlled the AAF procurement process from initial requirements to completed system. AMC favored rigid directives establishing specific AAF-determined goals for contractors without involving civilians in the planning process. Critics complained that research fell victim to production priorities at AMC. The new arrangement reflected Arnold's flexible approach to research and development whereby Rand would conduct broad investigations to see what could be accomplished and recommend courses of action and the new Air Staff office would provide central direction. AMC never reconciled itself to the new Air Staff position, while the Air Staff remained unwilling to assign it specific responsibility for the satellite and guided missiles programs. These would remain subjects of intra-Air Force organizational squabbles throughout the pre-Sputnik period. Nevertheless, initial prospects for achieving Arnold's goals appeared bright when he selected as his first Deputy Chief of Staff for Research and Development the hard-driving combat veteran, Major General Curtis E. LeMay.[20]

In November 1945, General Arnold became the first prominent military figure to address future warfare in terms of missile and satellite potential. In a report to Secretary of War Robert Patterson on 12 November, the air chief described the future importance of missiles and satellites as a means of preventing another Pearl Harbor-like surprise attack on the United States, and he outlined his vision for the nation's air arm. Strongly opposing shortsighted focus on present day forces, he cautioned that national safety would be endangered by an Air Force whose doctrines and techniques are tied solely to the equipment and processes of the moment. Present equipment is but a step in progress, and any Air Force which does not keep its doctrines ahead of its equipment, and its vision far into the future, can only delude the nation into a false sense of security.[21]

For Arnold, the forces of the future must never be sacrificed for the forces of the present. While the current state of technology convinced him to support manned aircraft, he envisioned a pilotless air force and supported developing intercontinental ballistic missiles (ICBMs) for the future Air Force. Profoundly affected by the German V-2 (A-4) rocket missile, he called for a similar weapon for the American arsenal, one "having greatly improved range and precision, and launched from great distances. [Such a weapon] is ideally suited to deliver atomic explosives, because effective defense against it would prove extremely difficult." In perhaps his most controversial prognostication, he proposed launching such "projectiles" from "true space ships, capable of operating outside the earth's atmosphere. The design of such a ship is all but practicable today; research will unquestionably bring it into being within the foreseeable future." Much of General Arnold's vision for his future Air Force received strong backing from Theodore von Kármán's monumental study on the state of air force technology—past, present, and future.[22]

After General Arnold suffered a massive heart attack in October 1945, von Kármán drove his group hard to conclude their work by the end of the year. In mid-December 1945 von Kármán's team produced the remarkable 33-volume report, *Toward New Horizons*. The first volume, "Science: The Key to Air Supremacy," set the tone by declaring that the Air Force should establish its policy, create new organizational alignments, and lay the "foundation of organized research" so that science would become an integral part of the Air Force. Von Kármán proceeded to discuss many specific means for providing technological training for service personnel and adequate research and development facilities, for disseminating scientific ideas at the staff and field levels, and promoting cooperation between the Air Force and science and industry. Regarding the latter, he noted that the Air Force preferred to sponsor research and development activities outside its own organizations, and this should be continued on a broader scale through extensive contacts with universities, research facilities, and scientists. As a means of providing continued scientific advice to Air Force leaders, he recommended that Arnold continue the Scientific Advisory Group as a permanent institution.[23]

Toward New Horizons expanded on issues discussed in von Kármán's "Science: The Key to Air Supremacy." The report's assessments of space issues are particularly interesting. Both jet and rocket propulsion received considerable attention, and von Kármán predicted the eventual operational success of ICBMs and declared the "satellite"…"a definite possibility." In his memoirs, von Kármán recounts that his group "examined the thrust capabilities of rockets and concluded that it was perfectly feasible to send up an artificial satellite, which would orbit the earth. We did not, however, give consideration to the military potential of such a satellite."[24] In fact, neither ICBMs nor satellites received more than passing mention because von Kármán and his colleagues believed that technological barriers would delay successful ballistic missiles for at least a decade. The report proceeded to emphasize what could be achieved within the atmosphere with jet propulsion. Indeed, von Kármán proposed that the Air Force implement a deliberate, step-by-step guided missile development program based on air-breathing missiles rather than ballistic rockets. The Air Force would accept von Kármán's argument and follow the air-breathing approach to missile development. Although von Kármán differed with other prominent scientists who dismissed the ICBM entirely, his recommendations served to chart an Air Force course that would delay development of the long-range ballistic missile.[25]

Nevertheless, *Toward New Horizons* proved to be a landmark because it established the importance of science and long-range forecasting in the Air Force. In staking out a role for military research, von Kármán differed fundamentally with colleagues like highly regarded Dr. Vannevar Bush, who believed that the military services should confine themselves to improving existing weapons and leave new scientific ideas to the civilian experts. *Toward New Horizons* helped ensure that the Air Force would reflect von Kármán's thinking. As his biographer aptly concludes, von Kármán's "detailed, highly technical blueprint set the agenda of research and development [in the Air Force] for decades to come."[26]

Arnold's and von Kármán's comments did not escape the attention of Dr. Bush, then the influential Director of the Office for Scientific Research and Development, and Chairman of the Joint Committee on New Weapons of the Joint Chiefs of Staff. Having sharply differed with von Kármán on military prerogatives in the research field, he turned his attention to the predictions of military officers on matters scientific. Appearing before a special Senate Committee on Atomic Energy in December 1945, Dr. Bush observed that

> We have plenty enough to think about that as [sic] very definite and very realistic-enough so that we don't need to step out into some of these borderlines, which seem to me more or less fantastic. Let me say this: There has been a great deal said about a 3,000-mile high angle rocket. In my opinion such a thing is impossible and will be impossible for many years. The people who have been writing these things that annoy me have been talking about a 3,000-mile high-angle rocket shot from one continent to another carrying an atomic bomb, and so directed as to be a precise weapon which would land on a certain target such as this city. I say technically I don't think anybody in the world knows how to do such a thing, and I feel confident it will not be done for a very long period of time to come. I think we can leave that out of our thinking. I wish the American public would leave that out of their thinking.[27]

When asked whether he was addressing his remarks to anyone in particular, he specifically identified General Arnold, whose report to Secretary of War Patterson had appeared the previous month.

Although Dr. von Kármán could characterize Vannevar Bush as "a good man... limited in vision,"[28] Bush and other prominent civilian scientists who expressed similar criticism had a major influence on the Air Force missile and space development programs. Their pessimism reflected current thinking in many postwar circles that contributed to stifling research by limiting it to the technical problems posed by ICBMs. In the postwar flush of victory and sense of American superiority, the American monopoly of seemingly scarce of fissionable uranium and the great weight of the first atomic bombs produced an air of complacency about the technological future. Atomic bombs of over five tons and relatively poor destructive capacity ("kill-radius") suggested that missiles could never be constructed with sufficient thrust and guidance accuracy to provide a credible operational weapon. Dr. Bush continued until his retirement in 1948 to manage research and development for the Defense Department, where he became known for his parsimonious funding of military programs that could not guarantee progress to his satisfaction.[29]

CHAPTER 1
Before Sputnik:
The Air Force Enters the Space Age, 1945–1957

In the aftermath of World War II Air Force leaders laid the foundation for future operations in space by establishing a clear research and development focus for the new service. Commanding General of the Army Air Forces Henry H. "Hap" Arnold and his eminent scientific advisor Theodore von Kármán set the course through their policy statements, organizational decisions, and comprehensive analysis of Air Force scientific requirements for a technological future. Their legacy appeared endangered in the late 1940s when tight budgets and higher priorities confined space and long-range missile development to low level studies at best. Air Force leaders seemed intent on establishing Air Force responsibility for the as-yet-to-be-determined space mission, but unwilling to promote the development of satellites and booster missiles that would make possible such a mission.

By the early 1950s, however, change was in the air. New concerns about Soviet political activity and ICBM development compelled leaders to reexamine the country's defense posture. In doing so, missiles and satellites received new attention. Larger defense budget outlays and successful testing of thermonuclear devices offered the promise of a feasible ballistic missile and space booster. A number of government officials and Air Force officers who shared Arnold's legacy acted as catalysts for change by creating new organizational structures and promoting greater awareness of spaceflight opportunities. They faced strong opposition every step of the way. Yet on the eve of Sputnik, their considerable efforts helped bring the Air Force and the nation to threshold of space.

Rand Proposes a World-Circling Spaceship

In a postwar America, with armed forces undergoing demobilization and reassertion of domestic priorities, Arnold and other Air Force innovators quickly realized that it was one thing to advocate an imaginative, liberally-funded research and development program for the Army Air Forces (AAF) and quite another to have it put into practice by a conservative military establishment. The Air Force's initial involvement with artificial earth satellites illustrates the difficulty of gaining approval for a system of the future rather than the present.

In early 1946, the AAF found itself about to be outmaneuvered by Naval officers who had been pursuing satellite feasibility studies since the end of the war. Captivated by a space study written in May 1945 by German space scientist Wernher von Braun, as well as the horde of captured V-2 rocket components, Dr. Harvey Hall of the Navy's Bureau of Aeronautics Electronics Division proposed a testing program to determine the feasibility of artificial satellites. Based on a current Naval hydrogen rocket motor development program, Commander Hall's team formed a Committee for Evaluation of the Feasibility of Space Rocketry and envisioned launching a liquid hydrogen-oxygen single-stage earth satellite to conduct scientific testing. Naval leaders agreed, and Hall called on four companies, including GALCIT, for technical assistance with fuels, electronic components and structural characteristics. Early in the new year, all four agreed that a satellite could be placed in earth orbit if the Navy proved willing to provide sufficient funding.[1]

Unable to gain the required Naval financial support, Commander Hall proposed a cooperative space venture to AAF representatives at a 7 March 1946 meeting of the War Department's Aeronautical Board. Although AAF Board members questioned the high costs involved, they expressed interest and promised to consult with Major General Curtis E. LeMay, Arnold's Deputy Chief of Staff for Research and Development, before the Board reconvened on 14 May. After discussions with General Carl A. Spaatz, who replaced General Arnold on 1 March as commanding general, LeMay informed Hall that the AAF could not support the Navy project but nevertheless would continue discussions on the subject. Already AAF leaders had decided that artificial earth satellite programs should be an AAF responsibility based on the argument that military satellites represented an extension of strategic air power. For the first time air leaders outlined the rationale for an Air Force space mission that would appear haphazardly over the next ten years, then surface prominently during the roles and missions debates after Sputnik.

To forestall the Navy's initiative in the spring of 1946 and help establish AAF primacy in the field, the service needed to demonstrate competence equal to the Navy's. In April LeMay turned to Project Rand for the necessary technical expertise. In just three weeks, the Rand team of sixteen experts completed their justly celebrated 250-page engineering analysis of a "World-Circling Spaceship."[2] Based on both the current state of technology and expected future engineering developments, the Rand team argued that "technology and experience have now reached the point where it is possible to design and construct craft which can penetrate the atmosphere and achieve sufficient velocity to become satellites of the earth." Indeed, the report predicted that the U.S. could launch a 500-pound satellite into a 300-mile orbit within five years at a cost of $150 million. Rand's analysts declared that even their most conservative engineers agreed, and they supported their prediction with a series of detailed studies in two chief areas.

One comprised technical feasibility studies dealing with such satellite-related issues as propulsion options, risks posed by potential meteor strikes, trajectory analyses, the important "re-entry" challenges posed by the intense heat objects would encounter returning through the earth's atmosphere, and, in contrast to the Navy's single-stage rocket, the use of a three-stage liquid hydrogen-oxygen rocket booster. The analysts argued that no technical challenge they investigated seemed overwhelming.

In a second area, noted radar expert Louis N. Ridenour examined a number of potential military satellite uses in a chapter titled "The Significance of a Satellite Vehicle." Focusing on defense support or "passive" military uses of satellites, he described satellites as nearly invulnerable observation platforms that could provide weather and bomb damage assessment data. He went on to describe the satellite as a communications relay station, in which satellites could be positioned at an altitude of approximately 25,000 miles so that "their rotational period would be the same as that of the earth." For the first time, a serious satellite proposal projected launching satellites into geosynchronous orbit for effective, worldwide communications.[3] Ridenour devoted most of his chapter to the satellites' scientific role in supplying important data unaffected by atmospheric conditions. As an aid to research, the satellite could facilitate the study of cosmic rays and provide precise gravitational measurements, as well as considerable astronomical data. Moreover, instrumented satellites could furnish important bio-astronomical information for medical scientists concerned with life in an acceleration-free environment.

Despite Ridenour's coverage of "passive" defensive military functions, he briefly raised the possibility of using satellites as offensive weapons. Given the onset of the missile age, he argued, satellites could provide both accurate guidance for missiles and serve as missiles themselves. He based his argument on the compatibility between missile and satellite technology as well as launch velocity requirements. As he explained,

> There is little difference in design and performance between an intercontinental rocket missile and a satellite. Thus a rocket missile with a free space trajectory of 6,000 miles requires a minimum energy of launching which corresponds to an initial velocity of 4.4 miles per second, while a satellite requires 5.4. Consequently, the development of a

satellite will be directly applicable to the development of an intercontinental missile.[4]

In short, if you produce an ICBM, you also have a satellite launcher. In the future,
however, the technical relationship between long-range missiles and satellites would
remain largely unexploited. Even missile advocates normally argued that satellite
development interfered with the greater need to accelerate missile programs. Later,
closer examination would show that satellite technology could be applied to missile
guidance systems and thereby contribute to missile development. The missions
identified by Louis Ridenour would become a part of the Air Force and the national
space program from the Eisenhower administration forward. Unfortunately, many
of the Rand study's predictions and analyses would be forgotten in the years ahead.
David Griggs, for example, in the report's introduction turned his vision to the
future:

> Though the crystal ball is cloudy, two things seem clear: 1. A satellite
> vehicle with appropriate instrumentation can be expected to be one of
> the most potent scientific tools of the Twentieth Century. 2. The
> achievement of a satellite craft by the United States would inflame the
> imagination of mankind, and would probably produce repercussions in
> the world comparable to the explosion of the atomic bomb.[5]

Armed with the Rand study, General LeMay formally declined the Navy offer of a
joint Navy-AAF program at the May meeting of the Aeronautical Board and staked
out the AAF's claim to potential satellite operations. At the same time board
members decided that the costs of developing and operating a satellite did not
justify a major effort on a project of questionable military utility. The Board agreed
to permit both services to continue their studies, with the jurisdictional assignment
of satellite responsibility left unresolved.

The 1946 Rand report established the technical feasibility of orbiting a satellite
but ruled out its likely use as an offensive weapon because available propulsion
systems could not launch heavy atomic weapons into earth orbit. Given this restriction, the problem became establishing a credible role for an orbiting satellite that
could justify the enormous cost and quiet the skeptics of "push-button warfare."
During the next several years, Rand's satellite proposals would continue to founder
on the criticism of cost and utility, while greater interest in developing guided missiles
served to retard satellite progress further.

The Air Force Shuns Ballistic Missiles

The analysts at Rand underscored the relationship between satellite and missile
development. Not only would progress with satellites promote greater interest in
missiles as boosters, but improvement in satellite technology could benefit missile
development as well. Yet, the Air Force establishment, which focused on its bomber
fleet, seemed unaware of the potential for mutual benefits, and later in the 1950s,

when missiles promised additional strategic firepower for the nation's arsenal, critics of a forceful space program argued that satellites must not be allowed to interfere with missile development. The Rand analysts also might have noted that the missile-satellite relationship meant that any progress with satellites would depend on developments in the higher priority missile field. In the years after the Second World War, however, neither subject drew significant attention from the Truman administration and the defense establishment. As with satellite proposals, initial postwar interest in long-range guided missiles soon succumbed to an Air Force policy that relied on strategic bombers carrying air-breathing missiles, interservice conflicts over roles and missions, and administration-imposed budget ceilings that compelled Air Force planners to focus on present needs.[6]

General Arnold was not the only military leader impressed by the German V-2 achievements during the war. In the flush of victory, all the services sought to build on the wartime experience by conducting rocket and guided-missile experiments based either on aerodynamic, jet-propelled "cruise" missile principles, or the German V-2 short-range liquid-propellant ballistic rocket technology. Operation Paperclip brought nearly 130 leading German rocket scientists, a vast array of data, and approximately 100 dismantled V-2s to White Sands, New Mexico. There, under Project Hermes, the Army Ordnance Division conducted upper atmospheric research into airborne telemetry, flight control, and two stage rocket capability with representatives from the Air Force, the Air Force Cambridge Research Center, the General Electric Company, the Naval Research Laboratory, and a number of scientific institutions, universities, and government agencies. From 1946 to 1951, participants received valuable data from 66 V-2s that first carried various scientific instruments then, later, primates.[7]

By early 1949 the Army, which viewed rockets as extensions of artillery, had successfully used a V-2 as the mother vehicle to launch the Jet Propulsion Laboratory's WAC Corporal second-stage rocket to an altitude of 250 miles. As Frank Malina noted, "the WAC Corporal thus became the first man-made object to enter extraterrestrial space."[8] These early V-2-based Bumper-WAC experiments set the stage for the Army's future missile and space program involving the Redstone, Jupiter, and Juno boosters developed by the von Braun team under Army supervision after it moved in 1950 from Fort Bliss, Texas, to the Redstone Arsenal at Huntsville, Alabama. Postwar naval rocket research led by the Applied Physics Laboratory of Johns Hopkins University and the Naval Research Laboratory in Washington, D.C., produced two reliable and effective sounding rockets: the fin-stabilized Aerobee, a larger version of the WAC Corporal modified for production as a sounding rocket, which achieved a height of 80 miles; and the more sophisticated Viking, which would reach an altitude of 158 miles in May 1954. A modified Viking eventually would provide the booster for the four-stage Project Vanguard, the nation's first "civilian" space program.[9]

Despite General Arnold's interest in developing long-range missiles of the V-2 type, the Air Force followed the path charted by Theodore von Kármán, which stayed within the atmosphere and the initial Air Force domain. Short-range jet-propulsion weapons seemed to offer faster development and better range and payload capabilities. They also directly complemented the strategic bomber fleet, the nation's intercontinental strike force of the day. In October 1945 the Army Air Forces solicited proposals from seventeen aircraft companies for a ten-year research and development program for pilotless aircraft, and the fiscal year 1946 budget included an impressive twenty-six different projects. Yet only two involved missiles in the 5,000 mile range, and one of these consisted of a Northrop Aircraft super-sonic turbojet vehicle. The other, a supersonic ballistic rocket design from the Consolidated Vultee Aircraft Corporation (Convair), would serve as the precursor of the future Atlas ICBM.[10]

If the Army Air Forces seemed devoted to shorter-range air-breathing missiles, it could not abandon long-range missile development to the Army or Navy. All three services jealously guarded their prerogatives and jockeyed fiercely with their rivals over roles and missions in the new postwar world. As it looked to a future as an indepen-dent service, the Army Air Forces proved particularly sensitive to new, unproved weapon fields such as rockets and missiles. While General LeMay in early 1946 staked out the AAF's claim to any prospective satellite mission, he also became embroiled with Army and Navy representatives over which service should be responsible for what types of missiles. Above all, the Army Air Forces took special interest in missiles it considered strategic.

Confusion and friction about missile development and operational control first emerged during the war in the competition within and among the services. A num-ber of Army Air Forces offices asserted their "special" interests, while attempting to ward off the Army Ordnance Command and various elements in the War Depart-ment. A directive issued by Lieutenant General Joseph T. McNarney, Deputy Chief of Staff of the Army, on 2 October 1944, attempted to clarify the situation by assigning the AAF responsibility for "all guided or homing missiles launched from the ground which depend for sustenance primarily on the lift of aerodynamic forces." Although this ruling appeared to award the Army Ordnance Department (the Army Service Force) the ballistic missile mission, the AAF, which sought primary responsibility for all missile "programs," continued to complain of Army encroachment into the aerodynamic field.[11]

Conflict persisted into the postwar era as each of the services pursued its own guided missile program while keeping a wary eye on its competitors. In a revealing memorandum in September 1946 to AAF chief General Spaatz, General LeMay expressed his concerns about the Air Force maintaining its rightful "strategic role." Admitting that the "long-range future of the AAF lies in the field of guided missiles," he cautioned that the Army's success in controlling guided missiles might embolden its leaders to seek control not only of close support but strategic aircraft as well. After all, he noted, the "stated opinion" of the Army Ground Forces is that guided missiles are extensions of artillery. LeMay saw the possibility of the Air Force losing control of a weapon system that might replace manned aircraft in the future. Yet control of the weapon did not necessarily mean that it should be developed, at least at the present time. Meanwhile, the Navy entered the contest for preeminence. Given AAF aspirations and its strategic mission, Naval leaders joined their Army counterparts in arguing that each service should have the freedom to develop missiles in response to its particular needs.

On 7 October 1946 the War Department's Assistant Secretary of War (Air), W. Stuart Symington, attempted to settle the dispute by awarding the AAF responsibility for research and development activities pertaining to all guided missiles. The directive remained silent, however, on the important question of ultimate operational assignment. The issue lay dormant until after September 1947, when the establishment of an independent Air Force reopened the competition. A year later the Defense Department achieved a modicum of peace when the Air Force relinquished its responsibility for conducting research and development work for the Army. In return, the Air Force received authority to develop strategic missiles, while the Army became responsible for tactical missiles. Meanwhile, the Air Force continued its pathbreaking ballistic missile defense studies, Projects WIZARD and THUMPER. Although the latter was cancelled in March 1948, Project WIZARD continued until early 1958, when then Secretary of Defense Neil H. McElroy reacted to persistent feuding over ballistic missile defense responsibilities by awarding the Army the mission of strategic defense and merging WIZARD with the Army's NIKE-ZEUS anti-ballistic missile system.[12]

The problem of interservice rivalry over missiles received little help from the defense committees most responsible for providing direction. With passage of the National Security Act of 1947, the Research and Development Board replaced the Joint Research and Development Board. Dr. Vannevar Bush continued as chairman until his retirement in October 1948. Neither he nor those active on subordinate committees, like the Committee on Guided Missiles, possessed the authority needed to provide the firm direction. Too often they allowed the complex committee system to work to their disadvantage and avoid decisive action.

Throughout the conflict over roles and missions, the Air Force demonstrated more interest in gaining and preserving its prerogatives than moving ahead with a strong missile research and development program. Paradoxically, as the Air Force's commitment to develop an ICBM diminished, its determination to be designated sole authority responsible for long-range missiles increased. Even with long-range cruise missiles, for which Air Force leaders sought exclusive control based on the service's strategic mission, it normally chose not to implement programs leading to operational missiles. Efforts to garner exclusive control of missiles would continue. In

September 1948, for example, the Defense Department awarded the Air Force operational control of surface-to-surface pilotless aircraft as well as strategic missiles. Two years later, in a very important March 1950 decision, the Air Force received official responsibility for developing long-range strategic missiles and short range tactical missiles. Later, near the end of the Truman administration, the Air Force successfully defeated the Army's bid to develop the Redstone rocket's range beyond 200 miles. The strategic mission would remain with the Air Force.[13]

In the late 1940s Air Force leaders signaled their attitude about research and development when forced to respond to the Truman administration's drastic economy drive that began in late 1946. Compelled to choose between supporting the forces of the present and those of the future, the Air Staff ignored the admonitions of General Arnold and Dr. von Kármán by focusing on manned aircraft to the detriment of guided missiles. As a result, Air Force research and development programs for missiles suffered drastically in the late 1940s.

Budget figures help tell the story of decline. Air Force leaders needed to show a firm commitment to research in terms of policy advocacy and budget allocation. As expressed by General Benjamin W. Chidlaw, commander of Air Materiel Command, "Many people have given lip-service to the magic phrase 'Research and Development.' Very few of us have really fought for it—and made sacrifices for it."[14] Without such a commitment, the Truman economy drive was bound to seriously erode research and development funding and projects. The fiscal year 1946 Army Air Forces budget allocated $28.8 million for research and development, with half earmarked to support the twenty-six guided missile programs sponsored in 1946. The initial fiscal year 1947 budget reflected the importance of research with a grant of $75.7 million, $29 million of which was dedicated to guided missiles research. Then the budget ax fell. Under pressure from the Defense Department, in December 1946 the Air Force cut the missile budget by $5 million and eleven missile projects. Additional funding cuts in May led planners to eliminate five more programs.[15]

Faced with drastic reductions in the guided missile program, the Air Materiel Command decided to protect those programs promising the earliest tactical operational availability, and in June 1947 General Hoyt S. Vandenberg approved the AMC recommendations. This criterion effectively eliminated the only long range guided missile project, the MX-774, and the Air Force terminated the Convair contract on 1 July. That same month the Air Staff established development priorities for managing the smaller budgets they expected in the future. The subsonic bomber and air-to-air and air-to-surface missiles received top priority. In the belief that long-range surface-to-surface missiles would be prohibitively expensive and require ten years to develop, build and launch, long-range ballistic missiles stood at fourth priority.[16]

By 1947 the pressure to downgrade the development priority of long-range missiles proved overwhelming. In the growing Cold War conflict the administration increasingly looked to strategic bombers, supported by cruise missiles, and the atomic bomb as the country's main line of retaliatory defense. Moreover, manned aircraft remained the heart of the Air Force, and advocates of a new, if potentially revolutionary, weapon and "push-button" warfare found themselves outmatched in competition for funding. Critics focused on the technological challenges of missile development. The budget slashers argued that putting scarce funds into a research program that might not be realized for a decade or possibly never could not be justified in light of current priorities. Therefore one must continue with a cautious step-by-step approach to any long-range missile program. Missile advocates found themselves victims of a circular argument: missiles seemed too challenging technologically, but no funds could be spent on solving the technological dilemmas; so the problems would go unresolved and the missile would remain "impossible." To questions about the logic of budgeting for missile programs, the answer always seemed to be the dogmatic response: "the time is not right" for an expanded program.

The Air Force's devotion to aerodynamic missiles like the intercontinental Navaho, with its combination ramjet-booster rocket propulsion, and the subsonic Snark and Matador missiles also must be seriously questioned. Planners consistently offered the rationale that aerodynamic research benefited ballistic missile research. This proved correct to a point, as shown by the transfer of the Navaho Rocketdyne engines for use in the Redstone and Atlas systems. Yet cruise missile guidance systems offered little commonality, while aerodynamic vehicles could provide no help with the ICBM's high-speed reentry from space into the upper atmosphere. Sadly, Air Force scientists never reexamined the assumptions so forcefully established in the 1945 von Kármán reports. Moreover, not one of the reports called for research and development to achieve strategic reconnaissance.[17] Although the Air Staff reassessed guided missile priorities in 1948 and 1949, it elected not to change them. Fortunately, Convair decided to use its own funds to continue the MX-774 project under imaginative Karel Bossart. Bossart's team persevered with their innovative experiments involving swiveling engines, internal fuel storage and tank design, and various means of separating the nose-cone warhead as a solution to the formidable reentry problem. All would prove important in designing the Atlas ICBM in the early 1950s. Meanwhile, in the late 1940s the outlook for the long-range guided missile project appeared bleak.

Ballistic Missiles Receive New Life

The first serious signs of a change in attitude toward research and development in general and guided missiles in particular appeared in 1949. Faced with growing criticism, General Vandenberg requested the Scientific Advisory Board to examine the state of Air Force research and development. It appointed a special committee chaired by widely respected Louis N. Ridenour. Throughout the summer of 1949 he and his committee examined research and development programs, then on 21 September submitted a highly critical report. The committee determined that "existing organizations, personnel policies, and budgetary practices do not allow the Air Force to secure the full and effective use of the scientific and technical resources of the nation." Its major recommendations included ensuring better assignment and promotion opportunities for technical officers and reorienting budget priorities because "if war is not imminent, then the Air Force of the future is far more important than the force-in-being and should, if necessary, be supported at its expense." The Ridenour Report is best remembered, however, for its organizational recommendations: the creation of a deputy chief of staff for research and development on the Air Staff, and a new major Air Force command for research and development.[18] This was von Kármán's wish, too.

Because of expected opposition within the Air Force to a "civilian" report that called for radical change, sympathetic officers like Major General Donald L. Putt, the Director of Research and Development in the office of the Deputy Chief of Staff for Materiel, helped create a parallel, senior-level military group that would undertake a review similar to the Ridenour study and thereby promote broader acceptance for its recommendations. Their efforts produced the Anderson Committee, named for its chairman, Air University's General O. A. Anderson, which conducted extensive interviews throughout the Air Force before issuing its report on 18 November 1949. The Anderson Report strongly supported the Ridenour Committee's findings, and used the effective argument that failing to implement the recommendations might easily lead the Army and Navy to "take over responsibilities abdicated by the USAF."

The powerful arguments for change convinced General Vandenberg to promptly implement the organizational recommendations. On 23 January 1950, the Air Force created the Office of the Deputy Chief of Staff, Development, and the Air Research and Development Command (ARDC) with headquarters at the Sun Building in Baltimore, Maryland. Yet it would take the "personal salesmanship" of Lieutenant General James H. ("Jimmy") Doolittle, acting as special assistant to General Vandenberg a year later in the spring of 1951, to end Air Materiel Command's foot-dragging. In late March General Vandenberg ordered the immediate transfer of AMC's Engineering Division and other designated responsibilities and functions to the new command, and reassignment of ARDC directly to Air Force headquarters rather than AMC. If the new arrangement divided responsibility for weapons acquisition between the two commands, it nevertheless served to highlight the importance of the research and development function in contrast to the heretofore production-oriented Air Materiel Command. Significantly, the Air Staff assigned the guided missiles program to the new command.[19]

While the Air Force made organizational changes in the early 1950s, events on the international scene contributed to major reassessments of the country's defensive posture. News that the Soviet Union had successfully detonated an atomic bomb in

August 1949, communism's triumph in China, and alarming reports of Soviet progress in missile development led to calls for increased military preparedness both in and outside the administration. The outbreak of the Korean War in June 1950 served to heighten the growing sense of national weakness. In the summer of 1950, for example, Under Secretary of the Air Force for Research and Development John A. McCone submitted reports on America's vulnerability to Soviet attack to Secretary of the Air Force Thomas K. Finletter, advocating a "Manhattan-type" program for missiles under the "most capable man who can be drafted." In late August 1950 President Truman responded to calls for action by appointing T. K. Keller, chairman of the Chrysler Corporation, "Director of Guided Missiles." Unfortunately, Keller approached his job as missile "czar" on a part-time basis, and focused largely on cruise-type missiles and the Army's tactical Redstone missile. Convair's low-priority Atlas ballistic missile project received little attention. Nevertheless, the McCone reports contributed to the movement for action on guided missiles.[20]

Other efforts to enhance defense proved more significant. President Truman early in 1950 authorized immediate development of the hydrogen or thermonuclear bomb, while after the outbreak of the Korean War, Congress authorized a 70-group Air Force and nearly doubled the administration's defense budget request from $14.4 to $25 billion. Armed with its new wealth, the Air Force reconsidered Convair's long-range rocket proposal. The company's presentations led to a contract in January 1951 for project MX-1593, whereby Convair would examine both the ballistic approach and the "glide" vehicles which use rocket power to reach the outer atmosphere then use their wings to glide through the atmosphere to their targets. The boost-glide approach reflected continued Air Force interest in the postwar "X"-series of high-altitude rocket-powered aircraft.[21]

Convair's six-month contract to conduct a "study and test program" for two types of missile propulsion hardly represented a ringing endorsement of the ICBM concept. Nevertheless, by late summer 1951 the Convair engineers had selected the ballistic-type rocket largely because it represented a weapon considered unstoppable for the foreseeable future, while they believed the formidable technical problems solvable by the early 1960s. ARDC, which had responsibility for the guided missiles program, agreed that the missile deserved greater support. Convincing Air Force headquarters to award it sufficient funding and project priority, however, proved next to impossible. In the fall of 1951, the Air Staff's Research and Development Directorate rejected ARDC's request for increased funding and directed a slowed-down five-year test program before considering further commitments. Convair continued to lobby Air Force headquarters in late 1951 and early 1952, while ARDC's new commander, General Putt, in a letter to his former office of Research and Development, argued that the ballistic missile project should be approved immediately because of its total "invulnerability to all presently known countermeasures and because of the relative simplicity of the entire weapons system." Putt also warned that the Soviets appeared to be pursuing development of such a weapon. In the spring of 1952 Air Force headquarters referred the ARDC request to the Guided Missiles Committee of the Defense Department's Research and Development Board. The Committee authorized only continued studies and component testing, not the complete Atlas system.[22]

Despite growing evidence to the contrary, skeptics on the Air Staff and in the Defense Department continued to view the intercontinental ballistic missile as a weapon system too complex and likely impossible ever to reach the operational stage. Much of the criticism focused on the old issue of warhead weight. Yet by 1950 the Atomic Energy Commission affirmed the existence of a sufficient number of atomic weapons small enough to be carried in guided missiles. Moreover, President Truman noted that in early 1950 his military service chiefs proceeded with elaborate plans to use the H-bomb on the assumption that the tests he had just authorized would be successful. Test results at Eniwetok in November 1952 proved the feasibility of thermonuclear technology and confirmed their optimism. Based on the test results, ARDC petitioned the Air Staff to reassess the overly restrictive weight and accuracy parameters for the Atlas. In response, a Scientific Advisory Board ad hoc committee chaired by Dr. Clark Millikan reviewed the technical issues. Although the Millikan Committee concluded that anticipated warhead yields called for reducing accuracy and guidance requirements, it saw no need to accelerate the program. Rather it recommended a "step-wise" project that would guarantee "a review of the project at appropriate intervals." A sense of urgency remained absent.[23]

At the end of the Truman presidency strategic bombers and cruise missiles represented the key elements in the nation's offensive arsenal, while the ICBM project moved painfully forward as a cautious, low-funded, phased study and test program that reflected the traditional skepticism of the Air Staff. Given the fate of ballistic missile development over the course of the Truman years, satellite proposals could be expected to garner even less support. Most decision-makers remained blissfully unaware that missile propulsion, guidance, and reentry technologies could be useful for early stages of space exploration, while the response to guided missiles suggests that such knowledge would have had little bearing on satellite developments.

The Air Force Studies Satellites

In postwar America satellite development followed a pattern similar to that of guided missiles. Initial interest faded under budget austerity, and serious government action only began to appear in the early 1950s. Although critics of ICBMs could stress their technological challenges, the German experience of World War II had demonstrated their potential military worth. Satellites, however, not only suffered from association with the "fantastic," but left many unconvinced of their military utility.

Back in 1946, while service jurisdiction over satellites remained undecided following the May meeting of the Aeronautical Board, Rand continued with its remarkable series of satellite studies. In February 1947, the "think tank" produced a second, multi-volume study that expanded on the initial 1946 report.[24] Led by James E. Lipp, head of Project Rand's Missiles Division, it provided detailed specifications for a reconnaissance satellite comprised of a three-stage rocket booster with a gross weight of 82,000 pounds, orbiting at 350 miles, and costing $82 million. Accompanying documents covering a variety of technical subjects from "Flight Mechanics of a Satellite Rocket" to "Communication and Observation Problems of a Satellite" offered contractors guidance for their own design work. The Rand analysis also identified for further development various component areas such as guidance control, orbital control, communications equipment and procedures, and reliable auxiliary power sources. Solar power and miniaturized electronic equipment had yet to be developed.

Two reference papers provided particularly insightful comments on the potential importance of reconnaissance satellites. In one, Yale astronomer Lyman Spitzer, Jr., addressed tactical satellite support of naval operations and the vulnerability of satellites to attack. Most interesting, he proposed satellites as communications relay stations and the application of astronomical telescopic principles to space reconnaissance. His work would contribute later to experiments using long-focal-length panoramic camera systems for surveillance purposes.

James Lipp's "The Time Factor in the Satellite Program" proved especially significant in light of future developments. He described the importance of satellites for scientific research, for military operations, for encouraging development of long-range rockets, and for providing the nation psychological and political benefits. Among his observations, he discussed polar orbits for recurring surveillance, geostationary orbits to compensate for the earth's rotation, and the use of television equipment and special telescopes for transmitting electro-optical images to ground stations. Several of Lipp's perceptive political and psychological assessments would prove hauntingly accurate. Noting that other nations would likely pursue satellite development, he argued that satellite feasibility had been proven, and "the decision to carry through a satellite development is a matter of timing, depending upon whether this country can afford to wait an appreciable length of time before launching definite activity."

Fully aware of the danger in waiting too long to develop satellites, he echoed the warning of David Griggs the year before by declaring: "The psychological effect of a satellite will in less dramatic fashion parallel that of the atom bomb. It will make possible an unspoken threat to every other nation that we can send a guided missile to any spot on earth." The importance of orbiting satellites outweighed the expense, he argued, and a "satellite development program should be put in motion at the earliest time."

Air Force leaders did not share James Lipp's sense of urgency. Six months passed before they requested Air Materiel Command to evaluate the Rand reports. In its late December 1947 evaluation, AMC officers offered a judgment that became commonplace in the years ahead. While they affirmed the technical feasibility of the reconnaissance satellite, they questioned both the high costs and lack of clear military utility. Constrained by "scarce funds and limited component scientific talent," the Air Force should not risk supporting a satellite development program when guided missiles deserved research funding priority. Characteristically, the Air Staff called for more studies on requirements and desired design specifications. In view of the severe missile program cuts in the fiscal year 1947 and fiscal year 1948 budgets, satellite advocates had no reason for optimism. With the only ICBM research program eliminated in July 1947, satellite studies represented the most proponents could expect and the least skeptical Air Staff planners needed to offer. Even so, during the next three years defenders of satellite utility studies needed to work hard to protect the "fantastic" elements from the budget ax.

Even though Air Force leaders proved unwilling to promote satellite development, they were not averse to campaigning for "exclusive rights in space." In January 1948, Chief of Staff Vandenberg became the first service chief to issue a policy statement on space interest when he declared that

> The USAF, as the service dealing primarily with air weapons—especially strategic—has logical responsibility for the satellite. Research and Development will be pursued as rapidly as progress in the guided missiles art justifies and requirements dictate. To this end, the program will be continually studied with a view to keeping an optimum design abreast of the art, to determine the military worth of the vehicle— considering its utility and probable cost—to insure development in critical components, if indicated, and to recommend initiation of the development phases of the project at the proper time."[25]

Although Vandenberg's statement might be faulted for its lack of clarity, clearly, once again, progress in the satellite field would depend on advances in missile technology without recognition that satellite technology might benefit the missile engineers. At the same time, funding would remain a major determinant. With sufficient money available, the Air Force, like the other services, would likely pursue a new mission to increase its share of the budget.

General Vandenberg's declaration appeared at a most opportune time for Air Force interests because Defense Department officials had decided once again to address the organizational squabble over roles and missions. Since September 1947 responsibility for satellite issues in the Defense Department belonged to the Research and Development Board's Committee on Guided Missiles. In December 1947, the latter formed a Technical Evaluation Group to assess satellite feasibility. Two months following Vandenberg's policy dictum, the Committee issued a report that verified the technical feasibility of satellites, but proceeded to assert that "neither the Navy nor the USAF has as yet established either a military or a scientific utility commensurate with the presently expected cost of a satellite vehicle." In hindsight it seems difficult to appreciate the question about military use, especially after the comprehensive, technical Rand report of 1947. At the same time, the satellite represented a "passive" weapon system that seldom elicited the interest of planners worried about supporting conventional strategic weapons. After all, they argued, what could the satellite do that aircraft could not, and at lower cost? Several years of analysis and promotion seemed to be required to establish military satellite utility. Significantly, the Committee recommended continuing with utility studies at Rand and allowing the research agency permission to consult with industry on system and component designs for a reconnaissance satellite.[26]

Satellites Receive New Life

While Air Force leaders might have been disappointed that the committee did not endorse Vandenberg's policy statement, at least the Rand studies continued to receive Defense Department funding. The Navy attempted to join the Air Force as joint sponsor of the Rand project but failed to overcome the opposition of LeMay and other Air Force leaders. By the end of 1948, the Navy had "suspended" its satellite work. The Army, meanwhile, would not reenter the satellite arena until its Redstone rocket team proposed Project Orbiter in 1954. This left the Air Force alone on the satellite field, such as it was. Based on the findings of the Technological Capabilities Committee, Rand proceeded to develop a satellite project with component analyses for "eventual construction and operation of a satellite vehicle." Rand's research and study subcontracts would be subject to AMC's approval and the availability of funds. The key question involved utility. Rand's 1947 study had shown the serious complications associated with designing a recoverable space vehicle. This drew their attention in the years ahead almost entirely to instrumented satellites rather than manned spaceships. The issue for instrumented satellites then became what equipment would be necessary and what military purposes would they serve?

Rand analysts addressed these questions in several 1949 studies, including one entitled "Utility of a Satellite Vehicle for Reconnaissance," and in a study conference in 1949 it sponsored on the military usefulness of satellites. The conference produced an unusually convincing argument for developing a reconnaissance satellite. Noting that technology did not yet permit satellites to operate as destructive weapons, conferees emphasized the passive satellite roles of communications and reconnaissance— especially as political and psychological weapons designed to alter Soviet political behavior. After establishing a list of eight basic satellite characteristics, the analysts assessed the possible functions such a satellite would likely perform. They concluded that as a surveillance instrument it could serve as a major element of political strategy. As a vehicle capable of penetrating the secrets behind the Iron Curtain, it could provide intelligence that might be used in various ways to modify Soviet actions. As the conferees concluded, "no other weapon or technique known today offers comparable promise as an instrument for influencing Soviet political behavior." The study group recommended that Rand impress the Air Force with the surveillance potential of such a satellite.[27]

This Rand study, too, produced few immediate results. As one more study, however, it helped foster growing awareness of reconnaissance satellite capabilities and helped lay the groundwork for passive surveillance applications when Rand commenced its component studies and designs in 1950 after the Air Force received authority to develop booster rockets. Advocates hoped the new concern with Soviet missile advances and the Korean War would generate increased interest in strategic satellites as it seemed to do for missiles.

In late November 1950 Rand recommended the Air Force authorize extension of Rand's research into specific areas of the reconnaissance satellite mission. With Air Force approval, Rand investigators produced two reports in April 1951: "Utility of a Satellite Vehicle for Reconnaissance" and "Inquiring into the Feasibility of Weather Reconnaissance from a Satellite Vehicle." The reconnaissance portion drew the most attention from the Air Force. Based on detailed analysis, it advocated "pioneer reconnaissance," or extensive coverage using television with a resolution of between 40 and 200 feet, in a 1,000-pound payload with a space vehicle weight of 74,000 pounds. With improvements in television technology, the researchers expected to achieve the 40-foot dimension in the near future. They hoped this would permit satellites to conduct all military reconnaissance and finally satisfy the skeptics.

The newly activated Air Research and Development Command enthusiastically supported the Rand findings and authorized Rand to recommend measures needed to begin development work in the reconnaissance program. Eventually this research would lead to the milestone Project Feed Back report of 1954. Rand began in 1951 by subcontracting key subsystems such as orbital sensing and control to North American Aviation, and optical systems, television cameras, and recording equipment to the Radio Corporation of America (RCA). In November 1951 the Air Force contracted with the Atomic Energy Commission to study small nuclear reactors as satellite power sources. By June 1952 the Commission reported encouraging results from preliminary testing, and Rand moved forward with its Feed Back research, which focused on designing and evaluating satellite components.[28]

The findings of the Air Force Beacon Hill Study reflected the state of these efforts at the close of the Truman era. In early 1952, the Air Staff authorized a study group, chaired by Eastman Kodak's Carl Overhage, and consisting of fifteen prominent reconnaissance specialists, including Polaroid's Edwin Land, Louis Ridenour, and Lieutenant Colonel Richard Leghorn, USAFR, considered by Rand one of the few "integrative" thinkers concerned with so-called pre–D-Day reconnaissance. The report called for various improvements to obtain strategic intelligence, and specified refinements to sensors lofted in high-altitude aircraft and balloons, sounding rockets, as well as long-range air-breathing missiles like the Navaho. The group also recognized the need for high-level approval for any overflight of foreign territory, an issue that would dominate political space policy debates during the Eisenhower administration. Although the Study addressed important issues, Rand officials referred to the Beacon Hill Report as "Reconnaissance without Satellites," and considered it a setback for reconnaissance satellites. Not a single Beacon Hill briefing or study addressed either weather reconnaissance or electro-optical reconnaissance, important applications Rand had been considering for years.[29]

On the eve of the Eisenhower administration, satellite advocates had cause for both hope and dismay. The Air Force-sponsored Rand studies had identified a mission, strategic reconnaissance, and produced increased technical justification for developing a military satellite. Feed Back research involving several hundred scientists and engineers seemed well underway by the end of 1952 and promised at long last to set the stage for satellite development. Renewed Air Force interest in the Convair long-range ballistic missile also indicated that large satellite booster rockets might soon be available. Yet the Rand reconnaissance proposal remained a planning project, and the ICBM program moved forward at a very leisurely pace. At the beginning of 1953, it remained to be seen how strongly the new Eisenhower regime would support both satellites and missiles.

Reviewing the course of missile and satellite development in the Truman years, clearly both satellites and missiles fell victim to skepticism about their practical, military use and to economic retrenchment that grew unabated through the 1940s. In a sense, General Arnold's retirement in March 1946 left no one of his stature in either the Air Force or the defense establishment willing to challenge national policy that favored strengthening the forces in being at the expense of future capabilities. Nor did Air Force leaders in the late 1940s question seriously the service's gradualist approach to guided missile development or the priority accorded aerodynamic, cruise missiles rather than long-range ballistic missiles. By the 1950s, however, heightened security concerns and technological change offered the prospect of breaking with the past and accelerating the satellite and missile programs.

Eisenhower Faces the Threat of Surprise Attack

President Dwight D. Eisenhower took office in January 1953 determined to implement a "New Look" defense policy that stressed strategic nuclear striking power at the expense of conventional forces.[30] In order to do this and roll back the Truman administration's Korean War budget from nearly $45 billion to $35 billion, he charged his Defense Department to end waste and duplication throughout the services. Missile and space programs could be expected to absorb their share of Defense Department cutbacks. Indeed, in early 1953 the administration expressed no particular interest in accelerating either program. Yet in the space of just four years, the regime would come to preside over a costly expansion of both military missile and satellite programs and a civilian satellite project that together represented the birth of the American space program. These events have left their mark on the nation ever since.

The rapid growth in space activities under Eisenhower, however, became lost in the wake of the Sputnik launches of October 1957. Critics contended that the administration had allowed the nation to be humiliated and endangered by failing to appreciate the political and psychological importance of being first into space and the demonstration of Soviet leadership in large operational boosters and ICBM technology. The public sensed a directionless program.

In fact, on the road to a national space policy, Eisenhower and his advisors followed a far more sophisticated, secretive, and complex path than many at the time appreciated. Early in the administration, they decided to follow what amounted to a dual space program that focused on launching a civilian scientific satellite to establish the principle of unimpeded overflight in space for the military satellites to follow. The administration had no intention of "racing" the Soviets in space affairs and gambled that the low priority and modestly funded civilian satellite venture could be completed in time for launch of the International Geophysical Year (IGY). Meanwhile, the major defense effort would be devoted to developing ICBMs for the "New Look" doctrine of "massive retaliation" as soon as possible. Given these priorities, the military reconnaissance satellite momentarily represented the odd man out in the space program.

The Eisenhower space program remains an impressive achievement, if not entirely preplanned. Early in the administration, three developments served to propel the nation to the threshold of space. One involved the President's determination to take all possible measures to forestall another "Pearl Harbor." Another concerned the technological "thermonuclear breakthrough" that solved much of the ICBM payload weight dilemma. Finally, several determined government officials risked violating bureaucratic routine to energize the decision-making process. Throughout the period, the Air Force remained divided between reform minded individuals who favored accelerated growth of missile and space programs, and more conservative officials who preferred a cautious, step-by-step approach leading to commitment well into the future. Although the reform group proved victorious, their members had to bypass traditional Air Force bureaucratic structure and procedures to achieve their goals.

Like General Arnold, World War II veteran General Eisenhower could never forget Pearl Harbor. As president, his scientific advisor, James Killian, remarked that Eisenhower remained "haunted"…"throughout his presidency" by the threat of surprise nuclear attack on the United States.[31] To avoid this horror, intelligence data on Soviet military capabilities became essential. Yet, neither news of Soviet advances in long-range bombers like the TU-4, or reports on Soviet long-range missile progress could be verified. At the same time, the development of a thermonuclear device and its testing in both the United States and the Soviet Union raised alarms about a potentially devastating surprise attack. A number of Rand studies in 1952 and 1953 heightened awareness by describing the vulnerability of strategic air bases to attack. The Rand assessments complimented the Central Intelligence Agency's (CIA) national intelligence estimates that forecasted imminent Soviet atomic weapons production and delivery capabilities.[32]

But reports remained confusing or contradictory, and the administration quickly realized that current intelligence methods could not provide meaningful data. Pre-hostilities intelligence information became increasingly essential, and all parties realized that aerial reconnaissance offered the most effective means to solve the dilemma. The near-term answer became the U-2 high-altitude reconnaissance plane, while the long-term solution would prove to be the military reconnaissance satellite. Meanwhile, the best potential satellite boosters also represented the best weapons to prevent surprise nuclear attack.

Trevor Gardner Energizes the Missile Program

While Eisenhower and his advisors worried about intelligence data, Trevor Gardner, the "technologically evangelical" Assistant Secretary of the Air Force for Research and Development, made it his mission in public life to convince the government that the nation must pursue a crash program to develop an operational Air Force ICBM or face nuclear disaster.[33] Ironically, he assumed his office with the mandate to implement the expected economy program in the Defense Department by ending waste and duplication in the Air Force missile program. Assistant Air Force Secretary Gardner was to have a profound influence on the nation's missile program, but he and his allies felt compelled to go outside established Air Force and Defense Department structures to carry out their goals.

In April 1953 Gardner called for review of all Air Force missile programs. He instinctively rebelled against ARDC's cautious approach and the Air Staff's persistent delaying tactics. Their reasoning reflected the dilemma of the self-fulfilling prophecy: missiles represented too costly an investment for an "impossible" system. But no development money meant that the problems would continue unsolved and the missile remain "impossible." Gardner, who had heard reports of the "thermonuclear breakthrough," knew that, now, accuracy and guidance performance requirements could be relaxed and the missile no longer need be considered "impossible."[34]

Fortunately, Gardner found willing allies to accelerate missile development among middle echelon ARDC and Air Staff officers, as well as the Convair group promoting Atlas. At the same time, the Joint Chiefs of Staff, as part of its military posture review for the incoming administration, called for a broad-based reexamination of the entire Defense Department missile picture. Gardner received the assignment to review the country's missile programs based on Secretary of Defense Charles Wilson's drive to eliminate waste and duplication among the services.

At this point Gardner decided to bypass the Air Force bureaucracy and appoint a full-time group of experts on whom he would rely for advice. Late in the fall of 1953 he convened the Strategic Missiles Evaluation Committee (SMEC) under the chairmanship of renowned Princeton Institute for Advanced Study mathematician and activist John von Neumann. This group, which came to be referred to as the von Neumann Committee, comprised an impressive assemblage of scientists and engineers, all of whom had been handpicked by Gardner for their "progressive" views on ICBM requirements as well as their technical brilliance. Trevor Gardner charged von Neumann's committee to determine the measures necessary to accelerate development of the Atlas missile.[35]

While von Neumann committee members deliberated, a Rand Corporation group directed by Bruno W. Augenstein neared completion of a similar study on mounting thermonuclear weapons atop ICBMs. Responding to Air Force direction to investigate aerodynamic systems, Rand analysts had produced a number of reports on missiles in the early 1950s that favored ramjets and boost-glide rockets over ballistic missiles. When nuclear weapons were made smaller, Rand concluded that ICBMs represented the optimum surprise-attack weapon, which heightened the challenge to produce pre-hostilities strategic intelligence. At the same time, an accelerated ICBM program would mean having space boosters available at lower costs. Rand evaluators worked closely with the von Neumann team, and Augenstein briefed von Neumann Committee members personally in December 1953 on his findings. To no one's surprise, the two groups reached similar conclusions in their final reports, which appeared two days apart in early February 1954. These reports would help convince President Eisenhower to convene that spring the Surprise Attack Panel or, as it was soon renamed, the Technological Capabilities Panel, chaired by James Killian.[36]

The von Neumann report confirmed the Rand analysis by calling for a drastic revision of the Atlas ICBM program in light of Soviet missile progress and newly available thermonuclear technology. Referring to the recent Operation Castle tests in the spring of 1953, von Neumann predicted the advent of thermonuclear warheads weighing only 1,500 pounds with a yield of one megaton. This meant that performance criteria for the Atlas could be reduced, making its development more feasible within the state of the art.[37]

Critical of Convair's management practices and design, which envisioned an enormous five-engine rocket to boost the earlier, heavier warhead, the committee recommended a thorough study of various alternate design approaches and the establishment of a new development-management agency in the Air Force authorized to provide overall technical direction. Committee members considered this agency more important than all the technical guidance, warhead weight, and reentry problems yet to be solved. Finally, panel members urgently recommended the project be given high priority and substantial funding. The von Neumann report would stimulate the revision necessary to develop the large boosters required for military reconnaissance satellites.[38]

Armed with the findings of the Rand and von Neumann Committee studies, Gardner set off to win support throughout the Air Force hierarchy to expedite an expanded ballistic missile development effort. After gaining approval from Chief of Staff General Nathan Twining and Secretary of the Air Force Harold Talbott, Gardner could successfully counter any disapproval from key air staff agencies and Air Research and Development Command. The traditional Air Force bureaucracy did not favor this civilian-sponsored initiative that proposed creating a separate development-management agency that would bypass established administrative channels. In the end, the Air Staff supported the Gardner-engineered initiative, perhaps because disapproval might result in appointment of a new missile "czar" completely outside the Air Force framework. If not all that the Gardner group desired, the results nevertheless proved "revolutionary." In April 1954 the Air Staff proceeded to create a new Air Force headquarters position, an Assistant Chief of Staff for Guided Missiles, with responsibility for coordinating all Air Force guided missile activities. The following month, Air Force leaders took a more significant step by directing ARDC to form a West Coast project office at Inglewood, California. Organized as the Western Development Division (WDD), the latter represented the central von Neumann committee recommendation, and Gardner insured that the new organization's chief would be his ally, Brigadier General Bernard Schriever. Shortly after the Western Development Division began functioning in August, General Schriever arranged for the Air Force to contract with the Ramo-Wooldridge Corporation as full-time technical consultant to his command. Schriever proved to be a splendid choice to head a crash ICBM program. A young disciple of Hap Arnold, whom he considered "one of the most farsighted persons" he had ever known, he had joined Trevor Gardner's reform group in early 1953 while serving on the Air Staff as Assistant for Development Planning in the office of Deputy Chief of Staff for Development. He used his intelligence, patience, and superb negotiating skills with military, government and private industry leaders to become an effective advocate for missile and space systems causes.

In order to produce an operational missile by the end of the decade, Schriever's command adopted a number of managerial innovations that would become common practice for the Air Force in future years. Help again came from the von Neumann committee, which had been reconstituted in April 1954 as the Atlas Scientific Advisory Committee. Together with Ramo-Wooldridge, the committee convinced Convair and the Air Force to design a smaller missile capable of carrying the lighter, powerful hydrogen warhead. Given the time constraints, the planners chose to develop a "light-weight" three-engine rocket with a thin metal air frame skin housing the liquid-fuel and oxidizer tanks made rigid through overpressure. The crash program called for special management techniques, too. In the summer of 1954 the ICBM committee recommended that the Western Development Division award alternate subsystem contracts, whereby each Atlas component would be "backed up" by an alternate relying on different technology. This more costly parallel development approach meshed effectively with the new "concurrent" procedures pioneered by Schriever and his staff. Under concurrency, all measures necessary to construct and deploy the system would be completed simultaneously. Still skeptical of Convair's capabilities, however, Air Force officials applied the parallel development concept on a larger scale by producing at the same time a second, more-sophisticated "back-up" ICBM, the Titan. Designers configured the new Titan as a two-stage liquid-propellant missile, with a more advanced guidance system, and rigid frame to permit underground deployment. Parallel development allowed Atlas and Titan program managers to replace subsystems in case of failure or technological breakthrough, while advanced designs could be pursued without risk to the overall ICBM program. It served as an effective risk mitigation approach that proved its worth when the Air Force launched both Atlas and Titan missiles successfully by the end of the decade.[39]

General Schriever could hardly have expected such future success when he surveyed the state of his command in the spring of 1954. Indeed, he faced a major battle within the Air Force to retain control of his project. Despite his relatively independent status under ARDC with responsibility for system planning, technical direction, and budgeting, the Air Materiel Command continued to control the major funding areas of system production and procurement. To do the job assigned, General Schriever believed he needed authority over all aspects of missile acquisition, from design, research and development, through production. The Air Staff, however, refused to compromise on this issue, and AMC maintained its production prerogative by establishing a Special Aircraft Project Office at Western Development Division to handle ICBM procurement. According to General Schriever, initial friction soon gave way to a reasonable "partnership" arrangement after the general established good rapport with the AMC officers. This far from optimum division of system management responsibilities would continue until the creation of Air Force Systems Command during the organizational reform of 1961.[40]

Managerial problems with the Air Materiel Command proved only the tip of the iceberg. Even though the Secretary of Defense had declared Atlas of "critical importance" in early 1955, the bureaucratic labyrinth at the Air Staff and the Defense Department continued to cause bottlenecks and delays because of the multiple program review levels. Once again Trevor Gardner—encouraged by General Schriever's active support—decided to bypass the Air Force bureaucracy by going directly to Congress. Meetings with Senators Clinton Anderson and Henry M. Jackson, the two most influential members of the Joint Committee on Atomic Energy, and congressional visits to Schriever's suburban Los Angeles headquarters, convinced the congressmen to support streamlined management procedures to eliminate the bureaucratic obstacles. At the same time, additional reports of new Soviet long-range bombers and missile tests picked up by radars in Turkey raised fears that the United States might be falling behind in the ICBM race.[41]

The congressmen wrote President Eisenhower in late June 1955 about their concerns and recommended immediate action on the Atlas program to avoid funding delays, overcome interference from major Air Force commands, and bypass the multiple review levels. By fall the President had designated the Atlas ICBM the "highest national priority" weapon system. Still, procedures remained unchanged, prompting Trevor Gardner again to seize the initiative by directing Hyde Gillette, Air Force Deputy for Budget and Program Management, to form a committee to devise new, more effective procedures for the missile program. In October 1955 the Gillette Committee's recommendations led to the establishment of a ballistic missiles committee at both Air Staff and Defense Department levels to function as the sole reviewing authorities for Western Development Division programs. Gone were the various separate offices that Schriever had to consult individually. Now he submitted a yearly development plan to a single committee, made up of representatives from the offices concerned with the ICBM program. Although not entirely able to overcome all Air Staff skeptics and AMC opponents, the Gillette procedures removed many bureaucratic bottlenecks, and the ICBM program moved ahead rapidly.[42]

By 1955 the momentous procedural and organizational decisions for ICBM development proved to have a major impact on the military space "program" as well. Gardner and Schriever, given their focus on missile requirements, could not be expected to devote their energies to lower-priority satellite activities. In fact, they viewed the military satellite space program as a competitor for personnel, funds, and contractors. Nevertheless, the relationship between satellites and missiles had become better understood as rockets with sufficient thrust soon would be available to launch the heavier satellites preferred by the Air Force. If the Western Development Division were to gain responsibility for the Air Forces' advanced reconnaissance satellite project, advocates hoped that the Gillette procedures would benefit satellite development as they promised to do for the ICBM.[43]

The Air Force Commits to the First Military Satellite

While Secretary Gardner and General Schriever worked on missile issues in the spring of 1954, the military satellite project also cleared major hurdles. Now, with ICBMs representing a practical option, Rand's studies on satellite systems received new life, as the Eisenhower administration sought solutions to the intelligence dilemma of providing accurate data on Soviet offensive capabilities.

Rand studies had proceeded on the assumption that the Atlas ICBM would provide the space booster required to launch a reconnaissance satellite. Rand also assumed that spaced-based sensing systems offered the best means of quickly relaying important intelligence data to ground stations. By the spring of 1953, Rand satellite studies of the previous two years—now referred to as Project Feed Back—began to draw a wider audience in view of new high-level interest in Soviet missile advances. Promising results from Atomic Energy Commission tests on nuclear power for satellites encouraged the Air Force in May to direct further study of the matter and to have ARDC begin "active direction" of the reconnaissance satellite program advanced by Feed Back. In the fall of 1953 Rand officials discussed satellite issues with a number of important government officials and military officers and, based on realistic near-term operational feasibility, recommended the Air Force issue a design contract within a year leading to full system development. By year's end ARDC had published a management "Satellite Component Study," and assigned it weapon system [WS] number 117L. Project Feed Back would place the satellite on the sure path of development.[44]

Authored by analysts James E. Lipp and Robert M. Salter, Jr., Rand's Feed Back report appeared in March 1954 in the midst of deliberations about the optimum ICBM organization.[45] It drew together findings from the previous two years' intense study of reconnaissance satellites. The "milestone" Feed Back study proposed an electro-optical reconnaissance satellite with a television-type imaging system projected to achieve a resolution of 144 feet from an altitude of 300 miles. The report readily admitted that this resolution could not deliver the accurate intelligence required and encouraged the Air Force to foster a competition among industrial firms to develop a higher resolution system based on long-focal-length, panoramic camera technology. It also discussed newly analyzed operational issues dealing with subsystems, cost projections, likely international political reactions, and a host of additional engineering requirements. With this "blueprint" in hand, Rand encouraged the Air Force to proceed on a full-scale basis with this "vital strategic interest" by implementing a seven-year development program budgeted between $165 and $330 million. In the next few years, while Air Force scientists and project officers worked to develop techniques for safe reentry of space payloads through the atmosphere, Rand engineers would stress two types of non-recoverable reconnaissance systems: one relied on television technology and "immediate" data transmission to ground stations; the other used tape storage of sensed data that would be transmitted at a later time.

After some initial hesitation, the Air Force agreed to pursue the Feed Back recommendations further, and in May 1954 directed Air Research and Development Command to review the military applications of the Rand satellite concept. Meanwhile, Rand and ARDC met with various Air Force, Defense Department, and industry leaders to "sell" the Rand proposal. At the same time ARDC proceeded with analyses of intelligence processing options, solar-electrical energy converters, auxiliary power sources, and guidance and control mechanisms. Following approval from the Defense Department's Coordinating Committee on Guided Missiles, the command issued a system requirement on 27 November 1954. With this decision, the Air Force in late 1954 clearly signaled its intention to develop an operational reconnaissance satellite system.[46]

The command followed up in March 1955 with a formal General Operational Requirement.[47] Now referring to the WS-117L reconnaissance satellite as the Advanced Reconnaissance System (ARS), the requirement prescribed continuous surveillance of "preselected" areas, especially aircraft runways and missile launching sites. In contrast to the Rand study's target resolution parameters, specifications now called for providing visual coverage of objects no larger than 20 feet on a side, and specified electronic and weather coverage capability, too. With an eye to continued technological advances, the scheduled operational date of 1965 seemed achievable. By August, ARDC had named as system project officer Colonel William G. King, Jr. In November he awarded $500,000 contracts to three firms—the Radio Corporation of America, Lockheed, and Glenn L. Martin—for a one-year satellite design competition under the code name "Pied Piper."[48]

Although by late 1955 space advocates might rejoice that at long last a military satellite program seemed underway, a number of long-standing, troublesome issues remained to be solved. One of the most important involved potential competition between satellites and missiles for scarce resources. Trevor Gardner resolved to insure that the Atlas ICBM schedule would not be compromised by satellite requirements. Back in November 1954, he had taken his worries to von Neumann's ICBM Scientific Advisory Group. The members asked General Schriever to assess the challenge of developing simultaneously satellites, Intermediate-Range Ballistic Missiles (IRBMs), and the "high priority" ICBM programs. Meanwhile, in January 1955 von Neumann's group, in an attempt to ease pressure on the ICBM program and accelerate satellite development, recommended that satellite work be confined to the spacecraft and its likely components rather than include booster elements, too. ARDC commander General Thomas S. Power agreed that satellite development not involve booster integration for the present. Nevertheless, the ICBM Scientific Advisory Group continued to worry about potential satellite competition with the ICBM schedule and addressed the issue again in June 1955. It cautioned that conflict could not be avoided because of satellite dependence on components developed through the ICBM program.[49]

General Schriever's analysis of the missile program convinced him that only centralized management of all military satellite and missile programs could minimize the problem of competition for scarce resources and avoid schedule delays. During his investigation, Schriever relied on the advice of Simon Ramo of Ramo-Wooldridge, the Western Development Division's technical consulting firm. Dr. Ramo met with von Neumann's Scientific Advisory Group and the Air Staff's Lieutenant General Donald Putt, Deputy Chief of Staff for Development to warn that the satellite program competed with the ICBM program for the same personnel and launch capabilities. Ramo strongly advised relocating management of the satellite program from the Wright Development Center at Wright-Patterson Air Force Base in Ohio to Schriever's Inglewood, California, complex.[50]

By the fall of 1955, with work on the satellite underway at the Western Development Division, General Power agreed to the management transfer, although not until February 1956 would the actual move begin from Wright-Patterson Air Force Base to the suburban Los Angeles facility. That the reassignment took nearly a year and a half to complete from the time Trevor Gardner raised the alarm suggests the reluctance of those concerned. General Schriever, in particular, would have preferred to focus on the ICBM program and not deal with IRBM and satellite competitors, while ARDC understandably preferred to keep the development program at its primary research facility in Ohio. In the long run, centralized management under the Western Development Division seemed the best alternative. At least Schriever's team could provide better management of risk and program scheduling with its "concurrency" approach to systems development and streamlined administrative procedures with higher headquarters. At the same time, satellite development could be expected to benefit from transfer out from under a research facility largely devoted to aeronautics to a "space"-oriented command located in the heart of the missile and satellite environment.

During the course of their deliberations on the ICBM program, Air Force planners and consultants had ample justification for concern over the attention their program would receive from the administration. Not only did they face the challenge of managing their burgeoning satellite and missile programs with limited resources, a new competitor for funds and development priority emerged in the summer of 1955. For over a year, the government had been considering sponsoring a scientific earth satellite to be launched during the International Geophysical Year (IGY), which was scheduled to extend from July 1957 to December 1958. Trevor Gardner and his fellow Air Force advisors kept a wary eye on these discussions of proceeding with an additional satellite program, which certainly contributed to their own concerns about satellite-missile relationships. In July 1955, once the administration formally agreed to sponsor a civilian satellite development program, this potentially high-profile competitor threatened to interfere with Air Force efforts to focus sufficient Defense Department attention and funding on both ICBMs and the Advanced Reconnaissance System.

The Administration Commits to the First Civilian Satellite

The decision to support a civilian satellite program reflected a genuine interest in promoting science, strong advocacy from certain elements in the scientific community, and the administration's national security concerns—especially the challenge of eliminating the possibility of a surprise nuclear attack on the nation. For most of Eisenhower's advisors, the civilian scientific satellite never represented solely an altruistic, international scientific venture.

By early 1954 Eisenhower expressed grave concerns about inadequate intelligence to the National Security Council (NSC). The President also followed with great interest the work on the country's strategic missile program undertaken by Trevor Gardner and the civilian scientists serving on the Scientific Advisory Committee in the While House Office of Defense Mobilization. In late March he called to the White House a number of prominent scientists, including committee chairman Lee A. DuBridge, president of Cal Tech, and requested their help on the problem of surprise attack. They responded in August by establishing a Technological Capabilities Panel (TCP), chaired by James Killian, president of the Massachusetts Institute of Technology (MIT). After five months of deliberations, in February 1955 it issued a momentous report titled, "Meeting the Threat of Surprise Attack."[51]

The Killian Panel projected changes in the relative posture of American and Soviet strategic forces. Confirming the vital need for pre-hostilities strategic intelligence on Soviet military capabilities, the panel supported development of the Lockheed U-2 high-altitude reconnaissance plane, the solid-fueled Polaris sea-launched ballistic missile, and more rapid construction of the Distant Early Warning (DEW) line across northern Canada. The report advocated an accelerated ICBM program, and rapid development of IRBMs as a stopgap security measure until the ICBM force became operational. The President in September 1955 endorsed their findings together with those of von Neumann's Strategic Missiles Evaluation Committee and assigned to the Atlas and Thor and Jupiter programs the highest possible priority. To the initial consternation of Gardner and Schriever, in December President Eisenhower declared the IRBM programs to be coequal with the ICBM.[52]

As for satellites, the Killian report responded to the growing satellite interest in the scientific community and the panel's strategic intelligence concerns by recommending immediate development of a small scientific satellite that would establish the precedent of "freedom of space" for military satellites to follow. Although government officials and Rand analysts had worried about satellite overflight in international law earlier, here, for the first time, advocates identified the requirement for a "civilian" satellite to establish the overflight precedent. Focused on Project Aquatone, the U-2 project that promised immediate results, the military satellite program received little interest or support from Killian and his experts. At that time, he considered the Air Force's reconnaissance satellite a "peripheral project." This attitude from one so influential helps explain the less than enthusiastic administration support of the Air Force's Advanced Reconnaissance Satellite in the two years preceding Sputnik. Despite the growing need for strategic intelligence and awareness that the U-2 represented a temporary solution, Killian declined to actively support the military satellite until after the launch of the first Sputnik. He believed an American scientific satellite had to precede the launch of a military vehicle to provide the overflight precedent for military satellites to operate with minimum international criticism.[53]

That spring of 1955 Eisenhower and his advisors acted further on the Panel's satellite and overflight recommendations by outlining a policy for outer space analogous to that of the high seas, whereby flight in space would be available to all without legal restriction. At the same time, the President attempted to redefine the legal regime already established for airspace when, on 21 July 1955 at the Geneva summit conference, he called on the Soviet leaders to join him in providing "facilities for aerial photography to the other country" and mutually monitored reconnaissance overflights. Although the Soviets rejected his offer, he continued to advocate his "Open Skies" doctrine, while moving forward to assure the nation of sufficient intelligence to avert surprise attack. The emerging Eisenhower policy on space seemed to accord nicely with the scientists' proposal for launching an experimental scientific satellite during the International Geophysical Year.[54]

Interest in experimental satellite research originated from several sources. Although the satellite studies done by the Navy in 1945 and Rand in 1946 focused more on scientific than military characteristics, only in 1948 did the larger scientific community become aware of this research when portions of the Rand analyses appeared in the so-called "Grimminger Report" in the October issue of the *Journal of Applied Physics.* The report generated widespread interest among various small national rocket societies as well as upper atmosphere research scientists who increasingly worried about continuing their work once the wartime stock of captured V-2 rockets had been used. Another interested group involved space enthusiasts who found a wider audience at proceedings like the Second Congress of the International Astronautical Federation and the First Symposium on Space Flight held in the fall of 1951. By the early 1950s a number of activists offered specific satellite proposals, too. Dr. Fred Singer, University of Maryland physicist, and members of the British Interplanetary Society, for example, proposed the launching of a "Mouse" (Minimum Orbital Unmanned Satellite, Earth) which attracted attention on both sides of the Atlantic. More important for subsequent developments, Wernher von Braun, the chief of the Army's Guided Missile Development Division at the Redstone Arsenal in Huntsville, Alabama, had mounted a campaign in and outside military circles for an experimental satellite using the Army's Redstone rocket as a first-stage booster. He also offered visions of a manned future space station in a series of articles in *Collier's* magazine, which attracted considerable attention. Eventually, interest in von Braun's proposals led the military services to offer their own satellite projects for the International Geophysical Year.[55]

Growing support for launching a scientific satellite led a group of prominent scientists in 1954 to discuss the idea with leading government and congressional leaders. In August of that year, Congress authorized IGY participation by the United States and proposed $10 million to support the American satellite entry. In early 1955, the various scientific satellite proposals arrived at the office of the Assistant Secretary of Defense for Research and Development, Donald Quarles. These included a formal proposal from the United States National Committee for the IGY, appointed by the National Academy of Sciences, along with the Air Force's WS-117L program and the Army's Project Orbiter. Quarles referred all the IGY proposals to his Advisory Group on Special Capabilities for review and recommendations. In early May, the director of the U-2 project, Richard M. Bissell, Jr., met with the director of the Central Intelligence Agency, Alan Dulles, and the director of the National Science Foundation, Alan Waterman, to decide how the scientific satellite initiative could best meet the Killian Report's "freedom of space" objective. Acting on their advice, on 20 May Quarles submitted a draft space policy to the National Security Council for review. The decisions reached at the NSC's 26 May 1955 meeting, issued in the form of NSC Directive 5520, rank among the most important of the early Eisenhower presidency for space policy. Affirming Quarles' recommendations, the NSC declared that an IGY satellite must not interfere with the "high priority" ICBM and IRBM programs then underway, and that the satellite launched for "peaceful purposes" should help establish the "freedom of space" principle and the corresponding right of unimpeded overflight in outer space. The NSC also agreed that the scientific satellites would serve as precursors of later, military satellites. Finally, the NSC showed itself fully aware of the prestige and psychological benefits likely to accrue to the first nation to launch a satellite into orbit. As Nelson Rockefeller, Eisenhower's Special Assistant for Foreign Affairs, noted in a forceful appendix to the directive, "The stake of prestige that is involved makes this a race that we cannot afford to lose."[56]

During the post-Sputnik hysteria, in late 1957, the administration publicly attempted to distinguish between its so-called peaceful satellite project and that of the military-oriented Soviet counterpart by emphasizing the separation of the civilian scientific satellite project from the country's long-range missile program. Yet, the deliberations of the National Security Council clearly show that separation of the satellite and missile program hardly occurred as part of an internationalist, altruistic policy of promoting "pure science." The administration's declaratory policy of "peaceful purposes" purposely obscured its real intentions. When the President publicly announced on 29 July America's participation in the International Geophysical Year effort, he pledged that this scientific venture would remain unconnected to the current military missile development programs. The National Science Foundation would direct the project, with the National Committee for the IGY responsible for the satellite. The Defense Department would furnish the rocket booster and provide logistic and technical support. He gave no hint of the underlying purpose of his emerging space policy for the civilian and military satellite projects then underway. The civilian satellite would serve as a stalking horse to establish the precedent of "freedom of space" for the military satellite, but the administration maintained great secrecy on the latter so that attention would remain focused on the former.[57]

In late 1954, Congressional approval of funding for the IGY project had opened the way for von Braun and others to submit competing satellite proposals. The Defense Department favored a joint service-IGY effort to avoid interservice rivalry, but only the Army's Ordnance Department and the Office of Naval Research could agree to cooperate. Led by the Redstone team of Major General John B. Medaris and von Braun, Project Orbiter envisioned launching a small inert satellite "slug" using a Jupiter IRBM booster with three Loki upper-stage solid-fueled rockets. While the Army developed the booster , the Navy assumed responsibility for satellite, tracking facilities, and data analysis. The Project Orbiter team had vigorously lobbied the Defense Department for their project since early 1955. The Naval Research Laboratory, on the other hand, countered in late spring with Project Vanguard, which specified adapting a Viking sounding rocket as booster for three new upper stages. The Vanguard project included an impressive Minitrack radio-tracking and telemetry system, which would support the scientific focus of the IGY proposal.[58]

The Air Force initially had declined to participate in the IGY competition because it might conflict with its long-range goal of developing heavier, military reconnaissance satellites. However, after Quarles had directed all three services to offer proposals, the Air Force in July submitted its own "World Series" project—an Atlas C booster and a modified Aerobee-Hi space probe rocket. Faced with the dilemma of selecting from among the three rival entries, Quarles appointed an Advisory Group on Special Capabilities in May 1955 under the chairmanship of Homer J. Stewart of JPL. Following a contentious assessment process, the Stewart Committee ultimately selected the Navy's Vanguard proposal, and the Secretary of Defense confirmed this decision on 9 September 1955, just over a month after the White House publicly committed the nation to launching a satellite during the IGY. Although the Air Force entry showed great promise, committee members realized use of the Atlas as booster could conflict with the ICBM schedule. The Army cried foul, claiming that the Vanguard selection represented a major development effort, while von Braun asserted that his Redstone rocket team could launch an 18-pound payload as soon as January 1957, and well under the Vanguard's budget.[59]

Critics of the Vanguard decision argued that the Committee's concern that its choice not "materially delay other major Defense programs" tilted the balance against Project Orbiter.[60] Perhaps so, but the selection issue proved more complex. While the criterion served to rule out the Air Force Atlas ICBM booster, Orbiter's Redstone did not cause similar consternation. Chrysler Corporation was about to begin production of the missile, and the Huntsville group did not receive the Jupiter IRBM assignment until well after the end of the IGY satellite selection process. Von Braun's Redstone team was available. In fact, the non-interference criterion ranked only sixth among nine criteria used by the Stewart Committee. The Committee clearly questioned Orbiter's reliability and its limited potential for future scientific space exploration, and found attractive Vanguard's "maximum scientific utility" and superior tracking system and satellite instrumentation. Rather than merely a question of selecting a "non-military" Navy system over a "military" Army one, the choice reflected efforts to combine the best scientific applications with a launch system that could not avoid military connections in any case.[61]

By the fall of 1955 the administration was supporting two satellite programs, WS-117L and the civilian Project Vanguard. Nevertheless, the Air Force's experience following the decision suggests that the door remained open for a possible military-oriented alternative regardless of the desire to maintain a civilian focus. Problems experienced by Project Vanguard most likely account for the Defense Department's extended review of an alternative Air Force proposal for a scientific satellite.

The Air Force Reconsiders a "Civilian" Satellite

For well over a year following the Vanguard award, the Air Force became involved with alternative "civilian" satellite proposals while challenged to develop an operational military satellite. The process reveals ambivalent attitudes about accepting a civilian project that threatened to compete for resources not only with the military satellite but the ICBM program as well. Throughout the course of events, Air Force planners seem to have operated without full knowledge of the ground rules, that had effectively eliminated a "military" project from the start, and the degree of seriousness the Defense Department attached to their proposals.[62]

From the start of their involvement in the IGY competition, Air Force planners worried that any Air Force scientific satellite that used an Atlas could interfere directly with the Atlas weapon ICBM schedule, while the competition's ground rules did not seem to exclude military criteria. Although ARDC might have thought the subject closed when the Stewart Committee selected Vanguard in August 1955, on 31 August Air Force headquarters directed ARDC to prepare another proposal that would integrate a scientific satellite with the WS-117L military reconnaissance satellite program. Then, on 14 October, with the new proposal still unfinished, ARDC halted that work, explaining it lacked sufficient funding and, in any case, the decision had been made in Vanguard's favor. In another reversal, the command resumed planning for a scientific satellite on 1 November, and the Western Development Division on 14 January 1956 submitted a scientific satellite variant of WS-117L, a 3500-pound satellite made from ARS components that could be launched by August 1958 atop an Atlas C at a cost of $95.5 million. General Schriever's proposal also specified a number of scientific experiments dealing with atmospheric density, solar radiation, and the upper atmosphere-near space effects on communications. The Schriever proposal reflected a consistent Air Force view that any satellite should serve a specific scientific or military purpose rather than merely serve as a public demonstration of the capability of launching a satellite into orbit.[63]

Most importantly, General Schriever advised that the scientific satellite could be developed without "significant compromise" to the military satellite program—

provided the operation be accorded sufficient funding, personnel, and resolve. He also established criteria to preclude interference with the ICBM program, but warned that any small delay in the Atlas schedule might mean a satellite launch beyond the IGY "window of opportunity." The general's caveats notwithstanding, ARDC forwarded the proposal to the Air Staff in January 1956. After a number of briefings on the subject, the proposal languished at Air Force headquarters in Washington throughout the remainder of 1956. Then, in early 1957, the Air Force notified General Schriever that the Defense Department had decided not to submit it to the Stewart Committee, which continued to oversee Project Vanguard. Evidently, by 1957 the Committee had decided to forego the luxury of a "back-up" satellite for Vanguard.[64]

General Schriever always maintained that his command could have handled development of both missiles and satellites. Yet the Western Development Division, which was redesignated the Air Force Ballistic Missile Division on 1 June 1957, would have required considerable additional resources from a parsimonious Defense Department to support three major long-range missile programs as well as two satellite projects. As for the civilian satellite planning effort during late 1955 and throughout 1956, Schriever and his staff affirmed that it did not significantly interfere with planning or funding for the military satellite. Once again, the so-called "non-military" criterion for an IGY satellite did not seem important enough to rule out lengthy consideration of the most "military" of satellite proposals. If concern for possible delays in the sensitive Atlas ICBM program again proved decisive, the story of the Air Force scientific satellite proposal also suggests that Air Force leaders felt compelled to remain involved in a potential program of questionable value. In view of the Air Force's aspirations to dominate the space mission, it could not remain uninvolved.[65]

Retrenchment on the Eve of Sputnik

While some Air Force planners labored on proposals for a civilian variant of the WS-117L reconnaissance satellite, work continued on the technical requirements for the military satellite. Following the program's transfer to the Western Development Division in early 1956, General Schriever appointed as project officer Colonel Otto J. Glasser, who directed preparation of a full-scale system development plan based on the winning design entry submitted from the three "Pied Piper" firms. By April the plan had been completed and approved by General Schriever and ARDC commander General Power. In June 1956, the Air Force selected the design from Lockheed's Missile Systems Division and awarded the firm a formal contract in October. The Lockheed choice surprised no one because the company had hired the majority of existing space research specialists, including former Rand analyst James Lipp.[66] Relying on the Atlas C booster, Lockheed proposed building a huge second-stage booster satellite, initially termed Hustler, that could provide high pointing accuracy from its stabilized orbit position. Eventually, this booster satellite would become the workhorse "Agena" that, together with its Atlas booster, would launch the heavier Air Force payloads. Lockheed's winning payload entry also included a unique feature proposed by engineer Joseph J. Knopow for an infrared radiometer and telescope capable of detecting missiles and aircraft from their hot exhaust "signatures." Offering the potential for "real time" data, the infrared system element of the Advanced Reconnaissance System would emerge as a separate missile launch detection alarm system (MIDAS) satellite project designed to provide early warning of missile launches. The Air Force plan predicted an initial orbit date of May 1959, with the complete system, including ground installations, expected to be operational in the summer of 1963.[67]

The advance of technology in 1956 and 1957 served to emphasize the Lockheed proposal's merits. Research on wider and slower reentry vehicles with ablative surfaces offered a solution to the old problem of aerodynamic heating when objects reenter the atmosphere and fostered renewed interest in retrievable reconnaissance systems. Recoverable film containers held the prospect of avoiding image degradation that might occur through TV sensing and transmission through the atmosphere. At the same time, current experiments with panoramic cameras with long-focal-length lenses offered both broad-area coverage and high ground resolution.[68]

Research in new technology promised to be costly, and funding for WS-117L had been a sensitive subject from the start. From General Schriever's perspective, the Advanced Reconnaissance System suffered from guilt by association with the troubled Vanguard project. Although neither satellite program received adequate support, Vanguard's priority status brought it the lion's share of satellite monies. When the civilian satellite experienced management and budgeting problems, the military reconnaissance satellite also encountered difficulties receiving the attention and funding its supporters believed it deserved.

The scientists themselves were largely to blame for Vanguard's problems. Their interest in maximizing the scientific output not only led to additional costly instrumented experiments which Eisenhower criticized as "gold plating," but served to make secondary the essential requirement to establish basic vehicle technology before adding sophisticated payload experiments. The scientists also wanted to increase the number of test launches from six to twelve, which drove up costs and contributed to schedule delays that poor management practices only exacerbated.[69]

In fact, Vanguard had been underfunded from the beginning. Even before the Stewart Committee selected Vanguard, Secretary Quarles indicated his skepticism about the initial Vanguard budget figure of $10 million by raising the satellite budget to $20 million. Even so, the Vanguard budget rose continually from an initial $28.8 million in September 1955 to $110 million by May 1957. Had the Vanguard team payed attention to the early Rand studies, they might have developed a more realistic budget. Nine years earlier, James Lipp proposed a similar satellite at a cost

of between $50 million and $150 million, depending on the payload. More attention
paid to the Rand reports would also have alerted the scientists to the importance of
international prestige associated with the country first into space. Instead, the
scientists focused on costly experiments that played havoc with the development
schedule. As for the experiments, Vanguard scientists did not clarify for President
Eisenhower the important contributions their satellite work offered for ICBM
development. Had the scientists done so, they might have been able to convince
the administration to elevate satellite priorities in the name of missile progress.
Ike, after all, listened to "his" scientists. A higher priority for Project Vanguard
might correspondingly have benefited the military satellite and ICBM programs.[70]

Without strong intervention from the scientists, Project Vanguard's managers
proved unable to stifle the concerns of an administration that was becoming
increasingly exasperated with the spiraling costs. What made matters worse, the
administration faced a larger problem brought on by the enormous costs and
excessive duplication associated with building two ICBMs and three IRBMs simultaneously. With all five missiles in development in fiscal years 1957 and 1958, the budget for guided missiles reached more than $1.3 billion, an enormous increase from
$515 million in fiscal year 1956, $161 million in fiscal year 1955, and only $14 million
in fiscal year 1954. In 1957 Eisenhower feared that the spiraling missile costs would
force defense spending beyond his fiscal year 1958 ceiling of $38 billion and directed
a budget review of all programs. By August Secretary of Defense Wilson had cut the
research and development budget by $170 million, reduced overtime work in the
Atlas program, accorded Titan a lower priority than Atlas, cut spending on the
Navy's Polaris project, and called a temporary halt to Jupiter and Thor production.[71]

The WS-117L satellite program did not prove immune to the budget slasher. Air
Force satellite program officers had hoped to obtain $39.1 million of the estimated
$114.7 research and development budget for use in fiscal year 1957. In August 1956,
however, ARDC received only $3 million to launch the project. On 17 November
1956, General Putt briefed Donald Quarles on the newly approved WS-117L program. If he expected to obtain additional funding from the Secretary, he was disappointed. Secretary Quarles directed the Air Staff's research and development chief
to ensure that the Air Force halt its military construction schedule and produce no
fabrication mock-up or initial satellite without his express permission.[72] General
Schriever felt compelled to vigorously lobby the Air Staff and the Defense Department for an additional $10 million:

> I can recall pounding the halls of the Pentagon in 1957, [he said later,]
> trying to get $10 million approved for our [USAF] space program. We
> finally got the $10 million, but it was spelled out that it would be just
> for component development. No system whatever.[73]

Even so, in July 1957 Secretary Quarles applied additional spending limits to the
WS-117L as part of the Defense Department-wide budget slashing exercise that

summer. This came after he had received intelligence information that spring predicting that the Soviets would be capable of launching a satellite before the end of the year. Quarles' actions should be viewed in terms of the administration's agenda for military satellites and space operations. The previous year, administration spokesmen had declared that no government officials were to speak publicly about spaceflight. General Schriever found to his dismay that the administration meant business after a February 1957 speech he gave in San Diego, California. Discussing the importance of studying potential military offensive functions in space, he declared the time ripe for the Air Force to "move forward rapidly into space." The following day Secretary of Defense Wilson instructed him to avoid the word "space" in all future speeches.[74]

The administration remained determined that the military satellite would under no circumstances precede the civilian satellite into space. It also opposed any discussion of military space operations that might generate a worldwide debate on the "freedom of space" for military spaceflight. This issue had to be avoided to maintain the declaratory policy of "peaceful purposes" as well as the action policy of having Vanguard provide the precedent for military space operations. As a result, before Sputnik the country supported two modestly funded space programs that did not interfere with ICBMs or other high-technology programs. Neither received the support its advocates sought. Yet neither permanently suffered from the 1957 budget cuts, which proved little more than an embarrassment after the launching of Sputnik on 4 October brought massive increases for satellite and missile programs. Although not of the administration's choosing, *Sputnik I* established the precedent, freedom of space, and underscored the administration's basic space policy.

Retrospective From the Threshold of Space

On the eve of the Sputnik flights, the Air Force and the nation finally had reached the threshold of space—a full decade after the intrepid Rand analysts first offered the Air Force the prospect of launching an observation earth satellite in five years' time. During the course of the decade Rand produced increasingly convincing analyses of solutions to technical problems, potential military functions, and the important political benefits and prestige that would accrue to the United States. Armed with the Rand studies, Air Force satellite champions repeatedly worked to convince their leaders and the Defense Department and administration officials of the wisdom of their cause. What they confronted, however, proved to be a decade-long pattern of disinterest, inaction, and dogmatic unwillingness to accept change. As a result, the Soviets became the first to launch an orbiting satellite.

President Eisenhower has received considerable criticism for allowing the country to be humiliated and its national security endangered. Yet, if the Eisenhower administration failed to launch the first satellite, Sputnik nonetheless established emerging United States space policy. With unimpeded overflight assured, a clandestine military space program could proceed apace with less scrutiny by domestic and international critics. Civilian spacecraft had set the precedent for the military satellites to follow—even if the pathbreaker proved to be Sputnik and not Vanguard. The failure of the administration's actions lies not in overlooking the importance of prestige, but in assuming that Vanguard, with all its problems, still would be first. Determined that the civilian scientific satellite would take precedence, administration officials remained unwilling to provide the WS-117L program the commitment its supporters desired and expected. Yet the real delay in the reconnaissance satellite program occurred during the Truman years, when the Russians began a serious program and the United States did not. The Soviets had an eight-year start by 1954, when Project Feed Back set the satellite on the sure path to development.

During the Truman era, satellite proposals continually fell victim to the logic of "realism" and higher priorities. The administration argued that national security in the postwar world could be best achieved by strengthening the forces in being, namely strategic bombers and subsonic cruise missiles. Budget austerity after 1946 meant that research programs for forces of the future suffered most. Problematical programs like missiles and satellites faced the severest cuts. The administration's argument received strong support from the scientific community. Experts like Dr. Vannevar Bush dismissed missiles as "fanciful" because they would require a decade of incredible expense to overcome technical problems of guidance, propulsion, and reentry. Bush was far from alone in his pessimism. After all, it would have taken a particularly insightful individual to foresee the incredible advances over the course of the pre-Sputnik decade in chemical fuels, rocket engine combustion technology, instrumentation, and missile frame construction.[75] Postwar America offered too few men of vision like Hap Arnold and Theodore von Kármán in positions of authority. And for all his achievements, von Kármán led the Air Force down the lesser, aerodynamic path of development.

Von Kármán's tenure as chief of the postwar Scientific Advisory Board suggests that Arnold, had he continued to lead the postwar Air Force, would have achieved only modest success against the forces of institutional inertia and intransigence. The newly independent Air Force benefited most from the Truman defense strategy. While General LeMay and others might admit that intercontinental ballistic missiles represented the strategic force of the future, the logic of the present seemed to favor the forces that could best ensure the survival of an independent Air Force and the security of the nation. When those forces happened to be manned aircraft and missiles supporting those aircraft, long-range guided missiles understandably became relegated to the distant future. Satellites suffered a similar fate. Satellites and missiles represented "new" and potentially "revolutionary" change for a service that traditionally viewed itself in terms of the airplane. Even in the early 1950s, when it became clear that technology had made ballistic missiles more feasible and Soviet actions precipitated a major budget increase for research and development,

Air Force decision-makers persisted in resisting the acceleration of satellite and missile projects.

On the other hand, the Air Force remained ever vigilant in protecting its authority over satellite and missile development. If it neglected its space programs, it nevertheless kept a wary eye on Army and Naval efforts to weaken the Air Force's claim to exclusive rights to these programs. The fierce contest for control of roles and missions proved to be a running theme throughout the pre-Sputnik decade, and clearly prevented faster progress. While the service squabble centered largely on missiles, on the eve of Sputnik the Air Force faced a competitor for its embryonic satellite program in the guise of Project Vanguard. Although Eisenhower's dual space program remained unclear to many in the mid-1950s, the "civilian" Vanguard satellite represented a future challenge for the Air Force in terms of civil-military roles and missions.

On the eve of Sputnik, the observer of early space events might be tempted to view the previous decade pessimistically as one of frustration and delay. The Air Force experience, however, also suggests a much more positive assessment. Many central characteristics of the future Air Force space mission emerged during the "dawn of the space age." For one, the Air Force made a strong bid for the preeminent military role in space matters, and by 1957 had carved out a relatively strong position with its Atlas and military satellite program. The emphasis on demonstrating military satellite utility served to intensify efforts to define and justify satellite operations in terms of providing better data more effectively than competing systems.

As for research and development, the Arnold and von Kármán legacy appeared far more secure after the downswing in the late 1940s. Reorganization provided research a greater focus, while General Schriever's command arrangements demonstrated impressive flexibility and effective improvisation. To be sure, it took activists like Trevor Gardner and his band of reformers working "outside" the system to facilitate change. Yet, in a sense, the Air Force established a tradition of going outside—to industry, scientists, research laboratories—that von Kármán's *Toward New Horizons* recognized and supported.

Considering both the failures and the achievements, the pre-Sputnik decade is best viewed as the conceptual period of the "New Ocean," during which the new ideas of space had to be tested and inertia and opposition overcome. After all, only a generation separated the upper-atmosphere explorers of the NACA and the early rocketeers from the Atlas and WS-117L teams on the eve of Sputnik. Few are blessed with the vision of Arnold and his disciples or the perceptiveness of the Rand analysts of 1946, who consulted their crystal ball and guessed correctly. Even by late 1957 the path ahead for most seemed unclear. With the nation on the threshold of space, the challenges for the Air Force remained formidable.

CHAPTER 2

From Eisenhower to Kennedy:

The National Space Program and the Air Force's Quest for the Military Space Mission, 1958–1961

The period from late 1957 to the spring of 1961 represents the watershed years of the national space program and the Air Force's place within it. In the wake of the Sputnik crisis, the Eisenhower administration implemented organizational and policy measures that provided the foundation of the nation's space program. Buffeted by pressure and counsel from an alarmed public and congressional and military spokesmen, President Eisenhower found himself fighting a rearguard action to hold to his view of civilian, military, and budget priorities for space activities. His dual military and civilian space program reflected his "space for peace" focus, one that fostered "open skies" for the free passage of future military reconnaissance satellites. Given the sensitivity of overflying the Soviet Union, during the formative years of his administration the civilian space program held center stage, while administration officials consciously downplayed the military space role and service initiatives.

Space advocates in all three military services and their supporters chafed at the government's refusal to sanction a broadly-based military space initiative in response to the Soviet menace. With visions of leading the nation into the space era, Air Force leaders found the situation especially frustrating. Relying on its "aerospace" rationale, they initially argued that the Air Force represented the logical service to head a unified, Defense Department-oriented national space program that would serve both military and civilian requirements. When it became clear that national policy preferred two programs, one a civilian-led effort dependent on military support, the Air Force sought to become the "executive agent" for military space.

The challenge proved formidable. Shortly after Sputnik, concerns with inter-service rivalry and duplication, among other reasons, compelled administration officials to create the Advanced Research Projects Agency (ARPA) for all Defense Department space research and development activities. Although the services retained their missile programs, they temporarily lost their independent space programs to the new agency. Moreover, the creation of the National Aeronautics and Space Administration (NASA) in the fall of 1958 further divided the space mission and raised thorny issues of civil-military authority that persisted well beyond Sputnik. Despite repeated government statements to the contrary, for many, a civilian NASA conducted "peaceful" space ventures, while the Defense Department and the military services, by implication, engaged in warlike or non-peaceful space activities. Air Force leaders found "space for peaceful purposes" an albatross that prevented them from pursuing a space program they believed necessary to provide the nation with the security it required. The latter involved not only recognized defense support functions such as satellite communications, reconnaissance, and navigation activities, but potentially offensive functions in space through space-borne antisatellite and antimissile defense measures. The Eisenhower administration believed otherwise, and permitted nothing more than studies of weapons in space.

Constrained by administration policy and the prerogatives of NASA and ARPA, and without a space "mission" to call its own, the Air Force also faced stiff competition from its service counterparts. Indeed, by early 1958, the Army and Navy had far more experience in space than the Air Force. Their success in orbiting the nation's first satellites (Explorer and Vanguard) seemed destined to propel one of them to victory in the quest for future space missions. Yet, by the spring of 1961, NASA had its sights on manned flight to the moon, ARPA had been relegated to obscurity, and the Army and Navy had been removed from any major role in space. The Air Force found itself effectively designated the executive agent for all military space development programs and projects. If Air Force leaders considered the victory incomplete, it nonetheless represented an impressive achievement that established the Air Force as the nation's primary military service for space.

Sputnik Creates a "National Crisis"

The administration's efforts prior to the Sputnik launches to downplay the American space program through a deliberately-paced civil and military research and development effort came to an abrupt halt following the electrifying news on the morning of 4 October 1957 that the Soviets had launched a 184-pound instrumented satellite into orbit atop a rocket booster weighing nearly 4 tons. By contrast, America's yet-to-be launched Vanguard weighed only 3.5 pounds.[1] *Sputnik I* dramatically demonstrated that the Soviets possessed both a highly advanced satellite program and booster technology sufficient to field an intercontinental ballistic missile force. For the first time, America seemed at risk of an intercontinental attack. Despite warnings of the psychological shock value of satellites repeated through various Rand studies and affirmed by the National Security Council a few years earlier, the administration found itself unprepared for Sputnik's "Pearl Harbor" effect on public opinion.[2]

President Eisenhower sought to reassure the American public and quell the press furor at home and abroad in his first news conference held five days after the Russian launch. On 9 October he downplayed the impact of Sputnik by declaring that, "so far as the satellite itself is concerned, that does not raise my apprehensions, not one iota." People had no reason to panic, and he would not involve the country in a needless space race or accelerate the launch schedule of the civilian Vanguard satellite. But neither the President's soothing words nor unfortunate public comments belittling the importance of the Russian effort by high-ranking administration officials proved able to silence a growing national debate over space and defense policies. They had a national crisis on their hands.[3]

At the same time, Eisenhower and his staff quickly perceived one important benefit from the Sputnik launch. Meeting with the President the day before his post-Sputnik press conference, Deputy Secretary of Defense Donald Quarles observed that "the Russians might have done us a good turn, unintentionally, in establishing the concept of freedom of international space." Eisenhower then requested that his advisors look five years into the future and provide an update of the Air Force's effort to develop a reconnaissance satellite. Clearly, the President intended to continue his public focus on civilian spaceflight and unrestricted satellite overflight to protect the viability of future military satellite operations.[4]

Throughout October administration officials reevaluated the entire missile program and discussed various courses of action. Then, nearly a month later, on 3 November, the Soviets successfully launched the 1,120-pound *Sputnik II* with its passenger, the dog "Laika." Although once again officials tried to calm troubled waters by claiming that the Soviet feat came as "no surprise to the President," the administration rapidly moved to gain control of the debate and reestablish confidence and prestige. On 7 November the President took one of his most important steps, appointing Dr. James R. Killian, his close confidant and chairman of the earlier Killian Committee, as Special Assistant to the President for Science and Technology and Chairman of the President's Science Advisory Committee. He immediately became the administration's "point man" for planning future space organization and policy.[5]

The day after Killian's appointment, the Defense Department authorized the Army Ballistic Missile Agency (ABMA) to proceed with preparations to launch its scientific Explorer satellite during the IGY under Project Orbiter as backup to Project Vanguard. Conveniently, incoming Secretary of Defense Neil McElroy had been visiting the Hunstville, Alabama, complex when the Soviets launched *Sputnik I*. Project director Brigadier General John B. Medaris and Wernher von Braun seized the opportunity to promise a successful Jupiter launch within ninety days. When they received official approval on 8 November, Medaris' team had been hard at work on the Orbiter booster since 5 October. Their hard work would pay off on 31 January 1958, when *Explorer 1* became the first U.S. satellite to achieve successful orbit. Although its miniaturized electronics relayed important scientific data, including discovery of the Van Allen radiation belts surrounding the earth, its 10 ½-pound payload seemed less impressive to the American public than the far larger and heavier, if less scientifically valuable, Sputniks.[6]

Secretary of Defense McElroy followed the Project Orbiter decision by announcing on 20 November his intention to create a new defense agency to control and direct "all our effort in the satellite and space research field." Representing the first step in reorganizing the government for space, Secretary McElroy planned to establish the Advanced Research Projects Agency (ARPA) in early February 1958 at a level above the three military services.[7]

The Air Force Seizes the Initiative

Meanwhile, the Air Force had been far from idle in the aftermath of *Sputnik I*. While Army and Navy teams continued preparations for Projects Orbiter and Vanguard, respectively, Air Force leaders in late 1957 initiated their own sweeping assessment of the nation's space activities and prospects. They hoped to develop a program of action with the Air Force playing the central role. The wide-ranging post-Sputnik debate on the national space course ahead seemed to present Air Force leaders with a golden opportunity to claim for their service the nation's space mission.

On 21 October 1957, Secretary of the Air Force James H. Douglas convened a committee of distinguished scientists and senior Air Force officers chaired by Dr. Edward Teller to evaluate the nation's missile and space programs. Completed in just two days, the Teller Report chastised the government for administrative and management practices that, it said, prevented either civilian or armed services agencies from achieving a stable and imaginative research and development program. It recommended a unified, closely integrated national space program—under Air Force leadership. A centralized program, the committee argued, would provide focus for an expanded national space program and avoid the divided effort likely to result from a fragmented program. Although the report received attention at high levels of the government, in the unsettled post-Sputnik period it failed to convince government officials to adopt a unified program either under military or civilian direction. Ultimately, the President would commit the nation to a dual program with separate military and civilian elements.[8]

On 7 November 1957, the Air Force's legislative liaison team, alarmed by what seemed to be a preference among congressmen for the Army's space initiatives, described the challenge confronting the Air Force. To avoid defeat in the race for the space mission, the Air Force must base its legitimate case on the position staked out in 1948 by General Vandenberg, that flight in the upper atmosphere and space represent logical extensions of the traditional Air Force realm of operations and the natural evolution of its responsibilities. The officers urged Air Force spokesmen to "emphasize and re-emphasize the logic of this evolution until no doubt exists in the minds of Congress or the public that the Air Force mission lies in space as the mission of the Army is on the ground and the mission of the Navy is on the seas."[9]

On 29 November 1957, Chief of Staff General Thomas D. White made this theme the focus of an important address to the National Press Club. As airpower had provided the means to control operations on land and sea, so in future "whoever has the capability to control space will likewise possess the capability to exert control of the surface of the earth." For the Air Force, he said, "I want to stress that there is no division, per se, between air and space. Air and space are an indivisible field of operations." By implication, an Air Force role in space must embrace offensive operations to provide proper national security. Publicly, Air Force leaders would seldom admit that the atmosphere and space represented fundamentally different mediums. In his talk, General White also addressed another basic institutional theme, affirming the service's traditional research and development focus. The Air Force still depended, he said, on the "skills, talent, ingenuity and cooperativeness of…science and industry to provide us the technological lead we need in the future." This future would be in space.[10]

In public addresses and Congressional testimony, General White and other Air Force spokesmen, including Under Secretary of the Air Force Malcolm A. MacIntyre, Lieutenant General Donald Putt, Deputy Chief of Staff for Development (DCS/D), and Major General Bernard A. Schriever, commander of the Air Force Ballistic Missile Division (AFBMD), would focus on the concept of space as a continuum of the atmosphere, a place for potential military-related operations rather than a function or mission in itself, and the logical arena for Air Force activities. Early in the new year, Air Force leaders coined a new term, "aerospace," to describe their service's legitimate role in space, and the following year "aerospace" officially entered the Air Force lexicon when it appeared in the revised Air Force Manual 1-2, *United States Air Force Basic Doctrine*, issued on 1 December 1959. According to the manual,

> aerospace is an operationally indivisible medium consisting of the total expanse beyond the earth's surface. The forces of the Air Force comprise a family of operating systems—air systems, ballistic missiles, and space vehicle systems. These are the fundamental aerospace forces of the nation.[11]

Along with policy and planning issues, the Air Force also addressed internal organizational concerns for space. To provide better focus for future Air Force space activities, on 10 December General Putt revealed the formation within his office of a Directorate of Astronautics, headed by Brigadier General Homer A. Boushey, whose long career in the "space" field included early rocket-assisted flight experiments with the von Kármán team during World War II. However, having created the new office without consulting Defense Department officials, General Putt and other Air Force leaders were chagrined by the vehement opposition from senior defense officials like William Holaday, newly-appointed Defense Director of Guided Missiles, who accused the Air Force of wanting "to grab the limelight and establish a position." This, of course, is precisely what the Air Force intended to do. When Defense Secretary McElroy objected to the term "astronautics" and criticized the Air Force for seeking public support, Air Force leaders realized they had overstepped military boundaries. The firestorm of protest convinced General Putt to rescind his memorandum three days later. Although Air Staff leaders remained committed to strong centralized headquarters direction of space projects, they continued to face roadblocks from administration officials.[12]

Unable to establish the Air Staff directorate in late 1957, the service's space supporters during the first six months of the new year followed the temporary expedient of coordinating USAF space activities through the Assistant Chief of Staff for Guided Missiles. Only in late July 1958, after the proposed civilian space agency received congressional approval and the National Security Council revised space policy, could the Air Force create a central Air Staff office for space. Even then, the term "astronautics" could not be used, and General Boushey's new office under the DCS/Development became the Directorate of Advanced Technology. Sharing space responsibilities with the Assistant Chief of Staff for Guided Missiles, General Boushey would have to wait another year before his office could be upgraded to assume direction of all headquarters space activities.[13]

In retrospect, given the administration's emphasis on strategic reconnaissance, of which he was well aware, General Putt should have been sensitive to any suggestion of an expanded military role in space. Four days after Sputnik, he and Vice Chief of Staff General Curtis E. LeMay met with Deputy Secretary of Defense Quarles to apprise him of the state of the military reconnaissance program and potential for satellite offensive operations. Quarles readily supported the Advanced Reconnaissance Program, which would become the government's most important space project. Yet, when the two officers advocated an offensive space role to forestall potential Soviet satellite weapon carriers, Quarles in no uncertain terms directed the Air Force not to consider satellites as weapon platforms and to entirely eliminate satellite offensive applications from future Air Force space planning. Air Force leaders would continue to find that the policy of "peaceful uses of outer space" embraced the development of reconnaissance systems but never offensive weapon systems. Weapons in space threatened the reconnaissance assets judged vital to national security.[14]

By the end of the year, the Air Force's initial foray into the space contest had produced mixed results. Its leaders had established the service's policy position for a legitimate space role, yet the lack of a Defense Department response to an Air Force-led space plan for the nation and Air Staff's rebuff suggested the need for a more cautious strategy to achieve Air Force space objectives. In future efforts, the Air Force would develop policy, planning, and organizational proposals as part of a well-organized quest for the military space mission.

The Government Organizes for Space

Beginning in early 1958, the administration took action to create a national space program. Its focus centered first on organizational measures, then embraced policy issues. By late summer, the National Space Act confirmed a dual civilian-military program designed to pursue a policy of space for peaceful development and exploration. Along the way, the administration and Congress considered various options in their attempt to create the optimum civilian-military balance. Although their decisions would prove enduring, they left unclear the precise relationship between military and civilian space responsibilities.

During the first week of the new year, the Defense Department requested a list of proposed space projects from each of the three services. Air Force leaders viewed this request as an open door for approval of a USAF space program. It had devoted considerable thought to the future space needs of the country ever since the first Sputnik flight and the Teller Committee's deliberations. In early December 1957, the Scientific Advisory Board reported on the subject of space technology. Pointing out that Sputnik and Soviet ICBM capability had produced "a national emergency," the board focused on the rocket field as the area which provided the Air Force the best means of contributing to "a proper national response." Its six-point program also included an accelerated reconnaissance satellite effort and a "vigorous" space initiative with an "immediate goal of landings on the moon." Both military manned space-flight and the WS-117L Advanced Reconnaissance System would remain centerpieces of future Air Force space proposals, while Air Force leaders would quickly realize that the relationship between missiles and space systems would prove the most effective key to achieving Air Force preeminence in military space.[15]

The result of the Air Force's post-Sputnik deliberations appeared on 24 January 1958, when the Air Staff submitted to the Director of Guided Missiles its "Air Force Astronautics Development Program." It comprised five major space systems: Ballistic Test and Related Systems, a lunar military base system, manned hypersonic (Mach 5 and above) research, the Dyna-Soar orbital glider, and the WS-117L Satellite System. Planners further divided the five proposals into twenty-one major projects that embraced a variety of military missions deemed "essential to the maintenance of our national position and prestige." The planners urged that special emphasis be accorded getting man into space at the earliest time.* Testifying before Congress in early January, Major General Bernard A. Schriever, Commander of the Air Force Ballistic Missile Division, emphasized that the Air Force possessed the means of developing an astronautics program with no detriment to ballistic missile programs. Much to its disappointment, the Air Staff received no reply from Mr. Holaday's office, and Air Force efforts to lead a national space effort proved fruitless. Statements by General Putt and his deputy, General Boushey, advocating missile-firing bases on the moon and eventually militarizing the planets alarmed rather than reassured their audience of civilian leaders in Congress and the Defense Department. By late February, the Air Force initiative had been "overtaken by events," and the Assistant Secretary of Defense assumed responsibility for coordinating military inputs for a national policy on outer space. When the Secretary of Defense created the Advanced Research Projects Agency on 7 February, frustrated Air Force officials realized that the Defense Department's request to the services represented little more than an effort to gain information that would assist the new Defense Department agency in assigning space development responsibilities among the Army, Navy, and Air Force.[16]

The comments by Generals Putt and Boushey reflected the uncertainty of the period and the great unknowns of space in the aftermath of Sputnik. After the demise of the Air Force initiative, Air Force leaders responded to the Defense Department's attempt to coordinate a policy input for the administration by calling for more basic knowledge to determine the potential and limitations of manned and unmanned spaceflight before formulating a national policy covering all available and contemplated space programs. Air Force thinking in the months ahead would be characterized by an emphasis on a "building blocks" approach to space development rather than on advancing fanciful ideas for military bases on the moon and planets from which to attack countries on Earth.[17]

ARPA Takes Control

ARPA began operations amid a flurry of great expectations from its admirers and dire warnings from its detractors. Secretary of Defense McElroy declared that the new agency would provide a "single control…of our most advanced development projects," while the services would continue with research and development of weapon systems that clearly fell within the "missions of any one of the military departments." ARPA, in fact, gained control over all U.S. space projects, military and civilian, until the National Aeronautics and Space Administration (NASA) commenced operations in the fall of 1958. For another year thereafter, the Defense Department agency retained control, including funding, of all military space

---

* See Appendix 2-1.

projects. The initial delineation of responsibilities between ARPA and the services proved difficult to maintain. Yet ARPA fulfilled two important administration objectives. For one, it ended the low military priorities heretofore accorded space technology in the absence of clearly defined military applications. For another, it offered the laudable prospect of avoiding interservice rivalry and wasteful duplication by transferring service decision-making power on space projects to the Defense Department agency.[18]

The congressional committees charged with military oversight viewed with suspicion any increase in the powers of the Secretary of Defense at the expense of the military services. In early January 1958, General Schriever and other military spokesmen testified against the creation of any agency with authorization to go beyond policy formulation and program approval to perform development and contractual responsibilities. Research and development, they argued, should be left to the services. Secretary McElroy promised Congress that ARPA's initiative would "be developed in coordination with the military departments to the point of operational use, so that…[weapon systems]…may be phased into the operation of one or more of the military services with a minimum loss of time or interruption of development and production."

The Air Force was not entirely reassured. Roy Johnson, ARPA's aggressive director, seemed too independent of service wishes and possessed too much authority over service space programs. Moreover, the President made ARPA responsible for civilian space projects as well until the proposed civilian space agency became operational. Nevertheless, until ARPA assumed control of most Air Force programs in late June, Air Staff planners, perhaps guilty of wishful thinking, continued to advocate an independent Air Force space program. As the historian of the Air Research and Development Command pointed out, the "classic and foreboding example of things to come…proved to be the reconnaissance satellite program, perhaps the most important single Air Force space program to light upon ARPA." Initially the Air Force applauded ARPA's focus on accelerating the WS-117L program on "the highest national priority basis." In response to Sputnik, by September 1958 ARPA had reprogrammed the Advanced Reconnaissance System into separate component projects with revised designations. The reconnaissance element received the name, Sentry, while MIDAS referred to the infrared sensing system. Under the designation "Discoverer," a cover for the covert CORONA project, ARPA grouped "vehicle tests, biomedical flights, and recovery experiments." In the fall of 1958, ARPA assigned all three projects to different Air Force organizations.

Operating on a project basis, ARPA direction signaled the end of "concurrency," the centralized systems management practice that had proven so successful in the crash ICBM program. In October 1958 ARPA also terminated the Weapon System (WS) designation altogether, declaring that the "system approach employed by the Air Force would be altered in such a way that all other items of the former 117L system would be budgeted as subsystems or components…for reasons of budget justification and program management." Omitting the weapon system designation contributed to the administration's low-profile approach to military space activities.

The other Air Force space programs received similar treatment from ARPA following their transfer in late spring.* The Defense Department agency organized its newly-acquired space activities into four broad programs: Military Reconnaissance Satellites, Missile Defense Against ICBM, Advanced Research for Scientific Purposes, and Developments for Application to Space Technology. Although ARPA redistributed most programs back to the Air Force and the other services, it did so under contract, thereby retaining technical and fiscal control and receiving credit for "its" programs. The Air Staff might set requirements, but ARPA made the decisions, directed the efforts, and dealt directly with other agencies and with private industry.

Air Force leaders also found ARPA's operating procedures highly unsettling. In late March Johnson informed the service secretaries that he intended to "cut red tape" and deal directly with subordinate agencies like the ARDC, AFBMD and other space and missile centers, bypassing established chains of command. At the Air Research and Development Command, for example, ARPA personnel frequently approached individuals and offices directly, which led ARDC commander Lieutenant General Samuel E. Anderson to establish a "focal point" to coordinate ARPA-ARDC activities. Even so, the "focal point" officer and his small staff faced considerable opposition from within the command and criticism from General Boushey's Directorate of Advanced Technology before they succeeded in keeping all parties informed on a consistent basis.

Yet, if the novel Defense Department agency acted high-handedly and pursued management practices that alarmed the services, the intrusion of ARPA could have been far more disruptive. In fact, dire warnings that ARPA might evolve into a "fourth service" proved false. Roy Johnson, much to the dismay of his staff, proved unwilling either to create and operate his own facilities and laboratories or to establish an in-house contracting capability with the armed services functioning as ARPA's contracting agents. In fact, for its expanded space program, ARPA remained dependent on the services for qualified personnel, necessary experience, and resources that included laboratories, launch complexes, rocket boosters, test facilities, and tracking networks. As a result, ARPA designated the military services its executive agents for most projects, with the Air Force receiving the lion's share of eighty percent. Along with the former Advanced Reconnaissance System, these represented the Air Force's most cherished space programs, including lunar probes, the 1.5 million-pound rocket booster, and a variety of measures designed to launch a

---

* See Appendix 2-2.

military man in space. ARPA, in fact, consistently supported the need for a military manned space mission, and already in late February 1958 had awarded the Air Force development responsibility for military manned spaceflight. Although the Air Force remained unhappy with its subordination to ARPA on space matters, Air Force leaders quickly realized that cooperation with ARPA would prove the best means of gaining development responsibility for space projects and, later, operational responsibility as well.

ARPA's rise to prominence reflected the country's alarm following Sputnik and the need to act rapidly to counter the Soviet advantage. As a result, ARPA became a prime mover for a variety of space projects, some of which, such as the lunar probe program, had no direct military requirement. Specifically authorized by the President, this effort would use available military resources, most notably the Army's Jupiter and the Air Force's Thor IRBMs as boosters. In short, ARPA served as the national space agency through much of 1958. Yet it remained clear from the spring of 1958, when the President submitted his proposal for a National Aeronautics and Space Administration (NASA), that the new civilian space agency would directly challenge ARPA's broad jurisdiction in the space arena and become an additional competitor for traditional Air Force space interests.

NASA Joins the Competition

Like ARPA, NASA represented an intervening space agency that challenged the Air Force for space responsibilities and program funding. NASA's civilian focus also raised the contentious issue of civilian-military space relationships. Despite the apparent logic in assuming that NASA would be responsible for civilian space activities and the Defense Department would handle military interests, the demarcation line between civilian and military space concerns often proved artificial and unattainable. On the other hand, if the Air Force found NASA an unwanted competitor for the space mission, it quickly perceived the benefits to be gained by cooperating with the civilian agency. For the foreseeable future, NASA would depend heavily on Air Force assistance, while its absorption of Army and Navy space assets would help propel the Air Force toward the military space mission.[19]

The "Sputnik crisis" produced demands by congressmen, scientists, and other civilian leaders for a more sweeping national organizational space effort than ARPA seemed to promise. The hearings begun in late November 1957 by Senate Majority Leader Lyndon Johnson's Preparedness Investigating Subcommittee of the Senate Committee on Armed Services focused on long-term space research and development requirements "from a broad national point of view." This could best be done, the committee's final report suggested, by either improved control and administration within the Defense Department or the establishment of an independent agency.

An independent space agency for long-term research and development outside the Defense Department gained increasing support in early 1958 from scientists concerned that centering space research in the Defense Department would likely alter and reduce the scale of scientific programs. While various individuals and groups proposed organizational alternatives, the National Advisory Committee for Aeronautics (NACA), which had considerably expanded its missile research under Chairman Jimmy H. Doolittle and Director Hugh L. Dryden, took an increasingly active role in the space debate.[20] In late 1957 it convened a special committee on space technology under MIT's D. G. Stever to examine space-age research and development requirements and determine the best role for the NACA to play. On 14 January 1958 the committee's report proposed an interagency cooperative space program that would involve the NACA, the Defense Department and the military services, the National Science Foundation, and the National Academy of Sciences. But just two days later the NACA's main committee passed a strong resolution on spaceflight proposing that fundamental scientific research in the upper atmosphere and space be conducted by the NACA rather than the military.[21]

Meanwhile, in early February 1958, the congressional leadership called for the formation of an independent civilian space agency, and, to address the "national crisis," Congress created two important committees: a Senate Special Committee on Space and Astronautics under Majority Leader Lyndon B. Johnson, and a House Select Committee on Astronautics and Space Exploration chaired by Speaker John W. McCormack. Yet, congressional hearings on the space agency itself began only after the administration submitted its own bill on 2 April 1958. The administration's delay in submitting its proposal is explained by the last ditch disarmament discussions Eisenhower carried out in January and the deliberations over the place of the military in the space program.[22]

In early February, the President charged his science advisor James Killian to proceed with specific recommendations for government organization for space activities. Recalling this early formative period, Killian admitted that he took on the assignment with a clear idea about what should be done.

> From the beginning, it has been my view that the Federal Government had...only two acceptable alternatives in creating its organization for space research, development, and operation. One was to concentrate the entire responsibility, military and nonmilitary, in a single civilian agency. The other was to have dual programs.... A possible third alternative, that of putting our entire space program under the management of the Defense Department always seemed to me to have so many defects as to be practically excluded as a solution. [23]

Because of his concerns for national security, in which strategic reconnaissance loomed large, Eisenhower did not share Killian's views. In fact, shortly after the new year, he thought simply of having the military direct the entire space research and development effort under ARPA's direction. He soon abandoned this idea because of congressional and scientific opposition, and because of the arguments of Killian.[24]

Nevertheless, the President always opposed creating an entirely civilian national space program or of diluting the Defense Department's overall responsibility for space research and applications. During the drafting of the bill, the administration's dilemma involved how much and what kind of military participation to authorize rather than choosing between military and civilian alternatives.

Once the administration accepted a civilian agency based on the NACA, it solicited comments from the Defense Department. Initially, defense officials thought little would change because of the traditionally cooperative military arrangement with the NACA. Commenting on the draft bill prior to its submission to Congress, Deputy Secretary of Defense Quarles reminded budget chief Maurice H. Stans that "it is assumed the operation of the new agency would bear the same relationship to the Defense Department in the field of space and aeronautics as the NACA now does in the aeronautics field." As it was, Quarles objected to a number of passages in the legislation, including one that he perceived as preventing the services from carrying out basic scientific research that had military mission applications. This issue would continue to cause tension long after passage of the Space Act.[25]

The administration's bill, drafted by the NACA general counsel Paul Dembling and sent to Congress on 2 April 1958, proposed that the nation's aeronautical and space science activities be directed by a civilian agency "except insofar as such activities may be peculiar to or primarily associated with weapons systems or military operations, in which case the agency may act in cooperation with, or on behalf of, the Department of Defense." Referred to as the "exception clause," this passage suggested a variety of interpretations. Would the new agency be the prime mover in government space activities, with the military playing a minor role? Did acting on behalf of the Defense Department mean that NASA would undertake military projects? Above all, as Donald Quarles suggested, did the narrowly constructed military mission preclude the Defense Department from performing basic space research closely related to defense missions?[26]

In congressional hearings, witnesses and committee members attempted to determine precise organizational relationships and functions. Defense Department witnesses strongly objected to the "exception clause." ARPA director Roy Johnson also criticized any implication that the law would give NASA veto power over military activities and restrict the Defense Department to operating space systems. His chief scientist, Herbert F. York, agreed and presented the Air Force's argument that space is not a program to be administered by a single civilian agency, but a place of civilian and military applications. From his reading of the bill, it seemed that the Bureau of the Budget and NASA would be responsible for programs either entirely civilian or jointly civilian, leaving the military with only the narrowly defined military agenda. The problem, declared military officials, centered on space requirements that could not be precisely known in advance, but often required identifying and refining during the course of development or research. Therefore, the Defense

Department must be permitted to conduct fundamental exploration of space technology in order to determine if particular defense tasks could be done more effectively in space. The administration's bill pointedly did not provide a clear, fixed division of labor between the military and the new civilian agency. But as an early House of Representatives staff paper concluded, "practically every peaceful use of outer space appears to have a military application."[27] In the bill's final language, Congress approved giving the Defense Department and NASA wide-ranging prerogatives in the space field, yet agreed that the Defense Department had authority both to develop systems and conduct any kind of space research and development "necessary to make effective provision for the defense of the United States." Even so, the gray area would remain.

To overcome the jurisdictional problem and permit basically separate activities without expensive duplication, Congress created two coordinating bodies: a cabinet-level National Aeronautics and Space Council (NASC) and a sub-cabinet-level Civilian-Military Liaison Committee (CMLC). During the remainder of the Eisenhower administration, neither would function effectively. The CMLC met often but had insufficient authority to resolve issues, while the NASC, which possessed the requisite decision-making capability, seldom met. The President refused to be constrained in his management of the space program.[28]

The establishment of NASA reflected the administration's determination to give the space program a civilian focus through a policy of "space for peaceful purposes" that encompassed scientific exploration as well as a less-publicized but far more important national security element. President Eisenhower signed the National Aeronautics and Space Act on 29 July 1958. Along with prescribing organizational and functional responsibilities of the National Aeronautics and Space Administration, the space act addressed policy in unmistakable terms. "The Congress hereby declares that it is the policy of the United States that activities in space should be devoted to peaceful purposes for the benefit of all mankind." [Sec 102(a)] Although the statement reflected Eisenhower's policy statements prior to Sputnik, before inclusion in the space bill James Killian and the Presidential Scientific Advisory Committee (PSAC) conducted a comprehensive examination of broad policy objectives as part of their assessment of organizational requirements.

At the request of the President back in early February 1958, Killian established a panel under the auspices of the PSAC to develop a national space program. Chaired by Nobel laureate Edward Purcell, the panel's deliberations focused on nonmilitary space programs and activities. Arguing that "even the more sober proposals…about space as a future theater of war…do not hold up well on close examination or appear to be achievable at an early date," the Purcell Panel strongly recommended passive military support applications while rejecting any use of military weapons in space. With the President's blessing, Purcell and panel member Herbert York briefed the Cabinet and other groups within the administration, and in late March issued a public version of their report. The brochure, "Introduction to Outer Space," stressed the peaceful, scientific objectives of spaceflight and the administration's cautious approach to the space age. The PSAC report would provide the basic guidelines for the military role in space. Despite strong objections from Air Force officers in the months ahead, the administration would confine offensive military space applications to studies only.[29]

With military satellite launches on the horizon, Eisenhower refined national space policy with two National Security Council directives that closely bracketed the signing of the Space Act. In June NSC Directive 5814, "U.S. Policy on Outer Space," advocated a "political framework which will place the uses of U.S. reconnaissance satellites in a political and psychological context most favorable to the United States." The NSC followed this on 18 August 1958 with a more definitive directive, NSC 5814/1, "Preliminary U.S. Policy on Outer Space," a broad statement which emphasized denying Soviet space superiority. Echoing the early Rand Corporation studies on satellite feasibility, the administration would seek to achieve this objective by "'opening up' the Soviet Bloc through improved intelligence and programs of scientific cooperation." This would be accomplished by the military reconnaissance satellites, whose mission, the directive asserted, fell squarely within the "peaceful purposes" guidelines and represented an asset of "critical importance to U.S. national security."[30] In effect, although NSC 5814/1 advocated an open, cooperative scientific exploration program, it also established the foundation for a national security reconnaissance space capability immune from international inspection or control. The latter received the highest priority from an administration that saw no contradiction in space for peace combined with space for national security.

With the 1958 Space Act, the government formally established a dual space program comprising separate civilian scientific and military applications projects. Both were directed to "peaceful," or scientific, defensive, and nonaggressive purposes. This accorded precisely with Eisenhower's commitment to insure unrestricted overflight in outer space of military reconnaissance satellites that the President so eagerly awaited to replace the increasingly vulnerable U-2 surveillance aircraft that violated national sovereignty in airspace overflight.

Although the framers of the Space Act did not equate "peaceful" with civilian or nonmilitary activities, government officials in the future often found themselves required to explain that both NASA and the Defense Department conducted peaceful space work, one primarily engaged in space exploration and the other in various military support activities devoted to keeping the peace. Air Force space leaders like General Schriever repeatedly criticized this policy which many interpreted as implying that NASA engaged in "peaceful" work while the military, pursued "nonpeaceful" activities. Such inaccuracies, he believed, along with policy restrictions limiting offensive space weapons to the drawing board, prevented the military from providing necessary security through an expanded space program. Air Force advocates of a dynamic, military-oriented national space endeavor remained frustrated by national space policy and organizational constraints that seemed to rule out anything except passive military space applications.[31]

NASA Takes Shape

With an organization in place by midsummer that provided for dual military and civilian programs, officials turned to the complex mission and project assignments remaining before NASA could commence operations on 1 October 1958. Lines of demarcation remained vague, while competition for prestige and funding promised to be severe. The initial question centered on facilities and infrastructure. During the congressional debate it became clear that the new agency would absorb NACA's existing aeronautical research facilities and personnel. These included nearly 7,000 personnel and the Langley and Ames Aeronautical Laboratories, the Lewis Flight Propulsion Laboratory, the High-Speed Flight Station at Edwards Air Force Base, and the Pilotless Aircraft Research Station at Wallops Island, Virginia.

To achieve space capability quickly, NASA needed an infusion of space programs, facilities, and funding from the military services. In the NASA raid on service assets, the Air Force emerged the clear victor. With little objection from the Navy, NASA received Project Vanguard's personnel and facilities, including its Minitrack satellite tracking network, and more than 400 scientists and engineers from the Naval Research Laboratory. Potential Army losses, however, proved far more sweeping and contentious. Newly-appointed NASA administrator, Keith Glennan, considered the Army space program most important for providing the agency credible space design, engineering, and in-house resources. He initially requested transfer of Cal Tech's contracted Jet Propulsion Laboratory (JPL), whose sympathetic director had visions of turning it into the "national space laboratory," and a portion of the Army Ballistic Missile Agency that included the von Braun team and its giant Saturn booster project. General Medaris, however, strongly objected and waged a public campaign to stall the process and reverse the decision. His effort produced a compromise. The JPL would be transferred to NASA by 3 December 1958, while the Huntsville complex would remain under the Army's jurisdiction and support NASA on a contractual basis. Medaris might postpone but he could not prevent a transfer. A year later the Army would lose to NASA its entire space operation at Huntsville, which would be renamed the Marshall Space Flight Center.[32]

As for the Advanced Research Projects Agency and its Air Force-related programs, the Defense Department agency intended to transfer only elements of its Advanced Research for Scientific Purposes program. In mid-August, however, Eisenhower awarded NASA overall responsibility for human spaceflight. As a result, ARPA relinquished all of its "man in space" projects, which NASA combined under the designation, Project Mercury. ARPA also relinquished its special engine research project, as well as satellite tracking and satellite communications, meteorological, and navigation satellite programs.* Air Force reaction proved mixed. While giving up what amounted to five space probes, three satellite projects, and some propulsion research represented largely scientific projects in early stages of research, the loss of the manned spaceflight mission created apprehension about the future of a military manned role in space. While $800 million for space in the fiscal year 1959 budget represented eight times the space portion before Sputnik, NASA's share outpaced ARPA's by more than $50 million and included $117 million transferred from ARPA. Of the latter, the Air Force gave up $58.8 million. In short order, NASA had acquired the missions of scientific space exploration, including the moon, as well as manned spaceflight and all civil applications satellites. To fund its new programs, NASA received a generous budget, which raised the specter of tough competition between civil and military sectors for space funds in future years.[33]

On the other hand, NASA's absorption of Army and Navy space programs had left the Air Force the front-runner for the military space mission. Air Force leaders quickly perceived the advantage of cooperating with the new agency and making the service indispensable to the national space program. An essential element involved the Air Force's dominance in available space boosters. In a 17 September 1958 memorandum, Under Secretary of the Air Force Malcolm A. MacIntyre offered guidelines for the Secretary of Defense to follow in his discussions with NASA over civil and military program jurisdiction. Under Secretary MacIntyre argued for continuing the Air Force man-in-space program in cooperation with NASA, and reminded the defense secretary that the Air Force possessed the booster engine capability to support manned spaceflight. Responding on 31 October 1958, ARPA Director Roy Johnson noted that the Defense Department and NASA were following the guidelines suggested, and the Space Council would decide jurisdiction in unclear cases. Moreover, he concluded, "the Air Force's foresight in anticipating the requirements of both agencies for booster vehicles is to be commended. The present outlook is that all that have been provided for will be greatly needed and well utilized." In the months ahead the Air Force would continue to work to gain approval of exclusive responsibility for space booster development.[34]

When NASA commenced operations on 1 October 1958, a year after Sputnik initiated the space age, its leaders recognized that it would remain in the Defense Department's shadow for the foreseeable future. The Defense Department continued to focus on system work and big projects. The Air Force, through ARPA, not only pursued space-related missile work on solid propulsion, launch facilities, and test ranges, it also combined space and missile activity through projects like MIDAS, Samos, and antisatellite identification. Its impressive list of projects involved work on a manned orbital glider/bomber, new boosters, a variety of satellites, studies for

---

* See Appendix 2-3.

developing manned satellites and space stations, and support for Project CORONA, the covert reconnaissance satellite program publicly known as "Discoverer," which planners readied for launch in January 1959. Meanwhile, NASA focused on scientific applications through its existing NACA laboratories, and depended on the Defense Department and the Air Force for assistance with a variety of responsibilities. Of its first eight space probe launches, for example, the Defense Department accepted responsibility for the initial five, with the Air Force launching the first two Pioneer lunar probes.[35]

By the end of 1958 the foundation to support American superiority in space had been laid. Policy prescribed space activities for peaceful, that is nonaggressive, purposes, while organizational arrangements promoted a dual effort with civilian scientific aspects centered in NASA and military research and applications directed by ARPA. Yet much remained unresolved, not only between the Defense Department and NASA, but within the military arena. While the Air Force continued to face challenges with ARPA over program development and operational responsibility, a new Defense Department office appeared in late 1958 to add to the confusion. In August, Congress passed the Defense Reorganization Act which, among other measures to centralize and clarify defense operations, created the office of Director for Defense Research and Engineering (DDR&E), whose chief reported directly to the Secretary of Defense. Subsuming the old responsibilities of the Assistant Secretary of Defense for Research and Development, the new office became the focal point for all defense research and development activities. However, it would be a number of months before the new agency would be able to build its staff, sort out jurisdictional arrangements, and exercise its authority. Meanwhile, ARPA would continue to function as the nation's centralized military space agency. Nevertheless, the fact that the new office received explicit recognition in Public Law, while ARPA had been established only by authority of the Secretary of Defense signaled the ultimate demise of Roy Johnson's space agency. Air Force leaders hoped that the new Defense Department office would allow the service more autonomy in the space arena.[36]

As NASA prepared to begin its operations on 1 October, the Air Force had clearly left the Army and Navy behind in the quest for sole control of the military space mission. Even so, the chief of the Air Force's Legislative Liaison Office perceptively described an Air Staff divided on whether the service should assert itself more directly. Some officers preferred a "wait and see" approach, because the Air Force had received from ARPA a share of the space mission. Others argued for a more active role given the Army's retention of its 1.5 million-pound Saturn booster project as well as signs the Army would be authorized to develop communication satellites and the Navy would proceed with its navigational satellite. By the end of the year, Air Force leaders decided that they could not stand on the sidelines and let events take their course.[37]

Renewing the Quest for the Military Space Mission

The Air Force decision to promote itself for the military space mission in early 1959 precipitated a wide-ranging review of its current space posture and available courses of action. In early February the Deputy Chief of Staff for Plans described the Air Force's weaknesses in space organization, operations, and research and development that resulted, he said, from its failure to develop a coordinated space program. Rather than formally requesting operating responsibility for space roles and missions, the Air Force should demonstrate successful stewardship, rely on available hardware [boosters], and establish "squatters rights." Despite the presence of ARPA, the Air Force should establish its own integrated space program while working to improve relationships with both ARPA and NASA. The Air Force, he said, "must assume the role of opportunist, aggressively taking advantage of each situation as it arises to assure that the Air Force is always predominate [sic] in any action that has a space connotation."[38]

The Air Force campaign focused on congressional hearings in the winter and spring. Beginning in February 1959 Air Force spokesmen repeatedly elaborated on the Air Force "aerospace" policy that viewed space as…an extension of the medium in which we are now operating in the accomplishment of assigned roles and missions." As General White testified before the House Armed Services Committee, "The missions that we foresee [in space] are largely an extension of the missions that are required in the atmosphere." He went on to argue for funding and program support in terms of three general requirements: first, to improve current forces; second, to develop new systems in areas with recognized military applications; and third, to study and develop systems in areas without clear military applications but with excellent potential for possible future military use. The Air Force's manned space program ranked high among the latter requirements. Unlike NASA, whose mandate encompassed manned spaceflight and exploration of the unknown in outer space, the military would find programs without known applications particularly difficult to justify to congressional budget overseers.[39]

The Air Force's campaign intensified with the convening in late March of Senator Stewart Symington's Subcommittee on Governmental Organization for Space Activities. Scheduled witnesses Under Secretary of the Air Force Malcolm A. MacIntyre and Major General Bernard A. Schriever could expect a sympathetic response to a strong Air Force argument from Senator Symington, who continued to criticize the administration's budgetary frugality in the area of space defense.[40] Aware that the Air Force witnesses appearing before Congress required well-coordinated statements, the Air Staff's Directorate of Technology (DAT) and Schriever's Ballistic Missile Division staff developed position papers that provided a comprehensive assessment of current service strengths and weaknesses as well as a strong case for an increased Air Force role in space.

The Air Staff analysis demonstrated that the Air Force had successfully identified thirteen major military uses of space, nine of which had been included in the important NSC directive, "Preliminary Outer Space Policy."[41] Five of these missions—photographic/visual reconnaissance, electronic reconnaissance, infrared reconnaissance, mapping and charting, and space environmental forecasting and observing—had received approval as Air Force General Operational Requirements (GOR) and represented missions previously identified and analyzed by Rand. At the same time, Air Force headquarters had underway seven important studies with industry or in-house agencies and offices. Moreover, the analysis asserted, Lieutenant General Roscoe C. Wilson's DCS/D had produced an important paper outlining "Priority Listings of Military Space Missions."* In every document cited by the Directorate of Technology's officers, satellites received top billing, with Samos and MIDAS heading the list, followed closely by a variety of manned spaceflight requirements. Despite NASA's human spaceflight mission responsibilities, Air Force space leaders clearly had not relinquished interest in military manned spaceflight.

The Air Staff's analysis focused on constraints that prohibited the Air Force from implementing its aerospace "policy" of performing the space missions formally identified in Air Staff documents and approved as General Operational Requirements. It noted that the Air Force retained authority for planning, budgeting, and development only in non-space areas because NASA's responsibility embraced the scientific space area and ARPA's the military space arena. In effect, the Air Force had no responsibilities for a space program of its own. Echoing long-held criticism of the Defense Department agency, the Air Staff paper faulted ARPA for its practice of assigning system development responsibility to a service on the basis of existing capability but without regard for "existing or likely [space] mission and support roles." ARPA, rather, should focus on policy decisions and forego the "project engineering" detail normally found only at the lowest Air Force operating levels.[42] As for NASA, the Air Staff critique noted that the Air Force, if prohibited from pursuing its own scientific space exploration and research might very well face dependence on the "fall-out and by-products" of the civilian, scientific agency. To avoid this, the Air Force rather than NASA should develop programs of common interest, such as space boosters and satellites, in order to meet the more stringent military specifications and priorities. This would leave NASA to apply its budget to "really scientific projects" like unmanned space probes. Ultimately, concluded the Air Staff directorate, Air Force leaders should lobby Congress for a greater role for the Air Force in space.

General Schriever's staff also agreed that "it is axiomatic that the Air Force has the prime military responsibility for operating in space. Yet the means for developing

---

* See Appendix 2-4.

this capability are denied by present NASA/ARPA policies and actions."[43] Given the command's responsibilities, the BMD analysts criticized NASA for assuming a major portion of the nation's booster development program, indicating interest in taking over guidance, control, and ground tracking communications programs, and showing signs of building up "a development, production, management and 'operational' capability which will duplicate that presently existing in the AF Ballistic Missile Program." ARPA appeared to acquiesce in NASA's objectives while continuing to pursue its own development activities without regard to the future military operational user. Both agencies appeared oblivious to the "systems" concept of development, leaving the Air Force unable to establish an "integrated Air Force space program with a logical stepwise progression towards stated goals."

General Schriever also found his command becoming overburdened with increased management responsibility for ARPA programs and NASA's requirements for boosters and launch support. In a letter to the chief of staff on 11 February 1959, the general described the critical shortage for the next eighteen months of six Atlas boosters and limited launch pad availability at both Atlantic and Pacific Missile Ranges. Without immediate Air Staff action, he predicted delays in either the ICBM or booster operational schedules. The booster issue proved especially sensitive in view of the new emphasis on using Air Force Thor IRBM and Atlas ICBM requirements as the wedge into an enlarged space arena. As Schriever's staff explained, the close connection between missiles and space vehicles represented the best means of achieving Air Force space objectives because "future booster development as well as subsystem development can be initiated against bona fide ballistic missile requirements." The Air Staff responded by programming for additional boosters and launchers.[44]

In their testimony before the Symington Committee in late April 1959, Under Secretary MacIntyre and General Schriever presented a strong defense of Air Force space projects and the case for a greater Air Force space role. General Schriever, in particular, argued that by 1970 the Air Force's responsibilities for strategic offense and strategic defense would be accomplished by an arsenal of space weapons consisting of "ballistic missiles, satellites and space craft." To help the Air Force move forward with its space missions, he recommended that ARPA be dissolved by 30 June 1959, DDR&E assume the role of providing policy guidance and assigning service operating responsibilities, and space research and development be returned to the military services.[45]

The testimony of General Schriever and other Air Force spokesmen before congressional committees in the spring of 1959 proved especially effective in light of the Air Force's growing involvement in space. They could cite an impressive array of their "own" projects as well as important support the Air Force provided ARPA and NASA on others.[46]

Heading the list of major Air Force projects appeared the three elements of the former WS-117L Advanced Reconnaissance System. Samos, formerly known as Sentry, represented the reconnaissance element. Consisting of the Atlas booster and Lockheed's second-stage spacecraft vehicle Agena, Samos involved collecting photographic and electromagnetic reconnaissance data and transmitting the information by means of a "readout" system or actual "recovery" of data packages by aircraft. In contrast to Project CORONA, which pursued the capsule recovery technique, the Air Force initially had elected the "readout" method, but eventually would attempt both methods of data retrieval. MIDAS (for Missile Defense Alarm System) also relied on the Atlas-Agena booster satellite combination. The MIDAS payload consisted of infrared sensors designed to detect missile exhaust plumes and be able to provide command centers a thirty-minute warning of an ICBM attack.[47]

Both Samos and MIDAS projects experienced technical and management problems not uncommon to projects on the leading edge of technology. For example, civilian and military officials continually differed over technical requirements and capabilities, funding, and operational arrangements. While the Air Force proposed assigning operational control of Samos and MIDAS to the Strategic Air Command (SAC) and the North American Air Defense Command (NORAD), respectively, the Army and Navy argued for a joint command that would operate all military space systems. Air Force officials also favored implementing a systems development approach that would achieve desired performance goals while development and testing proceeded. Solving problems "concurrently," they hoped, would result in achieving early operational capability. The Office of the Secretary of Defense, however, preferred a "fly before buy" arrangement, and focused on component subsystem performance and capabilities. As a result, MIDAS and Samos remained in flux with the Air Staff repeatedly defending and revising development plans, while looking ahead to initial test flights in 1961.

Although publicly Project Discoverer represented a third Air Force project of the former WS-117L program, it actually served as a cover for the covert Project CORONA. After President Eisenhower in February 1958 authorized a secret reconnaissance satellite as a joint CIA-ARPA-Air Force effort, it became known as Project CORONA, an experimental activity within the WS-117L program. However, alarmed by publicity identifying CORONA as a military reconnaissance system, administration officials in the late summer of 1958 decided to sever CORONA's public connection with WS-117L by creating two photo reconnaissance efforts. While the Air Force pursued its Sentry/Samos project using the Atlas booster, CORONA would continue as Project Discoverer and rely on the Thor booster. The Discoverer project embraced tests on satellite stabilization equipment, satellite internal environment, ground support equipment, and biomedical experiments using mice and primates and, most importantly, capsule recovery techniques. Officials had scheduled thirty-two polar orbit launches from Vandenberg Air Force Base using the Thor-Agena combination. Of the four launches attempted by the end of June 1959, the first two achieved orbit for brief periods and passed back useful experimental data despite loss of the capsules. The remaining two failed to achieve orbit.[48] Despite difficulties with the satellite systems during this early developmental phase, the Air Force could claim that it managed or supported the nation's most important satellite programs, and expected to be awarded greater operational responsibility in the near future.

In addition to its own Samos and MIDAS satellite projects, under ARPA's direction the Air Force provided launch support to the Navy's Transit navigational satellite, designed to support Polaris submarines, and the Army's Notus communications satellite effort.[49] The most important, however, proved to be the growing detection, tracking and satellite cataloguing project known as the Space Detection and Tracking System (SPADATS). Begun hurriedly under the name Project Shepherd by ARPA in response to Sputnik, all three services were to participate. The Air Force, under Project Harvest Moon (later Spacetrack), would provide the Interim National Space Surveillance and Control Center (INSSCC) data filtering and cataloguing center at its Cambridge Research Center in Massachusetts. Early efforts brought together radar data from MIT's Lincoln Laboratory's Millstone Hill radar at Westford, the Stanford Research Institute in Palo Alto, California, and an ARDC test radar at Laredo, Texas. Sensors included the new Smithsonian Astrophysical Observatory's Baker-Nunn satellite tracking cameras that it procured for tracking the IGY satellites and available observatory telescopes. The Air Force would also devise the development plan for the future operational system.[50]

ARPA assigned the Navy responsibility for developing and operating its east-west Minitrack radar fence and its data processing facility in Dahlgren, Virginia. Originally designed to support Project Vanguard, the Navy redesignated its sensor and control operation Spasur (Space Surveillance). The Army portion, termed Doploc, envisioned a doppler radar network to augment Spasur and, together, feed data to the INSSCC for cataloguing, trajectory prediction, and dissemination. ARPA and the three services realized the system's limited capability, but agreement on funding necessary improvements proved difficult to achieve. After the Army dropped out of the picture, the Air Force and Navy contested for operational control of the system. The Navy seemed to prefer operating a separate system, while the Air Force wanted its Air Defense Command (ADC) to acquire management responsibility and NORAD to possess operational control. By mid-1959, the controversy had reached the Joint Chiefs of Staff, where it became embroiled in a major roles and missions contest among the services.

As for NASA's requirements, the Air Force agreed to construct infrastructure facilities at Patrick Air Force Base for NASA's space probes and then provide booster support for the Pioneer lunar probes (Thor-Able) and Tiros cloud-cover satellite (Thor-Delta/Able). The Air Force also supported the Centaur high-energy upper stage based on hydrogen and oxygen as fuels, which it hoped to use in support of the Advent communications satellite project. Most importantly, the Air Force supported Project Mercury, NASA's man-in-space project, by furnishing Atlas boosters and launching services, along with considerable technical, biomedical, and personnel assistance. The issue of military manned spaceflight had always been a most sensitive subject for Air Force space enthusiasts. Like their German counterparts, early Air Force space pioneers looked to space as more than an arena for scientific exploration or simply a venue in which to pursue exciting new challenges. They considered a military man in space mission the logical extension and eventual goal of Air Force space operations. Not only did this objective correspond to Air Force thinking on "aerospace," but manned spaceflight seemed the next "logical" step in the chain of operational development from aviation medicine to space medicine. Indeed, by the time of Sputnik, Air Force medical personnel could look back on a wealth of aeromedical experience that put the service at the forefront of knowledge on conditions of flight in the upper atmosphere and near space. Space presented scientists the daunting challenge of mastering the complexity and weight problems in a space environment.[51]

At the close of the Second World War, the Air Force gained the services of a number of German scientists who had performed path-breaking medical research for the Luftwaffe. Although most joined the growing Aeromedical Laboratory at Wright-Patterson Air Force Base in Ohio, six received assignment as research physicians to the Air Force School of Aviation Medicine at Randolph Air Force Base near San Antonio, Texas. In February 1949, the latter established the world's first Department of Space Medicine, under the direction of Dr. Hubertus Strughold, who had coined the term "space medicine" at an important symposium the previous year. In November 1951 the Randolph school held another symposium, entitled "Physics and Medicine of the Upper Atmosphere," to avoid criticism of "Buck Rogers" projects within the Air Force. Nevertheless, at this meeting Strughold advanced the concept of the "aeropause," an area of "space-equivalent conditions" such as anoxia that begins much lower, about 50,000 feet, rather than at the 600-mile barrier normally cited by authorities as the boundary between the atmosphere and space. "What we call upper atmosphere in the physical sense," Strughold said, "must be considered— in terms of biology—as space in its total form." In effect, manned ballistic or orbital flight at the 100-mile altitude would be spaceflight. Strughold would come to be known as the "the father of space medicine" and go on to lead the Air Force's School of Aviation Medicine in exploring the space environment. Together with researchers at the Wright Air Development Center (Aeromedical Laboratory) and the Aeromedical Field Laboratory at Holloman Air Force Base, New Mexico, Air Force space medicine teams from San Antonio pursued a variety of experiments dealing with conditions of "zero g" or weightlessness in space, "g loads," or the effects of heavy acceleration and deceleration primarily through the upper atmosphere rocket plane flights and sounding rockets with animal passengers. Although the crash ICBM program in the 1950s interrupted animal flight research for a six-year period, other human factors experiments continued. By the time of the Sputnik launch, Air Force medical research specialists had accumulated a wealth of data on conditions of manned spaceflight and determined that the basic problems of weightless flight could be solved.[52]

While Air Force medical personnel continued their quest for data on conditions of manned spaceflight, scientists and engineers conducted research and development on space hardware systems that could eventually be powered through the upper atmosphere into earth orbit. Manned space vehicle concepts proceeded along two lines of thought based on the reentry technique used. One involved ballistic reentry using blunt-body capsules, the other aerodynamic reentry with winged vehicles. Although Air Force planners pursued both methods of spaceflight, initial interest centered on the winged suborbital vehicle later known as Dyna-Soar (from dynamic soaring).

Dyna-Soar evolved from the rocket plane studies and experiments of the early 1950s. By May 1955 hypersonic (Mach 5 and above) glide vehicle development had led to three related Air Force projects: Bomi, an acronym for Bomber Missile, but soon redesignated Robo, for Rocket Bomber; Brass Bell, a high altitude reconnaissance system; and Hywards, the actual boost-glide vehicle. Although designed for suborbital flight, the three could be launched into low earth orbits with adequate propulsion. After it became apparent that weapons in space would not proceed, on 30 April 1957 the Air Force merged the three programs under the name Dyna-Soar, and considered it the manned flight successor to turbojet bombers and reconnaissance aircraft. To reflect the requirements of the Air Force's first "aerospace" vehicle, engineers designed the Dyna-Soar as a manned, delta-wing aeronautical vehicle capable of being boosted into orbit while retaining reentry and controlled landing maneuverability. As such it filled a variety of accepted mission functions and could be supported by the vast network of existing ground facilities.

As early as the spring of 1956, the Air Force had discussed with several industrial firms its manned ballistic rocket research program. When the Air Force prepared its ambitious five-year astronautics plan in the heady weeks following the launch of Sputnik, it included projects for a manned capsule test system, manned space stations, and ultimately a manned lunar base. Although critics scoffed at such "fanciful" projects, ARPA director Roy Johnson did not. Shortly after his appointment in February 1958, he declared that "the Air Force has a long term development responsibility for manned space flight." With his blessing, Air Force leaders requested ARPA funds and directed Air Research and Development Command to prepare a development plan, called "Man-in-Space-Soonest" (MISS). It called for a four-phase capsule orbital process, which would first use instruments, to be followed by primates, then a man, with the final objective of landing men on the moon and returning them to earth.

The Army and Navy did not relinquish the field of manned spaceflight to NASA or the Air Force uncontested. In the spring of 1959 the Army unveiled its "Man Very High" proposal, later termed Project Adam, which called for lofting a man in a Jupiter nose-cone capsule on a steep ballistic trajectory that would produce a splashdown about 150 miles downrange from Cape Canaveral, Florida. Project Adam received no support from informed critics like NACA's Hugh Dryden, who explained that "tossing a man up in the air and letting him come back…is about the same technical value as the circus stunt of shooting a young lady from a cannon." The Defense Department rejected the Army plan from the start. The Navy's intriguing alternative, MER I (Manned Earth Reconnaissance), proposed orbiting a cylindrical vehicle with spherical ends. After achieving orbit, the ends would expand laterally to produce a delta-winged inflated glider. Although ARPA conducted studies on the proposal's feasibility, NASA's Project Mercury soon got underway and relegated the Navy plan to an interesting concept too bold for its day.

Although the Air Force MISS proposal came closest to "approval," ARPA balked at the high cost of $1.5 billion and the uncertainties surrounding the future direction of the civilian space agency. When NASA began operations on 1 October 1958, the Air Force had prepared seven Man-In-Space-Soonest development plans, each one dismissed by ARPA for cost, technical, or utility concerns. Fittingly, the last one omitted the word "soonest." When President Eisenhower assigned NASA the human spaceflight mission in August 1958, ARPA transferred its manned space programs and funds to the new civilian agency. Hampered by insufficient funding, the President's "space for peace" policy, and the inability to justify a military man in space, the Air Force had to abandon—at least for the time being—serious plans for a distinct and separate military man-in-space program.

NASA's assumption of the manned space mission left the Air Force with Dyna-Soar, a single-place vehicle, which the Air Force had protected from ARPA's grasp by stressing its suborbital, aeronautics phase of development. Although Dyna-Soar had received approval for development in 1958, by the spring of 1959 the Air Force still had not identified an adequate booster to fulfill the as yet undetermined aeronautical, missile and, especially, space requirements of Dyna-Soar. An initial proposal called for using a cluster of the yet-to-be-developed Minuteman solid-propellant rockets, but the problem of separating the rockets as they would be expended proved too challenging and costly. This opened the door to possible encroachment from the Army and NASA.

The Army's Saturn appeared as a logical candidate, and Wernher von Braun made several attempts to convince the Air Force to accept the Saturn–Dyna-Soar combination. But the Air Force demurred, preferring to continue with its own 1,500,000 lb-thrust engine project it had underway. Given NASA's interest in Saturn, however, the Air Force might very well lose Dyna-Soar to NASA if the civilian space agency acquired the Army's big booster. In the spring of 1959, the Air Force continued to move forward with the Dyna-Soar project and hoped that it could keep alive a military manned spaceflight mission. Meanwhile, it would continue its strong support of NASA's Project Mercury.

By the spring of 1959, the Air Force's expanding role in space led Air Staff leaders on 13 April 1959 to enhance the headquarters focus on space by providing General Boushey's year-old Directorate of Advanced Technology the authority to coordinate within Air Force headquarters all space issues. The new arrangement eliminated the space responsibilities of the Assistant Chief of Staff for Guided Missiles except for coordination activities involving boosters, test facilities, and range and launch complexes. Gone at last was the divided authority within the Air Staff for space requirements.[53]

The Air Force's 1959 campaign for the military space mission did not go unnoticed by the Army and Navy. They closely followed the Air Staff realignment, the growing Air Force responsibilities for space systems, and the coordinated testimony of its spokesmen before Congress. In fact, General Medaris seized his opportunity before Senator Symington's committee to accuse the Air Force of a long history of noncooperation with his Army Ballistic Missile Agency. Although General Schriever provided a lengthy, detailed rebuttal, Medaris refused to withdraw his charges. The dispute only served to reinforce the views of legislators already critical of interservice rivalry.[54]

In a move more threatening to Air Force interests, Admiral Arleigh Burke, Chief of Naval Operations, in late April 1959, made "a bold bid for a major share" of the space mission, by proposing to his Joint Chiefs of Staff colleagues the creation of a joint military space agency. In effect, he advocated a unified command for space based on the "very indivisibility of space," projected large-scale space operations in the near future, and the interests in space of all three services. Army Chief of Staff General Maxwell D. Taylor agreed, arguing that space activities transcended the particular interests of any one service. But Air Force Chief of Staff General White opposed the proposal because, he said, it violated the practice of treating space systems on a functional basis and integrating weapons within unified commands. He argued that space systems represent only a better means of performing existing missions and should be assigned to the appropriate unified or specified command.[55]

The Navy-Army initiative to gain a greater military space role by working through the Joint Chiefs of Staff to realize a joint command compelled General Schriever in mid-May to argue for an Air Force counter-campaign to acquire all or part of the military space mission "as soon as possible." In a letter to Lieutenant General Roscoe C. Wilson, DCS/D, the ARDC commander described his concerns and provided a draft letter for either Air Force Secretary James H. Douglas or Chief of Staff General White to forward to Secretary of Defense McElroy. His suggested letter asserted that "since its inception" the Air Force had been operating in aerospace through the mission areas of strategic attack, defense against attack, and supporting systems that enhanced both the strategic retaliatory and active defense forces. The Air Force had important requirements for earth satellites, which represent aerospace vehicles of the foreseeable future. Characteristically, Schriever criticized existing fragmented satellite program management and advocated a unified, systems development approach that would "achieve the most effective deterrent posture" by coordinating and integrating satellite systems within the broad Air Force strategic and air defense force. Moreover, Army and Navy requirements, the general asserted, would be best achieved by the Air Force acting as "prime operating agency of the military [national] satellite force."[56]

While the services argued over roles and missions, ARPA director Roy Johnson stoked the fire in June by recommending a tri-service Mercury Task Force to support NASA, while Defense Secretary McElroy requested advice from the Joint Chiefs of Staff on assigning the services operational responsibility for several important space projects, including MIDAS and Samos. Service views reflected the division over the unified command issue. While the Navy and Army favored a Mercury Task Force as well as a Defense Astronautical Agency to direct and control all military space systems, the Air Force opposed both for the reasons General White explained earlier in response to Admiral Burke's proposal.[57]

With no resolution of the differences by the fall of 1959, Secretary McElroy in September made three decisions that propelled the Air Force further forward in its quest for exclusive responsibility for military space activities. Differing with Admiral Burke's prediction, DDR&E director Herbert York had argued that the country could expect relatively few satellites in orbit in the foreseeable future, and thus the nation did not need a unified space command. The Secretary of Defense agreed, and sided with the Air Force position by declaring that "establishment of a joint military organization with control over operational space systems does not appear desirable at this time." He too disapproved both the proposed Defense Astronautical Agency and Mercury Task Force. In place of the latter, he designated Air Force Major General Donald N. Yates, Atlantic Missile Range commander, to "direct military support" for NASA's manned space project. Most significantly for the Air Force, the Defense Secretary assigned to it responsibility for "the development, production and launching of space boosters" as well as payload integration. Satellite operational responsibility, however, would continue to be assigned to the services on a case-by-case basis. Initially, the Air Force would receive Samos and MIDAS (in November) and, in a separate action, Discoverer (in December). The Navy acquired the Transit navigation satellite, and the Army four Notus communications satellites. In short, Secretary McElroy agreed with Dr. York and Air Force critics by reversing his established policy that favored ARPA and reassigning space projects among the three services. The Air Force received the major share. Admiral Burke's proposal for a unified command for space would prove twenty-five years too early.[58]

Secretary McElroy's directive in September represented the first fruits of the Air Force campaign of 1959 for the military space mission. Legitimately hailed as a landmark decision on the Air Force's road to space, it nevertheless provided the Air Force an incomplete victory over its protagonists. Pessimists pointed out that civilian control over development of military space systems remained unchanged at the secretarial level, and ARPA retained its authority to conduct project engineering supervision. Moreover, the Air Force received responsibility for space boosters but not for all space satellite systems. On the other hand, the Air Force had warded off a joint operational agency for space and received designation as the nation's "military space booster service"—a major objective of the spring campaign, and a further blow to the Army's space fortunes. The Air Force now found itself poised to assume command and control of operational space systems, while receiving operational control of Samos, MIDAS, and Dyna-Soar—all space systems with growth potential.

On balance, in the fall of 1959, Air Force leaders could express optimism about the space future, fully aware that much needed to be done to consolidate the September gains. At the Air Force major commanders' conference on 1 October 1959, the audience heard that "the Army and Navy can be expected to continue their efforts to neutralize this interim Air Force victory" by showing that missile range and tracking facilities as well as satellite payloads deserved unified command direction and control. Now that the Air Force had gained its first chance to issue plans for development and operation of particular space systems, it would need to make good use of this opportunity. "Future steps toward gaining the assignment of space responsibilities will be determined...by the manner in which the Air Force handles the responsibilities it has just been assigned."[59]

Before it discharged any of these responsibilities, however, the Air Force began lobbying for the Army's Saturn heavy-lift booster project. As Vice Chief of Staff General Curtis E. LeMay explained to the Secretary of the Air Force on 29 September 1959, "in view of this directive [18 September 1959] it appears that the in-house capability of the Department of the Army for the development of space boosters and systems, which is represented by the Army Ballistic Missile Agency at Huntsville, Alabama may now be available for transfer to the Air Force."[60] But Saturn was not a weapon system, and NASA, with funds available and manned spaceflight on the horizon, could make a far better case for the big booster than could the Defense Department. Try as they did, Air Force planners could not specifically justify the need for a 1,500,000-pound thrust engine. Apparently, Secretary of Defense McElroy offered the Saturn to NASA's Director Glennan, who contacted General Medaris. After DDR&E York publicly confirmed that the Air Force would develop all space boosters needed by the Defense Department, integrate space payloads and launch the combination, Medaris preferred to transfer to NASA the entire von Braun team and missile operation, rather than have the Redstone complex and personnel separated and parceled out to various agencies. Despite objections from the Joint Chiefs of Staff, President Eisenhower approved Saturn's transfer to NASA on 2 November 1959. The Air Force would have to await more favorable circumstances to gain authority to develop military superboosters.[61]

With the President's decision underscoring NASA's claim to human spaceflight, Air Force leaders realized that the Dyna-Soar project had become endangered. At the end of October 1959, General Boushey, chief of the Directorate of Advanced Technology, declared that the Saturn decision suggested that "the loss of the Dyna-Soar project to NASA appears imminent." He predicted such an action would effectively remove the Air Force from super booster development and nullify the 18 September 1959 memorandum assigning the Air Force space booster responsibilities. Events proved General Boushey's pessimism misplaced. York reaffirmed the Air Force's Dyna-Soar project and the service selected Boeing as contractor in November 1959. Yet Air Force leaders remained aware of the fragile state of the project's future.[62]

The end of the year also brought the official demise of ARPA as the central Defense Department agency for space activities. Following the transfer of most of its space projects to the services in the fall, a 30 December 1959 directive from Secretary McElroy designated ARPA as "an operating research and development agency of the DoD under the direction and supervision of the DDR&E." In the future, ARPA would manage only a limited number of advanced research programs. General Schriever and other Air Force leaders rejoiced at ARPA's demise and the return of development responsibilities to the user agencies. Yet it meant removing a high profile centralized space management agency close to the Defense Secretary. With the military spotlight on space now reduced, space projects faced competition from other worthy service requirements in the battle for funding, while greater service rivalry over space systems without clear service roles became a distinct possibility.[63]

DDR&E now became the dominant Defense Department reviewing office with far more authority over Air Force research and development proposals than ARPA possessed. In late 1959 Lieutenant General Roscoe C. Wilson, Deputy Chief of Staff for Development, expressed his concerns about the civilian technical influence that resulted in considerable wasted time and effort before decisions from "on-high" reached the Air Force. He also complained about civil-military relationships within the Air Force community. One involved Secretary of the Air Force James H. Douglas' initiative, in October 1959, to have all space decisions taken by the civilian-led Air Force Ballistic Missile Committee in the Office of the Secretary of the Air Force without significant Air Staff participation. Although Douglas' successor, Dudley C. Sharp, agreed to allow prior review of space issues by the Air Staff and increase its role in space development planning, the final decisions remained with his Ballistic Missile Committee.[64]

Despite these concerns, by end of the year the Air Force clearly had become recognized as the dominant military service in space. Lacking the boosters, facilities, and space experience of the Air Force, the Army and Navy found themselves on the periphery of the space picture, while ARPA had been reduced to insignificance. The changes in late 1959 affected the "space budget," too. The Air Force benefited the most from ARPA's loss of 80% of its funding. While NASA succeeded in nearly doubling its fiscal year 1961 budget from $535,6000,000 to $915,000,000, Air Force funding multiplied by nearly 120 times, from $2,200,000 to $249,700,000. Air Force leaders now could argue that the service had regained control of much of its "own" space program. Moreover, NASA remained dependent on the Air Force for launch boosters and range support and, Project Mercury notwithstanding, the Air Force's Dyna-Soar manned space program continued on the drawing board. If the Air Force had not achieved the complete victory sought by its leaders, it nonetheless seemed well on its way to gaining management responsibility for all service requirements as the Defense Department's executive agent for space.[65]

The Air Force Seeks to Consolidate Its Position

As the Eisenhower administration entered its final year, the President could take pride in the country's space program. In the spring of 1960, the number of American scientific and space probe launches totaled 24, of which 14 had achieved successful orbit. The Soviets had succeeded only in launching three such spacecraft, although they continued to garner world prestige from their spectacular "feat" of hitting the moon and photographing its far side. On the international front, the United Nations (UN) prepared to establish a permanent 24-nation Committee on Peaceful Uses of Outer Space, ten European nations discussed formation of a joint agency for scientific space exploration, and the administration continued its nuclear test ban and disarmament efforts by offering the Soviets use of America's global tracking network for its manned space experiments.[66]

Nevertheless, Air Force leaders continued to chafe at what they considered a policy that produced too modest a defense-support space program and prevented offensive space weapon system development altogether. They centered their criticism on the administration's National Security Council 18 August 1958 national space policy, "Preliminary Policy on Outer Space." If this directive represented a preliminary statement of policy, hopefully a more conclusive formulation of policy would provide specific recognition of military requirements. Back on 30 June 1959, President Eisenhower had charged the National Aeronautics and Space Council to review the preliminary policy. It took the group a full six months to prepare their report. Approved by the NSC as Directive 5918 on 26 January 1960, the "U.S. Policy on Outer Space" continued to emphasize a policy of civilian "peaceful" scientific exploration and development activity. It lauded the UN's approval of the "launching and flight of space vehicles...regardless of what territory they passed over"—as long as they involved the "peaceful uses of outer space." Although the directive accorded the military mission better recognition, it restricted military space functions to defense support and, once again, specifically limited offensive space weapon systems to study only.[67]

Although the revised space policy disappointed military leaders, Eisenhower's attempt to have the Space Act amended in early 1960 provided another opportunity to promote greater recognition of the military space role. The President believed that the single national space program implied in the act was impractical and undesirable. Dual civilian and military programs represented reality and should be formally recognized. Because NASA and the Defense Department cooperated effectively without what he considered inappropriate congressional mandates, the National Aeronautics and Space Council and Civilian-Military Liaison Committee should be abolished. Furthermore, he desired presidential relief from direct program planning responsibility but, to avoid duplication, sought specific authority to "assign responsibility for the development of each new launch vehicle, regardless of its intended use, to either NASA or the Department of Defense."[68]

The President sent his proposed amendments to Congress on 14 January 1960, where they received considerable scrutiny in hearings that winter and spring. Not only did many legislators remain unhappy that the country seemed to trail the Soviet Union in space progress despite administration statements to the contrary, the fact that 1960 was a presidential election year assured a lively and contentious debate on space in the months ahead. Overton Brooks, Democrat from Louisiana and Chairman of the House Committee on Science and Astronautics, predicted as much in late fall of 1959 when he warned that Congress early in the new year would "probe every facet of the [space] program." Brooks, in fact, had been trying since the spring of 1958 to convince the administration that the country should have an integrated space program.[69]

Representative Brooks and other congressional leaders convened a number of committees to examine the President's request and review the merits of whether the country had or should have one or two space programs under civilian and/or military control. Since there appeared no ready solution to the issue, the Eisenhower administration's preference of separate programs continued. As for the President's recommendations, the House agreed to eliminate both oversight bodies, but in so doing convinced the administration to accept a substitute, the Aeronautics and Astronautics Coordinating Board (AACB). Cochaired by the Defense Department's Director for Defense Research and Engineering (DDR&E) and NASA's Deputy Administrator, the new coordinating body, unlike the CMLC, possessed the authority to make binding decisions. The Senate, however, chose not to act on the President's request until a new administration could review the issue. As a result, the NASC and CMLC continued in law yet ceased to function, while the AACB began operating in September 1960.[70]

The hearings provided an opportunity for Defense Department witnesses to lobby for a wider military role in space. At the same time, pointed questions about space planning revealed the weaker side of the Defense Department and Air Force approach to space. NASA impressed committee members by presenting a "10-year plan" with funding milestones for research, development and exploratory space activities in pursuit of peaceful objectives. The NASA initiative placed the Defense Department on the defensive. The Defense Department had no such plan and, as DDR&E Herbert York explained, it saw no reason to prepare one. Testifying on 30 March 1960 before Senator John Stennis' NASA Authorization Subcommittee of the Committee on Aeronautical and Space Sciences, his argument reflected the logic of the Air Force concept for space planning and operations. Unlike NASA, he said, the Defense Department did not view space as a mission, with spaceflight and exploration as ends in themselves, but rather as a means for achieving better military space applications to improve existing terrestrial military mission capabilities. "Considering the nature of our space objectives, it is not logical to formulate a long-range military space program which is separate and distinct from the overall defense program." The Defense Department's planning process also served the administration's political agenda by highlighting the civilian program rather than the military.[71]

While DDR&E York presented a sound argument, the Defense Department's unwillingness to produce a space plan left it open to criticism from a Congress sensitive to duplication and effective development and coordination between NASA and the Defense Department. While NASA seemed to know where it wanted its space program to go in future, the Defense Department appeared less certain. Especially in the field of space exploration, which demanded initial funding for programs without definite military mission applications, the military found it difficult to convince Congress without benefit of an effective long-range plan. For the Air Force, this meant that its budget reflected space not as a program in itself, but as part of traditional mission areas. The Samos reconnaissance satellite, for example, appeared under strategic elements, while the MIDAS early warning system supported air defense mission requirements. Even after ARPA had transferred space projects to the Air Force, the scattering of space projects throughout the budget prevented a strong focus for advocacy of a military space program during the budget process.

At the same time, Air Force planners encountered difficulty in development and operational planning for space systems. While the so-called indivisibility of "aerospace" provided a conceptual approach to space that supported the service's quest for military space missions, it did not contribute effectively to a planning process that required consideration of space as a separate medium. Not only did space systems, in fact, involve different technical challenges, determined by orbital dynamics in a hard vacuum, but the lack of basic knowledge about many aspects of space contributed to the complexities of the planning process.

Nevertheless, the Defense Department's lack of a space plan per se did not mean that the Air Force conducted no long-range space planning. Air Staff planners had attempted since early 1958 to develop conceptual plans for space by means of an Air Force Objective Series (AFOS) paper. An agreed-upon AFOS paper would be complemented by a Required Operational Capability (ROC) document, which would identify the forces necessary to achieve the objectives (AFOS). Only by September 1959 could planners agree on a ten-year plan for peacetime and wartime operations that seemed to meet Air Force requirements without conflicting with national policy. Yet critics claimed that the draft document treated space as a separate "entity" in violation of the "aerospace" concept, and subsequent AFOS drafts failed to gain approval throughout the spring. Meanwhile, Air Staff officers working on the ROC also encountered roadblocks when they presented their "revolutionary" developmental program. Looking ahead to an operational date of 1975, they proposed a high-profile program with major funding increases to achieve innovations in propulsion, structural materials, and guidance, as well as "human factors development" as part of a future military man-in-space program. The ROC clearly treated space as a mission by calling for development of space weapons regardless of whether earth-based aeronautical systems might provide a more efficient and cost-effective alternative. Air Staff critics dismissed the plan as too "utopian" and risky. Without approval of these two planning documents, the 120-page qualitative force structure analysis that would logically follow in the form of a Research and Development Objectives (RDO) paper, remained a "dead letter."[72]

Not until the fall of 1960 could Air Force planners agree among themselves and gain the necessary approval for their ROC and RDO proposals. Another nine months would pass before the Air Force issued its first Objective Series statement depicting long-range concepts and its vision of military space activities. By then, Air Force leaders dealt with another administration that appeared to be far more sympathetic to their objectives. Much of the planning dilemma resulted from the unwillingness of General White and other Air Force leaders to issue official guidance for meeting national space policy and engage in an Air Force-wide educational campaign on space. The administration's "space for peace" policy tended to inhibit independent, high-profile Air Force military initiatives, while any official Air Force statement on space would prove of marginal value as long as space remained the preserve of ARPA or NASA for funding, management, and overall technical direction.[73]

While Air Force leaders might very well ballyhoo the concept of "aerospace" in public forums and argue that "aerospace power is peace power," current political and organizational constraints called for a more cautious approach to Air Force pronouncements on space. Back in July 1959 Air Staff planners initiated a formal space policy study, which received greater attention following ARPA's demise in the fall. By the end of the year, the Chief of Staff's "policy book" contained a number of statements for use in the 1960 congressional hearings. General White, however, desired a comprehensive space policy statement he could issue officially. After numerous reviewers on the Air Staff and in the Office of the Secretary of the Air

Force had their say, a final version seemed ready for publication in mid-March. Yet General White considered the timing "inappropriate." As the Air Force headquarters historian concluded, the chief of staff worried that "publication of an official [space] policy statement at a time when so many facets of the space program were still undecided would have unfavorable reverberations in Congress, the Office of the Secretary of Defense, and the other military services."[74]

General White's caution was not misplaced. In early May 1960, shortly after the Air Force had submitted its operational plans for MIDAS and Samos, Admiral Arleigh Burke, the Chief of Naval Operations, again challenged the Air Force position on space operations. He reaffirmed the need for a joint [unified] military space agency based on major technological developments of the last year and a half that propelled several systems to the "operational threshold." He also referred to the substantial interservice support for NASA's Project Mercury, and the joint agencies soon to be established for command, control, and communications functions. After dividing along the lines of the previous summer, the Joint Chiefs of Staff forwarded its divergent views to Secretary of Defense Thomas Gates, who had held the post since December 1959. On 16 June 1960, he reaffirmed the decision earlier taken by his predecessor on 18 September 1959.[75]

For a second time, the Air Force had deflected an Army-Navy challenge to its growing military role in space. Its prudent, cautious approach to asserting its prominence in the military space picture seemed vindicated. By late summer, however, the Air Force would lose control of one of its largest and most important space missions.

The downing of the U-2 reconnaissance aircraft piloted by Francis Gary Powers on 1 May 1960 destroyed plans for an East-West Summit Conference and limited reconnaissance flights exclusively to the periphery of the USSR. It also brought the troubled Samos and MIDAS satellite programs more funding from the administration and Congress, while compelling officials to reassess the reconnaissance satellite program at the highest government levels.[76]

Eisenhower's "peaceful purposes" space policy covered CIA as well as military involvement in a reconnaissance satellite program. Back in February 1958 the President authorized the CIA to develop a reconnaissance satellite, assisted by elements of the Air Force, after being told by his Board of Consultants on Foreign Intelligence Activities that Samos could not meet near-term requirements, because it used film readout and relied on the Atlas booster. While the Atlas would not be operational for several years, by using the Thor IRBM, the CIA might have a film recovery satellite launched by the spring of 1959. Using as a cover the Air Force's Discoverer project, the CIA designated its highly sensitive operation Project CORONA.

Satellites had to fill the intelligence gap created by the loss of the U-2. On 10 June 1960 Eisenhower directed Secretary of Defense Gates to reassess intelligence requirements and the prospects for fulfilling them using the Air Force Samos readout system. In turn, he appointed a three-man panel made up of the President's science advisor, George B. Kistiakowsky, John H. Rubel, Deputy Director for Defense Research and Engineering, and Joseph V. Charyk, Under Secretary of the Air Force. Apparently, over the summer Kistiakowski and the President's Scientific Advisory Committee performed most of the work, assisted by Richard Bissell and his CIA science advisory panel. CORONA, meanwhile, achieved its second success in four-teen attempts on 20 August, recovering the first film capsule. Kistiakowsky pre-sented the Samos findings to the President in a NSC meeting on 25 August. The report concluded that the Samos satellites, like CORONA and the U-2, represented a national asset. As such, the project should not be directed by a military service, but by a civilian agency in the Defense Department. The President agreed and autho-rized an accelerated program directed by the Secretary of the Air Force and report-ing to the Secretary of Defense.[77]

The new program arrangements took shape quickly. On 31 August Secretary of the Air Force Dudley Sharp created within his department the Office of Missile and Satellite Systems under the Assistant Secretary of the Air Force, who would be responsible for coordinating Air Force, CIA, and later Navy and National Security Agency (NSA) intelligence reconnaissance activities. Secretary Sharp named Brigadier General Robert E. Greer director of the Samos west coast development field office. At the same time, the Secretary established two advisory bodies: a Satellite Reconnaissance Advisory Group made up of four civilian technical spe-cialists, and a Satellite Reconnaissance Advisory Council. Chaired by the Under Secretary of the Air Force, the council included General Greer, the three Air Force assistant secretaries, the vice chief of staff of the Air Force and two senior Air Staff officers. Within months, the Office of Missile and Satellite Systems became the secret National Reconnaissance Office (NRO), directed by the Under Secretary of the Air Force, and responsible for all reconnaissance satellite projects, including CORONA. The Samos effort disappeared from public view as surely as it did from Air Force control.[78]

Although the new reconnaissance satellite offices remained within the Office of the Secretary of the Air Force and employed serving Air Force officers, Air Force headquarters was essentially excluded from the operations of this highly sensitive national project. As a result, the military satellite reconnaissance program would operate outside the Air Force area of responsibility. Moreover, when continued funding and technical problems led to cancellation of Samos in the early 1960s, only the equally troubled MIDAS missile early warning satellite and the Vela nuclear detection spacecraft remained in the Air Force satellite inventory.

While the Air Force lost control of the Samos satellite program, it took action to create The Aerospace Corporation to insure that it would have the technical competence to meet current and future space age challenges.[79] Although the new systems approach had proven successful during the crash missile program, the systems engineering role played by Ramo-Wooldridge Corporation generated criticism from aerospace firms and Congress about its privileged position. When, on 31 October 1958, it merged with Thompson Products, Inc., to become Thompson-Ramo-Wooldridge (TRW), Inc., its Space Technology Laboratory (STL) became an "independent" subsidiary of TRW. Nevertheless, conflict-of-interest charges and congressional scrutiny compelled General Schriever to seek an alternative based on a nonprofit, noncompetitive arrangement.

Secretary of the Air Force James H. Douglas and other Air Force leaders agreed. A special committee confirmed the nonprofit corporation approach, and in the spring of 1960 General Schriever and Under Secretary Joseph Charyk worked with an organizing committee to form a new corporation. By 3 June they had established The Aerospace Corporation on El Segundo Boulevard in Inglewood, California, adjacent to the Ballistic Missile Division headquarters. At a news conference on 25 June, Chairman of the Board Roswell L. Gilpatrick declared that his organization represented "a new approach on the part of the Air Force in the management of its missile and space programs." By the end of the year, the new corporation had acquired more than 1700 employees and responsibility for twelve major Air Force programs. Eventually, the Aerospace Corporation would provide general systems engineering and technical direction (GSE/TD) for every missile and space program undertaken by the Air Force.

Air Force leaders had good reason for optimism in the fall of 1960. They had beaten back space challenges from the Navy and Army and had created the Aerospace Corporation. Despite losing control of the Samos program, the Air Force continued to expand its space role in the Space Detection and Tracking System, in booster development, and in development of infrastructure to support national space policy. The Air Staff's Brigadier General Homer A. Boushey forecast in the fall of 1960: "We can go into space with our feet firmly planted on the ground." Yet, Air Force leaders soon threw caution to the wind. With the prospect of a new and potentially more space-oriented administration on the horizon after the November 1960 election, Air Force leaders decided to embark on a campaign to influence the thinking of the new administration on space issues.[80]

The Military Space Mission Goes to the Air Force

Senator John F. Kennedy made space an issue in the 1960 presidential election campaign. Referring to Soviet "firsts," he cautioned that "if the Soviets control space they can control the earth, as in past centuries the nation that controlled the seas dominated the continents....We cannot run second in this vital race. To insure peace and freedom, we must be first." He called for an accelerated space program.[81]

Shortly after his narrow victory over Vice President Richard M. Nixon, Kennedy appointed a committee to review the country's space program. Chaired by MIT's Jerome B. Wiesner, the "Wiesner Committee" included among its nine distinguished members Trevor Gardner, prime mover of the Air Force Atlas missile program. While serving on the Wiesner Committee, Gardner also accepted an invitation from General Schriever to chair a committee that would examine the status of Air Force space activities. Schriever hoped that Gardner would be able to produce a von Neumann Committee type of report that would lead to a "comprehensive, dynamic Air Force space development program" along the lines of the crash ICBM program.[82]

The Wiesner Report appeared on 10 January 1961.[83] It began by severely criticizing the organization and management of NASA and what it termed a "fractionated military space program." It recommended that one agency or military service be made responsible for all military space development and cited the Air Force as the logical choice. Already providing ninety percent of the support and resources for other military agencies, the Air Force, said the report's authors, represented the nation's "principal resource for the development and operation of future space systems, except those of a purely scientific nature assigned by law to NASA." Their recommendations also included more emphasis on booster development, manned space activities, and military applications in space. The Air Force could not have been happier with the Committee's report.

Meanwhile, early in 1961 the Air Force had to confront the unwanted fruits of its assertive late-fall campaign for a greater space role. Back in late November 1960, the Air Staff's Deputy Director of War Plans, Brigadier General J. D. Page, prepared a paper describing the Air Force position on space for use in briefing the new administration's officials. The paper restated the Air Force view of "aerospace," stressed the importance of space applications, and described seven such projects: Samos, MIDAS, a space-based antisatellite or missile system, a satellite inspector known as Saint, the Space Detection and Tracking System (SPADATS), the Advent communications satellite, and the Transit navigation satellite. Additionally, four more projects, Discoverer, Dyna-Soar, the Aerospace Plane, and HETS, a so-called hyper-environmental test system, were identified as "learning type" projects designed to determine the feasibility of new technology for space. General Page's rationale also assessed relations with NASA, suggesting that the Air Force work to have the Space Act be amended to provide clear recognition of the military's role in developing space systems.[84]

The Page paper seemed at variance with General White's efforts to promote a good relationship with the civilian space agency. In late 1959, for example, the chief of staff had circulated a letter to the Air Staff directing the fullest possible cooperation with NASA and had continued to foster good relations between NASA and the Air Force. General Page's paper of late 1960, however, suggested that less harmony existed between the two organizations than publicly admitted, and a more forceful effort might be needed to right the balance. The Page paper coincided with an intense public and internal information campaign to express Air Force views on space to congressmen, journalists, businessmen, and other influential people. The self-promotion effort immediately raised a storm of protest in the press over what it termed the Air Force's "political offensive to bring about changes in national space policy and law." Critics predicted an approaching contest with NASA for the country's major role in space.[85]

The outcry came to the attention of Congressman Overton Brooks, whose House Committee on Science and Astronautics planned to meet in February 1961 to examine the possible Air Force-industry "plot to undercut the space agency." Brooks' intentions prompted General White to write the congressman a letter, in which the Chief of Staff declared that "any action or statements by any Air Force individual or groups which tend to create such impressions [of unhealthy competition between the service and NASA] are in direct contradiction to the established beliefs and policies of the Air Force." General White requested Congressman Brooks to identify the "'pressure groups within the USAF'…and the specific actions taken by these groups toward 'degrading the position of NASA.'" Despite General White's assurances, the chairman reiterated his concerns in a 14 February 1961 letter to NASA director Glennan, who passed the letter on to General White. The Air Force Chief of Staff responded by assuring Glennan he was sending his key officers to meet with the new NASA leadership to determine how they could lay to rest the "ghost of this alleged NASA-Air Force dissension and duplication" once and for all.[86]

General White also appeared before the Brooks Committee in March to deny that his service had a plan "to take over NASA." During the congressional hearings in 1960, he had reassured his questioners that all was well between the two agencies, and that Air Force support to NASA had been extensive. This included providing the space agency sixteen Atlas D boosters modified for Mercury capsules and adapters, launch facilities at the Atlantic Missile Range Complex 14, and one-half of Hanger J with adaptations to accommodate telemetry, communications, and data transfer equipment. Along with normal base support and office space and equipment, Air Force infrastructure support also encompassed guidance sites and computers used for the Atlas, along with more than 400 Air Force military and civilian personnel. General White specifically referred to good working relations in evaluating requirements and preparing schedules, reaching agreements to share facilities on a priority basis, and cooperating on a demarcation of missions. As for the latter, he declared that the Air Force had no conflict with NASA handling space exploration and civilian uses and the Defense Department pursuing military applications. General White did, however, suggest the need for a single point of contact for Defense Department-NASA affairs and argued that the Air Force represented the logical Defense Department representative.[87]

Nevertheless, Congressman Brooks in March 1961 called on the new president to clarify the civilian and military roles and explain what seemed to be a tilt toward the military by the Wiesner Committee. In reply, President Kennedy declared that, while he never intended NASA to be subordinated to the Defense Department, there remained "legitimate missions in space for which the military services should assume responsibility."[88]

In fact, the President had already agreed to a new military directive that assigned remaining military space efforts and effectively awarded the Air Force the bulk of the military space "mission." Shortly after taking office, Secretary of Defense Robert McNamara directed his staff to review the military space program in light of the Wiesner Report's criticism of the "fractionated military space program." After studying the issue and soliciting comments from important Defense Department officials, the Defense Secretary decided to centralize space system development within the Defense Department by assigning the Air Force responsibility for "research, development, test, and engineering of Department of Defense space development programs or projects." Air Force enthusiasm remained tempered by other parts of the directive which authorized each service to conduct preliminary research and asserted that operational assignment of space systems would be done on a project-by-project basis. Nevertheless, by effectively making the service the executive agent for military space development projects and, thereby, the lead military service in space, the directive represented a major step in the Air Force's quest for the military space mission.[89]

On 17 March General White announced a major reorganization to better manage the missile and space programs. Although the timing suggests that the Defense Department directive precipitated the Air Force action, actually the reverse describes the course of events more accurately.[90] Apparently in early January 1961, Roswell Gilpatrick, the new Deputy Secretary of Defense, bolstered by the Wiesner Report's findings, contacted General White and offered the Air Force major responsibility for the military space mission if it "put its house in order." Gilpatrick and General Schriever had discussed the fragmented state of Air Force research and development activities when they worked together in forming the Aerospace Corporation the previous year. At that time, the main split in weapons systems responsibilities was between research and development, and procurement, the former function being assigned to Air Research and Development Command and the latter to Air Materiel Command. General Schriever had argued that the Air Force could not handle the military space mission unless one Air Force command held responsibility for research and development, system testing, and acquisition of space systems. The ARDC commander had advocated such a reorganization for a number of years. The problem had become more urgent by 1960. While the ARDC's Ballistic Missile Division in Los Angeles had retained research and development responsibility for space projects, its most important mission in 1960 involved close coordination with Air Materiel Command's collocated Ballistic Missiles Center to activate the new intercontinental ballistic missile (ICBM) force. As a result, two major, national programs—missiles and space—competed for resources and management focus within a single research and development organization. General Schriever expressed his concerns to General White in September 1960 and received authority to begin dividing the west coast space and missile functions by moving the latter to Norton Air Force Base, California, and retaining all space responsibilities at the Los Angeles site. Yet the ARDC commander remained convinced that the Air Force required more sweeping organizational reform. Deputy Secretary of Defense Gilpatrick agreed.

Following Gilpatrick's offer, General White asked Schriever to form a small task force to prepare an acceptable plan for centralizing weapon system development and procurement. Only Secretary of the Air Force Eugene Zuckert, Under Secretary of the Air Force Joseph V. Charyk, and Generals White and Schriever had been informed of Gilpatrick's offer, and General White preferred to keep the knowledge to a minimum. Although the Air Staff's Major General Howell M. Estes, Jr., Assistant Deputy Chief of Staff for Operations, chaired the small group, Schriever's chief appointee, Colonel Otto Glasser, actually formulated the plan and briefed it to senior officers and officials in the Air Force and to Defense Secretary McNamara. Afterward, General White informed the Air Council of what had transpired.

The centerpiece of the Air Force reorganization of the spring of 1961 involved creation of the Air Force Systems Command (AFSC), with responsibility for all research, development and acquisition of aerospace and missile systems. With the inactivation of the Air Materiel Command, a new Logistics Command was established to handle maintenance and supply only. To carry out this challenging assignment, AFSC received four subordinate divisions: an Electronics Division, an Aeronautical Systems Division, a Ballistic Missile Division, and a Space Systems Division. The new arrangement reflected the separation of missile and space management functions that General Schriever had favored for the past two years. The new Space Systems Division would be formed at the Los Angeles site from elements of ARDC's Ballistic Missile Division and AMC's Ballistic Missiles Center. The Ballistic Missile Division, also comprised of elements from ARDC's Ballistic Missile Division and AMC's Ballistic Missiles Center as well as the Army Corps of Engineers' Ballistic Missile Construction Office, would relocate to Norton Air Force Base. An additional measure involved establishment of an Office of Aerospace Research (OAR) on the Air Staff for basic research elements.

The Air Force reorganization represented a fitting complement to the Defense Department's directive assigning to the service future military space development responsibilities. With its own house in order, space activities promised to receive the management and research and development they would need in the years ahead. Fittingly, General Schriever received a promotion to four-star rank and became the first commander of Air Force Systems Command.

The Defense Department directive awarding the military space development mission to the Air Force could not be expected to please Army and Navy leaders.

Their grumblings reached the ears of Congressman Brooks, who held hearings beginning on 17 March, the day the Air Force announced its organizational changes. Before the committee, however, Army General Lyman Lemnitzer, Chairman of the Joint Chiefs of Staff, and other Army and Navy representatives denied opposing the directive. At the same time, Deputy Defense Secretary Gilpatrick assured the committee that centralization of space research and development would prevent duplication and prevent "misuse of resources," while General White declared that the Air Force would "bend over backward to meet the requirements of the Army and Navy as prescribed by the directive." The Chief of Staff also stressed that the new arrangement would improve cooperative relationships with NASA. The committee took no action, but promised "continuing close scrutiny" of the new directive's implementation.

Meanwhile, on 20 March 1961, three days after the public announcement of the Air Force reorganization, Trevor Gardner submitted his committee's report to General Schriever.[91] Although General Schriever had hoped to have Gardner's report by mid-January 1961, the former Assistant Secretary of the Air Force found it necessary to establish two study groups to provide the managerial and technical information needed. The report's conclusions proved alarming. The United States, it claimed, could not overtake the Soviet Union in space achievements for another three to five years without a major increase in the Defense Department's space effort. The report reserved particular criticism for the Eisenhower administration's emphasis on separate "military" and "peaceful" space programs, which had relegated the military program to a "stepchild" status with little participation in the scientific exploration of space, which was reserved to NASA. Above all, the report recommended that planners avoid prescribing detailed space requirements and operational systems in favor of first developing a firm technological basis, with the Defense Department and NASA focusing on fundamentals or "building blocks." Finally, like the Wiesner Report, the Gardner Report called for military participation in a comprehensive, lunar landing program that would land and return astronauts sometime between 1967 and 1970. The broad technological capabilities resulting from such a major national effort, the report predicted, would provide important "fallout" for both military and civilian space purposes.

While the Gardner Report underwent high-level review, on 12 April 1961, Soviet cosmonaut Yuri Gagarin became the first man to orbit the earth. Motivated in part by this Soviet space "spectacular," Secretary of Defense McNamara directed Herbert York, DDR&E, and Secretary of the Air Force Zuckert to assess the national space program in terms of defense interests and the Gardner Report's conclusions. The Defense Secretary's initiative led to an intense two-week study effort that centered on a special task force at the Space Systems Division led by Major General Joseph R. Holzapple, Air Force Systems Command's Assistant Deputy Commander for Aerospace Systems. On 1 May 1961, in forwarding the report to Secretary McNamara,

Secretary Zuckert reiterated the Air Force's concerns about "the inadequacy of our current National Space Program." Not surprisingly, the Air Force's "Holzapple Report" confirmed the conclusions reached by Trevor Gardner's committee. Following an analysis of military space objectives and current development efforts designed to meet them, the report focused on the large booster program as the most pressing problem and reason for Soviet supremacy. Like the Gardner Report, the Air Force proposal also called for a national lunar landing initiative, whose framework would provide an urgently needed comprehensive research and development "effort." Although the Air Force recognized that NASA would head the expedition, it looked forward to a close, cooperative effort that would enable it to reenter the field of superbooster research that had been a NASA preserve since it acquired the Army rocket team in October 1959.[92]

The Air Force recommendations ultimately were incorporated into the National Space Program announced by President Kennedy in May. Shortly after receiving the Air Force proposal, Secretary McNamara and newly-appointed NASA Administrator James E. Webb met to propose major initiatives and budget increases necessary "to establish and to direct an 'Integrated National Space Program.'" Although the lunar landing objective topped their list of essential projects, they also called for developing global space communications and meteorological networks and large boosters for both civilian and military programs.

After receiving public and congressional support for an expanded space program, President Kennedy on 25 May 1961 appeared before a joint session of Congress to challenge the nation to overtake the Russians in space.

> If we are to win the battle that is now going on around the world
> between freedom and tyranny, the dramatic achievements in space…
> should have made clear to us all…the impact of this adventure on the
> minds of men everywhere who are attempting to make a determination
> of which road they should take….It is time to take larger strides—time
> for a great new American enterprise—time for this nation to take a
> clearly leading role in space achievements."[93]

Echoing the agreement between McNamara and Webb on the nation's future course, the President listed the moon expedition as the first space goal, followed by development of nuclear rockets [big boosters] for interplanetary space exploration, and creation of global communication and meteorological satellite systems as soon as possible. Congress had already raised the funding of the Defense Department's large solid-fuel booster project from $3 to $15 million. As a result of the Kennedy-proposed space program, the Air Force, as the "space booster service," would receive $77 million to begin development of both an upper stage and a large solid-fuel booster to compete with NASA's liquid-fueled Nova engine.[94]

By May of 1961 President Kennedy realized the importance to national security of reconnaissance satellites. Although he did not alter the Eisenhower policy of "space for peaceful purposes," he clearly believed that the nation found itself in a race for space supremacy with the Soviets and should accept the challenge. The Air Force fully expected to play a central part in the ambitious space program that lay ahead and to benefit from the technological achievements along the way.

The Air Force Rise to Military Space Preeminence

The Eisenhower administration's space policy never wavered from its central objective of permitting the launch and operation of military reconnaissance satellites. The "spy satellites" would enable the country to guard against the President's old nemesis of surprise attack, while reinforcing the moral high ground of "space for peace" by providing the means to verify future arms agreements and nuclear test ban treaties. Relying on the "Sputnik precedent," he preferred to avoid direct confrontation with the Soviets by stressing civilian spaceflight and limiting military operations to defense support activities. This would best insure the success of clandestine satellite operations for the nation's defense.

Throughout the late 1950s Air Force leaders often failed to appreciate the subtleties of the Eisenhower space policy. For them, the policy of "space for peaceful purposes" served only to restrict military space activities to modest defense support projects and no offensive initiatives beyond the study phase. As military planners, they preferred defense preparations to combat potential enemy capabilities rather than prepare for operating in an "outer space sanctuary." Given their focus on space as the ultimate "high ground" and the extension of traditional Air Force operations, Air Force leaders believed that the country should achieve space "supremacy" in order to deny offensive space operations to the enemy. Because such activity might jeopardize space reconnaissance assets, the Eisenhower regime categorically refused to permit it.

Given these circumstances, the Air Force remained unable to conduct an independent space program. Prevented after Sputnik from leading a nationwide space effort to overtake the Soviets, it found itself forced to respond to ARPA's direction, then compete with NASA for funds and projects. Only with the demise of ARPA in late 1959 did the Air Force regain control of its "own" space program. Even then the future course with NASA and DDR&E seemed unclear, while key projects continued to experience growing pains. Moreover, much of the Air Force space responsibility involved supporting other agencies with booster and infrastructure assistance. Operational direction remained the responsibility of other services and agencies. This did not always seem to reflect the aspirations of the service that had been assigned, in the words of General Schriever, the "prime responsibility for the military space mission."

By the end of the Eisenhower presidency, the program had in place the five functional areas of defense support operations that would characterize Air Force space operations from then until the Reagan administration reopened the issue of weapons in space.* Of the five areas, only the missile detection and space defense functions remained largely under Air Force control. At this time, the Air Force supported others who had responsibility for communications, navigational, and meteorological satellites, while "observation of the earth" now encompassed highly sensitive "black" systems outside the Air Force's control. Looking back on the McNamara directive's impact on the Air Force following loss of reconnaissance assets to the National Reconnaissance Office, Air Force Secretary Eugene Zuckert declared, "It was like getting a franchise to run a bus line in the Sahara Desert." Yet, Secretary Zuckert's comment did not express pique at the service not getting what it wanted. The March 1961 decision, he explained, was jurisdictional and provided the Air Force all the jurisdiction it needed in the space field. How much support the service would get remained in doubt. In effect, the Air Force received the research and development franchise for space systems, including offensive space-based systems, but it awaited customers and support from the Defense Department in the course ahead.[95]

If the Air Force did not acquire all the military space missions it desired, it had much to celebrate in the spring of 1961. Of its space programs, the MIDAS early warning infrared satellite remained a high national priority, and the Air Force continued to develop its Samos reconnaissance project. At the same time, it provided important launch and infrastructure support to the national reconnaissance effort under Project Discoverer. By the end of June 1961, the Air Force had launched twenty-six Discoverer satellites in support of various projected space systems, and the program had been expanded from an original thirty-five planned vehicles to forty-four. The Air Force also played a major part in the Space Detection and Tracking System with overall planning responsibilities and development of its Spacetrack network, and it moved forward with an elaborate air and missile defense system that would provide collateral support for Spacetrack. Already, the Air Force programmed the Thor and Atlas boosters as standard launch vehicles of the future, with an improved Titan to follow. Boosters had been the foundation of the Air Force dominance in space and represented the best means to perpetuate that dominance. At the same time, despite Project Mercury, in these, the Dyna-Soar years, the military man-in-space mission remained a viable option.

In the aftermath of Secretary McNamara's directive and President Kennedy's lunar challenge, the Air Force could look back on the years since Sputnik with satisfaction. Its cautious, well-orchestrated, opportunistic three-and-one-half-year quest for the military space mission had succeeded. The losses of rival service assets to NASA had resulted in Air Force gains, and efforts to create a unified space command for space had been successfully thwarted. Along the way the Air Force

---

* See Appendix 2-5.

prepared itself for the space mission by demonstrating the flexibility to establish its own in-house technical expertise with the Aerospace Corporation, and implement a major reorganization to better handle the research and development challenges ahead. The Air Force had staked its claim to space through the "logic of aerospace," and it had been accepted. Most importantly, despite the difficulties with space program advocacy this often presented, Air Force leaders remained convinced that space must be approached in terms of its utility for traditional operations. This would be an important legacy for the future. In the years to come, space would become an increasingly important medium in support of both strategic and tactical military operations. That, in turn, would serve to institutionalize space within the Air Force. In 1961, the Air Force had garnered the bulk of the military space "mission." The challenge now would be to strengthen its position by developing a military space program vital to the nation's defense.

CHAPTER 3
The Air Force in the Era of Apollo:
A Dream Unfulfilled

In the spring of 1961 the Air Force appeared poised to play the dominant role in the nation's military space program and, hopefully, the national space effort for at least the next decade. In March, Secretary of Defense Robert McNamara designated the Air Force the military service for space research and development, thereby diminishing the prospects for disruptive interservice rivalry. In response, the Air Force reorganized its research and development elements to provide a stronger focus on space issues. Although the administration in May awarded NASA the lunar landing mission, the Air Force fully expected the civilian agency to remain dependent on the service for program management, key personnel, various launch vehicles, and ground support. Above all, Air Force leaders continued to believe that NASA's lunar landing agenda did not preclude its own aspirations for testing the usefulness of military manned spaceflight. Despite the promise of major advances by unmanned, artificial earth satellites in support of operational requirements, man-in-space remained the centerpiece of Air Force efforts during the 1960s to institutionalize space within the traditional airplane-oriented service.

Unlike its predecessor, the Kennedy administration promised the nation an integrated, national space program retooled to overtake the Soviet lead in space. The Air Force interpreted the new approach as a challenge to convince government leaders that national security requirements demanded an expanded military space program under Air Force control. For two years, the Air Force waged an aggressive campaign to achieve leadership of an "independent" space program. By 1963, however, its hopes and expectations ended in the wake of NASA's growing confidence, its success in Project Mercury, the formation of the National Reconnaissance Office, and the McNamara Defense Department's assertiveness and rigid criteria for space program approval.

The Air Force would find itself the loser in the tug-of-war between the civilian space agency and the Defense Department over priorities and responsibilities for space exploration, both manned and unmanned. Although the service would continue lobbying for an ambitious military space program, its efforts would prove fruitless. Ultimately, it failed to gain approval to establish an operational space-based anti-satellite and antimissile capability to thwart potential Soviet space dominance. It also encountered roadblocks to develop proposals it considered important for defense support functions. Above all, the service proved unsuccessful in retaining its man-in-space mission. From the Dyna-Soar orbital glider to the Blue Gemini space capsule to the Manned Orbiting Laboratory, the Air Force fought hard to convince skeptical Defense Department officials of the need for a military man-in-space role independent of NASA's responsibilities and capabilities. By mid-decade success seemed assured when President Johnson announced development of a military space research laboratory under Air Force management. But later in the 1960s, the growing financial and emotional demands of the Vietnam War and the Great Society, along with public disenchantment with space, doomed Air Force pretensions for manned spaceflight in the competitive battle over the defense budget.

With the advent of the Nixon administration, Air Force leaders readjusted their priorities from space requirements to other more pressing and achievable needs. Frustrated by failure to claim leadership of an expanded "independent" space program and stymied in realizing its main goal of manned spaceflight, Air Force leaders turned their attention to more traditional "flying" needs of the service. Represented by major Air Force commands, priorities for tactical and strategic weapons took precedence. While NASA basked in the glow of the historic Apollo lunar landing, the Air Force seemed confined to a secondary role in the national space program. Yet appearances proved deceiving, because the Air Force had quietly established a space applications satellite program that rapidly made space support routine and important to tactical as well as strategic commanders. At the same time, the Air Force found itself with a major voice in development of the Space Shuttle, the re-usable space launch and manned orbital system of the future. Space seemed ready to move from the arena of research and development to operations.

The Air Force Position in the Spring of 1961

With the advent of the Kennedy administration, Air Force leaders had every reason to believe that their service would play a larger role in an expanded national space program to achieve space leadership and thwart potential Soviet space threats to national security. The new President clearly recognized the requirement for both civilian and military space activities. In his 1961 report to Congress, President John F. Kennedy declared that "space competence is as essential for national security as it is for national growth."[1] While affirming the Eisenhower policy of space for peaceful purposes, he noted that his vision of an expanded national space program "included space projects to help keep the peace and space projects to increase man's well-being in peace."[2] His initial actions encouraged the Air Force to believe that military space proposals would receive new emphasis in the high-profile national space program. With the President's announcement on 25 May 1961 of the ambitious lunar landing initiative, the nation received a distinct, long-range objective, the pursuit of which promised to make space big in business and government. As the responsible agency, NASA's fiscal year 1962 budget request came to $1.8 billion, twice the previous year's appropriations. Administrator James E. Webb predicted that final costs for what became known as Project Apollo would reach between $20 and $40 billion.[3]

At the same time, the military also benefited from the new space priorities. The final fiscal year 1962 appropriations totaled $1.1472 billion, nearly $350 million higher than the previous year and only $0.7 billion less than NASA's final figure of $1.7968. Moreover, every major Air Force space program, whether approaching operational capability like Samos and Spacetrack, or still in the exploratory stage like MIDAS and Saint, the space-borne satellite detection and inspection proposal, received increased funding.* Beyond specific system development projects, the Defense Department received greater funding for basic research in some areas that had no clear military application at that time. Of the latter, the large solid-rocket-motor project represented an important achievement for Air Force space advocates who, during the Eisenhower administration, had repeatedly championed development of a military superbooster and the need to conduct basic space technology and exploratory research apart from the civilian agency.[4]

Indeed, the Kennedy administration's highly touted "national" and "integrated" space program encouraged the Air Force in its quest for a greater leadership role in space.[5] As Vice President and Space Council Chairman Lyndon B. Johnson asserted, "It is national policy to maintain a viable national space program, not a separate program for NASA and another for Defense and still another for each of several other agencies."[6] Although NASA could move forward with plans for big rockets, an operational communications satellite system, and manned orbiting spacecraft experiments, the agency's mushrooming requirements for facilities, equipment, bioastronautics data and personnel would encourage Air Force leaders, including Chief of Staff General Thomas D. White, Vice Chief of Staff General Curtis E. LeMay, and newly appointed commander of Air Force Systems Command Lieutenant General Bernard A. Schriever, to believe that NASA's dependence on the Air Force

---

* See Appendix 3-1.

would continue to allow the service a major voice in NASA's manned and unmanned spaceflight operations.

With responsibility for ninety percent of the military space effort in the spring of 1961, the dominant Air Force role in space had received acknowledgment that March from Secretary of Defense Robert S. McNamara, whose directive, "Development of Space Systems," accorded the Air Force what General Schriever referred to as "the prime responsibility for military space."[7] Although the Army and Navy would continue with their existing satellite projects and conduct preliminary space research, the Air Force became responsible for nearly all future defense space research and development, with exceptions authorized only by the Secretary of Defense. If the Air Force did not receive sole responsibility for the military space mission, the Defense Department directive for all intents and purposes made the Air Force the leading military space service and effectively muted the rivalry among the three services over space issues that had plagued the Eisenhower administration.

In response, the Air Force had reorganized internally to provide the desired focus for leadership of the military space program. General Schriever's newly formed Air Force Systems Command now controlled release of new weapon systems from research and development to operational status, while its subordinate Space Systems Division on the West Coast prepared to direct the service's space effort with strong technical support from the Aerospace Corporation. The service hoped and expected to lead a "crash" program for space similar to the high-powered ICBM effort of the 1950s. This had been General Schriever's purpose in charging Trevor Gardner's committee in late 1960 to perform a role for space similar to that of John von Neumann's earlier Strategic Missiles Evaluation Committee for missile development. The Gardner Committee's report of 20 March 1961 advocated an ambitious Air Force-led space program to overtake the Soviets and achieve military spaceflight dominance. In the spring of 1961 Air Force leaders believed that the McNamara directive and the national space agenda would provide such a mandate, and they considered the Air Force well-organized and prepared to lead the effort.

Despite the service's new prominence, Air Force leaders realized that a campaign for a greater Air Force role in space faced major challenges. The President's announcement of the lunar mission heightened NASA's prestige and responsibility in support of the nation's "space for peace" policy, while its new manned spaceflight mission threatened to eliminate the Air Force focus on a military man-in-space mission of its own. At the same time, the Air Force confronted a Defense Department intent on maintaining the precedent of "freedom of space" and, therefore, skeptical of earlier Air Force proposals for antisatellite and antimissile space capabilities as well as military manned space operations that might threaten it. Under its dynamic leader, Secretary Robert S. McNamara, the Defense Department advocated an integrated national space program in the name of cost effectiveness and the end to wasteful duplication.

Such a program meant emphasizing mutual cooperation, coordination, and support between NASA and the services. The Air Force found itself in an ambivalent position. As the military service for space, it could look forward to greater involvement with a civilian space agency still dependent for much of its hardware, infrastructure, and launch support on the Air Force. At the same time, a centralized space effort might very well find the Air Force overly dependent on the civilian space agency for scientific and technical data and hardware. Above all, it might be frozen out of manned spaceflight activities that NASA now claimed as its own, and compelled to rely on experience derived from NASA's near-earth orbital and lunar projects for military applications, if any.

Air Force leaders decided on an aggressive campaign to lead an expanded military space effort. In 1961 their "plan of action" would proceed on three discernible levels that often overlapped. First came policy concerns. Despite the President's acknowledgment of a major military role in national space policy, service spokesmen publicly assailed what they considered an artificial distinction between military and civilian space activities. This resulted in a narrowly-construed "space for peace" policy that prohibited development and deployment of offensive space systems that could deny the Soviets space superiority. Air Force spokesmen often took their argument public to convince sympathetic congressmen and a reluctant administration that only an offensive space capability and military manned spaceflight proficiency could ensure space for "peaceful purposes." On a second level, Air Force planners moved rapidly to shed the constraints of the Eisenhower administration and devise a formal Air Force space plan with related programming documents. These, they hoped, would serve to crystallize Air Force institutional thinking on space and win from the administration permission to lead an ambitious national space effort. The third element of the campaign involved establishing what Secretary of the Air Force Eugene M. Zuckert termed an "equal partnership" with NASA. This meant lobbying the Defense Department for formal designation of the Air Force as the executive agent for military support to NASA. While Air Force leaders expected to parley their pervasive support of NASA into a major voice in NASA's affairs, they also solicited NASA's help to overcome a growing Defense Department tendency to rely on the civilian agency for military space needs. The Air Force resorted to logic, cooperation, and pressure to convince NASA officials that, despite the policy of an integrated national space program, NASA alone could not satisfy military space requirements in the two vital areas of space exploration and man-in-space. In effect, NASA might serve as the wedge Air Force space leaders needed to maneuver Defense Department officials into approving a larger Air Force role in space.

Over the course of the 1960s, the Air Force would find itself in the middle of an ever-evolving saga of cooperation and competition between NASA and the Defense Department for leadership in space. In retrospect, the ambitious Air Force plan of action might seem doomed from the outset in view of Secretary McNamara's strong leadership and NASA's high-profile Project Apollo. Nevertheless, in the spring of 1961 the new administration's ambitious space goals, Air Force prominence in the space program, and sensitivity to Soviet manned space successes opened the door to an aggressive Air Force campaign for an expanded space program. Not until the end of 1962 did it become clear to Air Force leaders that their efforts had proven unsuccessful and that they would need to reassess the service's relationship with NASA and the Defense Department.

Seizing the Initiative

The Air Force opened its campaign for a greater space role by renewing its criticism of what it termed the "space for peace" policy.[8] In July 1961, newly-confirmed commander of Air Force Systems Command, General Bernard Schriever, the service's highly respected and most outspoken space advocate, appeared before Senator John Stennis' Senate Preparedness Investigating Subcommittee and testified that the military space program was inadequately supported. "I think we have been inhibited in the space business through the 'space for peace' slogan," Schriever declared. "I think that there has been too arbitrary a division made between the Department of Defense and NASA in this area."[9] Coming in the wake of Soviet Yuri A. Gagarin's historic first manned orbital flight on 12 April, Schriever found a congressional audience receptive to charges of neglect and artificial impediments to America's space potential. Impressed with the General's testimony, committee members requested that he provide them a written report on the problem.

By late summer the proverbial political winds seemed increasingly favorable to Air Force efforts to have the "space for peace" policy modified. The Soviets' second manned space spectacular, a 17-orbit flight on 6 August by Cosmonaut Gherman S. Titov, reaffirmed the specter of Soviet space superiority and compelled congressmen to deem the American situation "critical." Even NASA watchdog Representative Overton Brooks, chairman of the House Committee on Science and Astronautics, seemed to capture the public mood when he asserted that the Soviets "obviously now have the capability to send up manned satellites carrying bombs and other equipment for destroying other nations."[10]

General Schriever's statement, which received Air Force Secretary Zuckert's blessing, reached the Stennis committee on 11 September, soon after the Titov flight. His report described the potential threat posed by the cosmonaut's space flight and a Soviet space program unencumbered by its American counterpart's handicap: "an unnecessary, self-imposed restriction—namely, the artificial division into 'space for peaceful purposes' and 'space for military uses,' when in fact no technical and little other distinction between the two exists." The general focused on manned space-flight by stressing the findings of a recent Air Force Scientific Advisory Board study, which concluded that "the sense of urgency that exists across the whole front of space projects should be injected into the manned military space program."[11]

Alarmed by Schriever's argument, a sympathetic Senator Stennis took to the Senate floor in late September to warn his colleagues and the nation of a growing Soviet space threat. Afterward, he promised to study the issue over the congressional break in preparation for holding major hearings early in 1962 on the issue of "whether the present division of responsibility between the military and NASA is proper in light of international developments."[12]

Responding to congressional and public concern, Air Force leaders that fall spoke out more openly for a stronger military space program. In an address to the American Rocket Society on 12 October 1961, General Schriever reiterated his theme of artificial constraints on Air Force programs and the growing threat posed by Russian rockets equally as capable of carrying 100-megaton warheads as of launching cosmonauts.[13] Later that month, on 26 October, Chief of Staff General Curtis E. LeMay drew a parallel between airpower during the First World War and space in the early 1960s. Speaking to the American Ordnance Association in Detroit, Michigan, he described the evolution of early airpower operations from peaceful, chivalric, un-armed reconnaissance flights to combat efforts designed to deny the enemy air superiority. "I think we will be very naive," he declared, "if we don't expect and prepare for the same trends in space."[14] By late fall President Kennedy and his Space Council chairman, Vice President Johnson, publicly acknowledged the increasing Soviet space threat and expressed interest in a greater military space role. The Vice President cautioned against applying "arbitrary distinctions...between military and civilian space efforts," while the President asserted that America could not let the Soviets dominate space.[15]

At the end of 1961 Air Force leaders had good reason to believe their criticism of the nation's military space posture foreshadowed an expansion of the Air Force space role. The stage seemed set for a major congressional debate early in the new year, while administration leaders increasingly responded to public pressure and Air Force concerns. Even the troubled Dyna-Soar manned space glider program benefited from the changing climate when the Defense Department in December authorized the Air Force to eliminate the suborbital phase and proceed with an accelerated orbital flight program using the Titan III booster in place of the Titan II. Air Force leaders fully expected that the momentum established for an expanded space effort would lead to major Air Force-led space initiatives.

The Soviet Union's monopoly with respect to manned spaceflight and the new administration's commitment to a greater national space effort in 1961 also stimulated an internal   Air Force space planning and programming initiative to prepare the service for its expected leadership role in space. Gone were the Eisenhower admini-stration's proscriptions against publishing long-range Defense Department military space plans, which had stemmed from considering space as supporting traditional mission areas rather than as a distinct mission in itself. The Kennedy administration's focus on an integrated national space program provided the Air Force the necessary "green light" to undertake preparation of a comprehensive space plan. Such a plan could serve to clarify Air Force views on space objectives in a rapidly changing technological environment and help gain the Defense Department's support for Air Force goals.

At the suggestion of Major General William B. Keese, the Air Staff's Director of Development Planning, the Chief of Staff directed Keese to establish a task force made up of Air Staff and Air Force Systems Command representatives to prepare the plan. The group completed work on the Air Force's first formal Space Plan on 21 September 1961. In the tradition of Theodore von Kármán's post-World War II *New Horizons* study and subsequent service proposals like the Gardner Committee's report, the plan emphasized the importance of a basic research and development focus that would establish the technical foundation for enhanced military space operations.[16]

The space plan called for an "aggressive military space program" focused on "a vigorous applied research program…[conducted in the fields of guidance, propulsion, and sensors]…to insure that military potentials, when developed, will be promptly identified and vigorously pursued…[with operational systems]…to insure the security of the Nation."[17] Such an initiative would support an integrated national space program in which Air Force capabilities and facilities would support the entire national program. Consistent with earlier views on mission application, space capabilities would be used only when deemed the sole available recourse or most cost-effective operational solution to support existing mission areas, which the planners identified as reconnaissance and surveillance, defense, offense, command-control, and support.[18]

The space plan proceeded to recommend future action in specific Air Force space program areas.* Discoverer (Project CORONA), MIDAS, Samos, and the Blue Scout research vehicle, for example, should be continued at their present pace, while efforts to develop weapons in orbit, the antisatellite and antimissile defensive systems, should be accelerated. Planners recommended that Saint, the satellite inspector project, be revised and enhanced to include testing of unmanned techniques for rendezvous, inspection, docking, and "satellite neutralization," while Bambi, the space-based anti-ballistic missile concept, be shifted from ARPA to Air Force control and prepared for feasibility demonstrations. Authorized to develop a large heavy-lift booster, Air Force planners advocated acquisition of an economical and reliable military space booster capable of launching payloads of 10,000 to 50,000 pounds into a 300-mile low earth orbit.[19]

Military manned spaceflight requirements received special attention from the Air Force planners. Declaring that "it is…imperative for the United States to determine

---

* See Chapter 4 for discussion of specific unmanned Air Force space programs.

the military utility of man-in-space at the earliest possible time," the plan outlined tasks potentially handled better by manned systems, such as command and control decision-making, especially "placing man in a satellite inspection and neutralization system," as well as reconnaissance, and in-space maintenance and repair. Planners strongly supported an accelerated Dyna-Soar project designed to achieve manned orbital flight and emphasized the need for a close, cooperative relationship with NASA. The Air Force should expand and accelerate its bioastronautics program in conjunction with NASA, they said, while the civilian agency could share its experience in earth orbital programs "in order to provide for early multi-manned testing of military subsystems in space for duration up to two weeks." In addition, the space plan called for increased study and research efforts to develop "a manned, maneuverable, recoverable spacecraft" and, for the first time, declared the Air Force's strong interest in "a long-duration military test space station." The space plan indicated that the Air Force would continue to pursue both the aerodynamic and ballistic methods of reentry.[20]

After hearing a presentation on the space plan, Secretary Zuckert recommended updating the basic plan periodically and using it to develop "detailed implementing plans on major aspects of the program." The space plan's initial impact came at year's end with its use in preparing the space budget presentation in early 1962. On 4 December 1961 the Vice Chief of Staff appointed Lieutenant General James L. Ferguson, Deputy Chief of Staff for Research and Development, to develop programming documents depicting costs and schedules for use in defending the Air Force fiscal years 1963 and 1964 space program before Congress in February 1962. The Ferguson group consisted of eight separate panels of Air Staff and Air Force Systems Command space specialists, who laid the groundwork for the most comprehensive testimony of the decade describing the Air Force's position on space.[21]

On 19 February 1962 General Ferguson appeared before the House Armed Services Committee and testified in favor of an expanded military space program. Based on the September 1961 Air Force Space Plan, the Air Force space budget recommended raising the fiscal year 1963 figure of $826.2 million proposed by the Office of the Secretary of Defense to $1.31 billion, and the fiscal year 1964 total from $1.32 billion to $1.86 billion. General Ferguson argued that the nation must exploit space to achieve military superiority as the best means of insuring "the peaceful use of space." This meant a potential "offensive" military requirement to inspect non-U.S. satellites, perform surveillance and reconnaissance functions, and establish a defense against potential ballistic missile attack.[22]

Although he noted that an integrated national space program found both NASA and the Air Force pursuing mutually supportive rather than competitive programs, he strongly argued that:

> some operational and related technological needs are not common to
> both the civilian and military effort…military tasks frequently require
> routine and repetitive operations. We therefore need low-cost, high
> reliability and, if possible, reusability in our systems. Military tasks also
> may require quick reaction, positive control, and the ability to operate in
> a combat environment. These factors have different implications than
> those involving scientific, commercial, or prestige missions.[23]

As one example, he cited the importance of rendezvous in space with "non-cooperative" targets that demanded techniques different from a lunar landing mission involving "cooperative" targets in specific, controlled orbits. He next proceeded to describe eleven important areas of technology in which the Air Force worked to exploit earlier military space applications and broaden its knowledge and capability. In doing so, he declared against the increasingly restrictive Defense Department guidelines for approval of space projects. "We must not be restricted," he said, "from exploratory developments merely because a clear application is not yet evident."[24]

The attainment of manned military space operations represented the main theme of his presentation. He argued that including man-in-space operations would markedly improve system flexibility and the likelihood of mission success. After describing the various functions for man in space outlined in the September 1961 space plan, he asserted that "it is for these reasons that we believe that man is essential not only in operational space systems, but also in those programs designed primarily to further technological capabilities in space." To answer the basic question of military man's utility in space, the Air Force advocated a program, coordinated with NASA, to develop a manned military test station in space. An orbiting space station, he asserted, would answer the urgent question of special military concern: "Can man effectively perform specific military combat and non-combat functions in space?"[25] General Ferguson concluded his statement with a strong plea for an expanded space program. The Air Force, he said, believed that space systems could solve major national problems both then and in the future if military space technology was adequately supported as proposed in the 1963 Air Force Space Program. Moreover, "the program in future years will need to be even more vigorous and comprehensive."[26]

Ferguson's testimony seemed to elicit the desired reaction from congressmen and helped increase pressure on the administration to reassess its military space posture. In short, the Air Force sought to force a decision on weapons in orbit and a change in space policy on Kennedy and McNamara. The question became whether the Eisenhower space doctrine would prevail or be overturned, as the Air Force desired. That same month, on 23 February 1962, Secretary of Defense McNamara pleased Air Force leaders by formally approving the accelerated Dyna-Soar proposal and informing Air Force Secretary Zuckert that he recognized the importance for national security of an investigation of military manned space roles. He acknowledged that "performance specifications and design requirements for military space systems may differ substantially from those stipulated for non-military applications." For the first time the Defense Secretary appeared to agree with the Air Force position on military manned spaceflight and the need to establish a military technological base and operational capability even without clearly defined missions.[27]

Encouraged by congressional and administrative action, Air Force leaders continued to press their advantage. In late March 1962 General LeMay spoke at Assumption College in Worcester, Massachusetts, on the need to "develop military space systems as quickly as possible" to avoid a Soviet technological surprise in the 1970s. Commenting on LeMay's speech, a *Washington Post* article compared LeMay and other Air Force leaders of the current period with their predecessors prior to World War II. They both possessed "supreme faith in the overwhelming need for military aerospace power but [were] unable to demonstrate it."[28] On 2 April, when the *Post*'s comments appeared, McNamara met with the Chief of Staff and suggested the Air Force outline specific technological needs, increase its space allocation in the fiscal year 1963 budget, and prepare a five-year Air Force space program to complement the effort of the Office of the Secretary of Defense already underway. The Chief of Staff called on General Ferguson, who responded first by reassessing the programs he presented to Congress earlier, then adding $252.9 million to the Air Force supplemental proposal for approved programs like Dyna-Soar, MIDAS, and Titan III, and those in the advanced study and development stage dealing with satellite interception and missile defense. On 16 May the Chief of Staff submitted the supplemental budget request to the Defense Department and authorized work on a five-year space program.[29] By the spring of 1962 Air Force leaders optimistically expected success from their efforts to champion an effective space plan and program.

Achieving a "workable" relationship with NASA represented the third element in the Air Force campaign for a greater space role. Following President Kennedy's announcement of the manned lunar landing project, NASA and Defense Department officials met to coordinate their requirements for mutual support and delineate lines of responsibility in order to avoid duplication. Much of their work centered in the Aeronautics and Astronautics Coordinating Board (AACB), which was cochaired by Dr. Robert C. Seamans, Jr., NASA's Associate Administrator, and Dr. Harold Brown, Director of Defense Research and Engineering, together with its six subordinate panels. The Assistant Secretary of the Air Force chaired the Launch Vehicle Panel and served as vice chairman of the Manned Space Flight Panel, while senior Air Force officers and officials maintained a strong presence on every panel.[30]

Already in the Kennedy administration the Defense Department and NASA had established a pattern for future cooperative measures through an agreement reached on 23 February 1961, by which both parties agreed to seek the consent of the other before developing new launch vehicles. Discussions during the summer of 1961 resulted in agreements that placed the Air Force well on its way to a guarantee of parity with NASA in booster development. In July the Defense Department and NASA established a large launch vehicle planning group that led to a division of labor concerning long-term booster requirements for both agencies. According to several formal agreements signed in the fall, NASA would pursue development of large liquid-propellant rockets, in tandem with the Air Force's work on large solid-propellant rockets until it became clear which would better support the lunar mission. The Air Force project initially included a proposal for a 3,000,000-pound thrust motor, but eventually settled on development of two large motors, one a 156-inch diameter segmented motor and the other a monolithic (unsegmented) 240-inch diameter motor. At the same time, the panel approved Air Force plans to develop a large, standardized "workhorse" booster for potential future needs of both NASA and the Defense Department. By autumn, this proposed system had become the Titan III, a vehicle which would consist of a basic Titan II, modified by the addition of two strap-on solid rockets. The Titan III would be capable of orbiting near-earth payloads of 5,000 to 25,000 pounds.[31]

A second coordination effort involved facilities and resources needed to support the lunar landing program, which NASA had already designated Project Apollo back in the summer of 1960. Interest centered on a joint study of possible launch sites conducted by Major General Leighton I. Davis, who had succeeded Major General Donald N. Yates as commander of the Air Force Missile Test Center and the Defense Department's representative for coordinating range support for NASA, and NASA's Dr. Kurt H. Debus, chief of the agency's Cape Canaveral launch operations. In July they agreed on Cape Canaveral as the Apollo launch site, with the recommendation that NASA purchase 80,000 acres on Merritt Island just north of the already over-crowded missile and space launch complex. On 24 August 1961, NASA Administrator James E. Webb and Deputy Secretary of Defense Roswell Gilpatrick signed an arrangement that made NASA responsible for costs associated with the lunar project and "technical test control" of its launch operations, while designating the Air Force range manager for the Apollo program. As agent for NASA, the Air Force would direct facilities and land improvements subject to NASA's approval.[32]

The Air Force expected to parley its strong supporting role into a "full partnership" with NASA. With this objective in mind, on 4 August 1961 Air Force Secretary Zuckert formally requested the Defense Department to name the Air Force "executive agent" for NASA support. Expecting a positive response in the near future, General Schriever received permission to begin discussions with the agency's Associate Administrator Seamans to develop the necessary organizational and procedural requirements for Air Force Systems Command support of NASA. He also directed Dr. Brockway McMillan, Assistant Air Force Secretary for Research and Development, to prepare essential NASA-Defense Department directives and procedures following acknowledgment by Defense Department representatives that the

Air Force would continue to provide the vast majority of military resources necessary to support NASA. Based on the fall discussions involving cooperation and support between Air Force and NASA representatives, in late December 1961 Secretary Zuckert also proposed formation of a new Air Force Systems Command office, Deputy Commander for Manned Space Flight, to include members of all three services and be located at NASA headquarters.[33]

While the Office of the Secretary of Defense studied the Air Force's December proposal, on 24 February 1962 it granted the earlier Zuckert request by officially designating the Air Force the "executive agent" for NASA support. Under terms of Defense Department Directive 5030.18, titled "Department of Defense Support of National Aeronautics and Space Administration (NASA)," the Secretary of the Air Force became responsible for "research, development, test, and engineering of satellites, boosters, space probes, and associated systems necessary to support specific NASA projects and programs arising under basic agreements between NASA and DoD." Air Force responsibilities included "detailed project level planning" and contract and management arrangements.[34]

As the 24 August 1961 arrangement suggested, NASA remained heavily dependent on Defense Department support. The civilian agency relied on the Defense Department's experience with the Navy Transit navigational satellite in planning its own commercial or civilian satellite system and looked to the Defense Department for procurement procedures, contract management services, and cost and work scheduling methods. From civilian agency's beginning, the Defense Department, largely through the Air Force, had supplied personnel, rocket boosters, launch and range facilities, and communications and tracking networks, as well as experience gained from the ballistic missile program. By 1962, the Air Force and NASA had concluded ten major agreements and a host of implementing arrangements. For NASA's Project Mercury, the nation's first manned program, the Air Force provided most of the astronauts, launch facilities and vehicles, range support, and the necessary recovery forces. The Defense Department and NASA already had begun talks on Project Gemini, the low-earth orbital follow-on program to Mercury, in which the Air Force would play a similar supporting role. Beyond this, the Air Force supported fourteen specific NASA programs, assigned ninety-six R&D officers to various NASA offices, and assisted NASA with substantial Air Force funding. Moreover, NASA officials recognized Air Force pretensions for a military role in space exploration and manned spaceflight, and they sought to assuage Air Force concerns by pledging that NASA would continue to support military interests as required.[35]

To Air Force leaders, the tactics of cooperation and advocacy appeared to be achieving their objective of "full partnership" with NASA in the nation's space program. Indeed, by the spring 1962 it seemed that Air Force space advocates could point to success in all three areas of their campaign for an expanded Air Force-led space program. Then came the "firestorm." On 11 June 1962, the *New York Times* reported on its front page that the Defense Department was "embarking upon a man-in-space program to prevent [foreign] military control of space as well as its exploitation." In response to this threat, the report stated, the Air Force would develop a manned satellite designed to destroy hostile space vehicles. The newspaper went on to assert that the White House and Space Council had authorized the Defense Department to conduct a six-month study in order to prepare an expanded military space program because, officials had said, NASA could not be relied on exclusively. Apparently, an earlier speech by Deputy Defense Secretary Gilpatrick on 13 May precipitated the *Times* article. In that speech Gilpatrick argued in favor of having military insurance in space. For the first time, he publicly acknowledged that the Defense Department "has decided to develop the technology of manned orbital systems able to rendezvous with satellites [neutralize or destroy them] and then land at preset locations on earth." Such a system might combine the capabilities of both Dyna-Soar and Saint. The Air Force interpreted the deputy secretary's remarks as authorizing feasibility studies for Saint and, that same month, began negotiations with contractors on a three-month study.[36]

The *Times* report in June unexpectedly precipitated a public outcry from critics who worried that a military man-in-space program meant direct competition with NASA and an antisatellite system in violation of the administration's declared use of space for "peaceful purposes." The immediate political fallout proved disastrous to Air Force hopes of changing administration policy. Administration officials quickly reaffirmed the "space for peace" policy, while the Defense Department denied authorizing the Air Force to proceed with antisatellite system development. The Air Force System Command's Space Systems Division immediately canceled its contract negotiations on the Saint project.[37]

Later in June Deputy Secretary Gilpatrick and Harold Brown, the Director of Defense Research and Engineering, appeared before the Senate Committee on Aeronautical and Space Sciences to publicly deny that the Defense Department intended to preempt NASA's role in manned spaceflight. But in doing so, Brown raised doubts about the entire concept of military manned spaceflight. In response to a question on the subject, he asserted that "I cannot define a military requirement for them. I think there may, in the end, turn out not to be any." In effect, the director also implied that the Department's new "building block" approach to research and development also might be invalid. If so, the Air Force would be prohibited from conducting research on all programs without clear, defined missions. Moreover, during a news conference following the newspaper story, President Kennedy responded to a question about a larger role for the military in space by saying, "No, the military have [sic] an important and significant role, though the prime responsibility is held by NASA and is primarily peace." Such a remark did little to alleviate continued public confusion about military space activities. Moreover, the Air Force could do little to educate the public following the government's information blackout on all military space programs that became effective on 23 March 1962. With the secrecy ban in place, which administration officials refused to acknowledge, the sensitive reconnaissance programs begun under the Eisenhower administration disappeared from public view. The ban also applied to the Navy's Transit navigational satellite and Air Force sounding rockets and space probes. As a result, the Air Force found it difficult to promote and justify the results of its successful "peaceful" space efforts in areas like communications, navigation, advanced spacecraft techniques, guidance systems, and basic scientific research.[38]

The controversial events of May and June 1962 signaled the end to the year-long Air Force initiative to modify the "space for peace" policy and gain a larger Air Force leadership role in space. In all likelihood, the Air Force space campaign and the spring "firestorm" of publicity contributed to President Kennedy's decision on 26 May, in National Security Action Memorandum (NSAM) 156, directing the Secretary of State to form an interagency committee to review the political ramifications of satellite reconnaissance policy. The 156 Committee focused on the question of banning weapons of mass destruction from outer space. Efforts to prevent the arms race from adding space to its arena dated back to the Eisenhower administration's policy of freedom of space through "Open Skies." But any agreement on space seemed unachievable apart from a general disarmament scheme that ensured adequate inspection and verification. With the development of a satellite reconnaissance and other intelligence capabilities, what became known as "national technical means" of verification answered this requirement. Soviet criticism of American "spy" satellites diminished in 1963 following the Cuban Missile Crisis and their own progress in developing reconnaissance satellites. By the end of the year, the United Nations passed a resolution banning weapons of mass destruction from orbiting in space. Later, in 1967, fear of a nuclear arms race in space had diminished to the point where negotiators, using the 1963 resolution as a basis for concluding a more comprehensive arrangement, succeeded in reaching agreement on an Outer Space Treaty that prohibited weapons in space.[39].

Although the brouhaha in the spring of 1962 took administration and Air Force leaders by surprise, several warning signs suggested that earlier Air Force optimism might have been misplaced. For one thing, on 20 February 1962, Colonel John H. Glenn, Jr., became the first American to orbit the earth part of the NASA Mercury program. The largest television audience to that date watched his three-orbit *Friendship 7* flight, and on 1 March he and fellow astronauts Alan B. Shepard, Jr., and Virgil I. "Gus" Grissom received a ticker-tape parade in New York City attended by four million people.[40] In the acclaim and euphoria after the Glenn flight, NASA's star ascended, and Soviet space achievements seemed less threatening and insurmountable. With the end of a Soviet monopoly on manned spaceflight, Senator Stennis and his colleagues lost interest in pursuing their investigation of the "peaceful purposes" policy and separation of responsibilities between NASA and the Defense Department. The Glenn flight relieved pressure on NASA and dashed Air Force hopes for a larger voice in the national space program.[41]

As for Air Force space planning efforts, Secretary Zuckert and Air Staff planners encountered little more than faint praise from Defense Department officials like John H. Rubel, Deputy Director of Defense Research and Engineering, who had listened to an earlier Air Force presentation of the plan in the fall of 1961 yet declined to recommend approval to his superiors. As one might suppose, the Defense Department hewed to the President's space policy, but the Air Force held different views about space objectives and the direction of Air Force space programs. Even so, Air Force leaders initiated a major planning and programming analysis in the spring of 1962 without first clarifying and agreeing with the Defense Department on military space objectives.[42]

Another sign that the administration began having second thoughts about an expanded military space program came with the Defense Department's final decision on proposed increases in the fiscal year 1963 budget. Despite Secretary McNamara's offer to entertain budget increases for Air Force space initiatives, by late spring of 1962 General Ferguson's new list of space projects and cost figures drew charges of padding from Assistant Secretary of the Air Force for Research and Development Brockway McMillan, and in August the Secretary disapproved the supplemental request. In the wake of the Glenn flight and the June "firestorm," the administration felt much less inclined to accede to Air Force arguments.[43]

Finally, the Air Force-NASA relationship proved less harmonious than suggested by signed agreements and expressions of mutual cooperation from their leaders. Almost immediately after the signing of the 24 August 1961 "Agreement on Responsibilities at the Manned Lunar Landing Program Launch Site," the two sides became embroiled in disagreements over interpretation of the accord. The precipitating issue involved the Air Force's desire to locate the proposed Titan III launch site within NASA's area of operation at Cape Canaveral, to purchase an additional 11,000-acre buffer region to the north, and to establish overflight procedures. By the spring of 1962, on the eve of the public outcry against perceived military ursupation of NASA's responsibilities, differences over range use remained unresolved, and the Air Force also had raised the issue of reimbursable funding for support costs. Although these issues might appear minor and easily settled, they in fact represented larger, long-term questions of position and responsibility within the nation's space program.[44] At the same time, the Air Force and the Defense Department did not always agree on responsibilities and relationships toward the civilian agency. Indeed, Defense Department officials proved in no hurry to recognize a special role for the Air Force in support of NASA. It took six months before Secretary McNamara sanctioned Secretary Zuckert's request to have the Air Force designated "executive agent" for

NASA support. Likewise, Zuckert's December 1961 request for an AFSC liaison office at NASA headquarters did not receive approval until April 1962, and another month passed before the Air Force designee, Major General O. J. Ritland, assumed his new duties at NASA headquarters. Moreover, while the Air Force became the official military service for NASA support, decision-making responsibility for supporting NASA remained in the hands of the Defense Department's Director of Defense Research and Engineering. With its campaign for a larger space role in shambles in late spring of 1962, the Air Force clearly needed to establish a more effective working relationship with both the Defense Department and NASA if it expected to preserve the prerogatives it still held.

By the summer of 1962, the 156 Committee had reaffirmed the Eisenhower policy on space and decided against the Air Force on the issue of weapons in orbit. The Air Force also failed in its efforts to take over management of Project CORONA following cancellation of its Samos reconnaissance satellite program in the spring. Moreover, with the military man-in-space mission in question, the Air Force now faced the prospect of greater reliance on NASA for any involvement in manned spaceflight operations. The decisions taken in 1962 effectively ended Air Force efforts to lead an expanded effort that included weapons in space.[45]

Confronting the McNamara Defense Department

In the early months of the Kennedy administration, Air Force leaders had chosen to overlook signs that their position as the military space service faced potentially severe constraints. By the same 6 March 1961 directive assigning future space research and development to the Air Force, Secretary McNamara moved to restrict "the independent freedom of action of the three military services...by limiting the latitude of the military departments to increase emphasis and funding for various projects."[46] In the McNamara Defense Department, the office of the Director of Defense Research and Engineering (DDR&E), under Harold Brown, became more forceful as the Secretary's central staff reviewing agency for all military space research projects. The 1961 directive noted that DDR&E—not the Air Force—would define the parameters of military space research, select projects for development, and review all space proposals before sending them on to Secretary McNamara.[47]

The lunar landing decision masked the full impact of the Defense Department's approach as both Congress and the administration increased funding and support to a variety of space programs. At the same time, while the Defense Department directive had specified and tightened the basic rules for performing space research and development, it left open the question of the criteria for acceptable military space programs as well as their relation to NASA's agenda. Under pressure from the Air Force campaign for a greater military space role, the intention of the Defense Department to force the services to defend their programs by comparing costs and benefits emerged only gradually over the course of 1961.[48]

More than any other service or agency, the Air Force found itself increasingly on a collision course with the DDR&E review agency that the Defense Secretary relied on to control costly new space development proposals. Having reorganized in large part to perform as the "military space agency," the Air Force hoped for a repeat of the relatively "free hand" it had to build missiles without undue concern for cost overruns and duplication. At the same time, the Air Force found itself the service most heavily committed to expensive space programs, especially those like Dyna-Soar and others that involved manned spaceflight, without well-defined military operational missions. With decisions on funding these important and expensive new projects in the hands of the Defense Secretary and his civilian staff offices, prospects for disagreement between the Office of the Secretary of Defense and the Air Force proved unavoidable.[49] Indeed, when confronted with Air Force proposals, Director Brown and his staff increasingly demanded more precise requirements and "program definition" in terms of costs, schedules, and technical hurdles. Defense Department review officials applied rigorous cost analyses to programs from the development stage through full-scale production to deployment. The initial history of the Titan III space booster illustrated the Defense Department roadblocks facing Air Force space programs.[50]

The prospect of a standardized launch vehicle strongly appealed to the cost-conscious McNamara Defense Department. Initial discussions by AACB members led DDR&E's deputy director, John Rubel, to promote the idea as a "unified program concept" that would provide the model for future space program planning. In early August 1961 he and Assistant Air Force Secretary for Research and Development Brockway McMillan organized under the auspices of the AACB an Ad Hoc Committee for Standardized Workhorse Launch Vehicles to examine alternate approaches for a rugged booster capable of orbiting 10,000-pound payloads at 300-mile altitudes. Later the committee raised the booster performance requirement, calling for a capability of launching payloads between 5,000 and 25,000 pounds into low-earth orbit. By September the committee and the Air Staff had agreed on the combination of a Titan II upgraded with strap-on solid boosters and a high-energy upper stage for future, heavier satellites. Led by Space Systems Division, Air Force agencies immediately began intensive studies of roles, designs, performance capabilities and reliability, and a cost and development schedule. On 13 October 1961 the Air Force received permission from Deputy Director Rubel to start a "phase I" study for a system "package" comprising "a family of launch vehicles based on the Titan III."[51]

Although the Air Force favored the prospect of a standardized booster more powerful than either the Thor or Atlas, the Defense Department's micromanagement soon proved unwelcome. As Secretary McMillan recalled, the Titan project became the "most comprehensive advanced development planning effort ever undertaken by the Air Force."[52] In effect, Secretary McNamara saw in the Titan III booster the ideal test case for applying his innovative management procedures to reduce costs and accelerate development schedules. As a result, Defense Department officials accorded the booster project the closest scrutiny of any project heretofore developed by the Air Force. Project "definition" required more detail; a strong program office supervised every aspect; and the Air Force received direction to use new Program Evaluation Review Techniques and establish special accounting and auditing procedures. Director Rubel involved his office in initial study proposals, and he required use of a civilian consultant agency throughout the bidding period. When the Defense Department delayed the release of funds and continued to "refine" procurement procedures, the Air Force had to extend the study's due date from 1 February to 1 April 1962. Meanwhile, after Space Systems Division presented its findings on technical aspects of the project, Rubel requested a "white paper" assessing the program's philosophy and technical approach. Even after a thorough review of the phase I plan by Air Force officials, Rubel returned it a number of times for additional data and lower cost estimates to assist the Defense Department's review. By late spring the repeatedly revised schedule projected an initial Titan IIIA test flight in May 1964 and the first Titan IIIC flight in January 1965.[53]

The Defense Department's intensive scrutiny and persistent involvement drew the wrath of General Schriever. On 30 April 1962 he complained to Chief of Staff LeMay of "unprecedented...demands for large volumes of information and program data that is magnified at each succeeding organizational level. Decisions on matters that have never been previously reviewed are being withheld for inordinate lengths of time." He especially worried about the future impact of demands for detailed design specifications before the decision on program approval had been taken. "If we are to be held to this overly conservative approach, I fear the timid will replace the bold and we will not be able to provide the advanced weapons the future of the nation demands."[54]

The Defense Department's management procedures and system development criteria failed to convince Air Force leaders that space systems could reach maturity faster and cheaper. Defense Department practices also threatened to eliminate all Air Force programs that failed to convince the Office of the Secretary of Defense of ultimate mission success. As a result, under the new administration, the old dilemma posed by the "new ocean" of space became more acute for Air Force planners. While space continued to represent an unknown frontier that required exploration to determine its potential uses and missions, the Defense Department's rigid approach to requirements cast doubt on the service's ability to preserve both its hard-won fight to conduct basic research in space and pursue projects whether or not they could claim a viable mission in the end. But how to answer the military's argument that, in order to counter Soviet superiority in space and avoid a technological surprise, the nation must pursue military space research and development initiatives regardless of guaranteed mission success? The Defense Department's solution was the "building block" approach to military preparedness.

Secretary McNamara first described this concept during testimony before Congress on the fiscal year 1963 budget in early 1962. It subsequently appeared in the President's *Aeronautics and Space Activities Report* for 1962. As the Defense Secretary explained, space projects comprise two categories, those with "identifiable military needs and requirements," and those "designed to investigate promising military space capabilities…[to insure]…a broad flexible technological base" ready for adaptation and development for systems once future military requirements were identified. The latter category represented "building blocks" for future use, and the Titan III, which initially supported no operational requirement, exemplified this approach.[55] In this manner, the Defense Department continued to fund a variety of additional space projects, including space probes, large solid-propellant rocket engines, laser technology, ion propulsion, and bioastronautics, along with a host of related supporting research and development activities. On the other hand, the "building block" rationale provided the Defense Department more control over a growing number of expensive projects. Air Force leaders became increasingly alarmed at the shrinking research and development budgets for space.* In General Schriever's view the McNamara Defense Department's focus on cost effectiveness and the desire to accommodate the Soviet Union stifled the Air Force's efforts to move from exploratory to advanced research.[56]

Following the public furor in June 1962 about potential Air Force "offensive" systems in space, the Secretary and his staff showed less willingness to accommodate Air Force proposals. The new attitude became especially clear by fall in the remarks of the Deputy Director of Defense for Research and Engineering. In a speech on 9 October 1962, John Rubel asserted that the Defense Department's space spending was as high as it could go given the "uncertainties" of the military program. Therefore, although new space projects might seem potentially useful, they would undergo increased scrutiny for their contribution to the military mission. Most alarming to Air Force leaders, Rubel suggested that many Air Force proposals did not meet the required high research and development standards of his office but merely served abstract doctrines about the military space role. He pointedly referred to the now traditional Air Force concept of aerospace, by which space represented a mere continuum of the atmosphere and the logical area for Air Force operations. He saw no useful purpose in such theories that suggested the vacuum of outer space would become the next battleground, or that "control" of space, whatever that implied, meant control of the earth. An expanded Air Force space program had no place in the Deputy's view of the nation's current and future space posture.[57]

Although all Air Force space proposals received increased attention from the Defense Department, Rubel's remarks indicated that the Defense Department found

---

* See Appendices 2-2 and 2-3.

fault more with new proposals than existing programs and studies. The "building block" approach would allow continuation of a variety of carefully controlled research projects, while providing the means of avoiding commitment to costly new programs. In light of the Defense Department's rigid criteria and conservative research and development philosophy, Air Force space planners encountered major road-blocks in their efforts to develop credible long-range space planning and program-ming documents. Rubel's speech, in fact, occurred shortly after the Air Force had completed its most intensive space planning effort to date. The Air Force endeavor represented the era's "last hurrah" in the service's aggressive campaign for an ex-panded, Air Force-led space program.

The Air Force Plans and Programs for Space Leadership

In the spring and summer of 1962 Air Force leaders carried out three major space planning initiatives in response to perceived weaknesses in the national space program: the "West Coast" phase; the Five-Year Space Program Study; and an Air Staff-supervised revision of the Air Force Space Plan. The "West Coast" phase involved a technically oriented study conducted at Space Systems Division in Los Angeles under the direction of Lieutenant General Howell Estes, Jr., Deputy Com-mander of Air Force Systems Command for Aerospace. An "Executive Committee" phase represented a second space study effort led by Lieutenant General James L. Ferguson, Deputy Chief of Staff for Research and Development, who formed a joint Air Staff-major command task group to formulate a Five-Year Space Program. Finally, during the spring and summer the Air Staff's Deputy for Development Planning supervised a revision of the September 1961 Air Force Space Plan.[58]

The "West Coast" phase occurred in response to Secretary McNamara's 23 February 1962 letter to Secretary Zuckert, in which he emphasized the need to establish the "necessary technological base and experience," or building blocks, for possible manned space requirements at some future date.[59] In mid-April General Estes convened a Space Technical Objectives [planning] Group composed of a wide spectrum of the "best scientific and technical personnel available to AFSC." Its mission was to formulate long-range space program requirements centered around technical objectives. In a revealing initial address to the group on 14 April, Estes described the prevalent atmosphere of great skepticism at the Defense Department surrounding the project. He was "shocked," he said, to find that the Defense Depart-ment believed the Air Force developed technical justifications to support preconceived ideas and objectives; moreover, Defense officials considered that their technical work in coordinating Defense Department-NASA programs had left the Air Force with little of value to offer. The general expected his study group's work to convince the Defense Department otherwise. He also reminded his audience that the Defense Department intended to maintain control of all military space programs and, as a formal proce-dure, had required Air Force Systems Command to obtain clearance from DDR&E

through specific development plans before proceeding with any space research project in excess of $200,000. As a result, every aspect of the task force's findings had to be absolutely credible and integrated into the overall space program. Finally, the Defense Department remained "suspicious of our desires to run a military space program," and believed that the Air Force should focus on building a sound technical base rather than development of operational systems.[60]

General Estes formed several directing committees and twelve technical panels to assess important space research and development areas, including launch vehicles, space propulsion, spacelift support, space communication equipment, weapons, reentry vehicles, and spacecraft. On 14 June, after two months of study, the general and his Space Systems Division colleagues presented their analysis and findings on current programs and future requirements to Defense Department representatives, who suggested that the Air Force, like the Defense Department, move forward on preparing a Five-Year Space Program. Although on 25 June the "West Coast" group briefed its results at Air Force Systems Command and Air Force headquarters, their report never received approval or release authority, even within AFSC. By the end of June the Estes study had been superseded by the Executive Committee's Five-Year Space Program effort.[61]

The "Executive Committee" phase of the Air Force space effort, which lasted from 26 June to 16 September, brought together at Air Force headquarters representatives from the Air Staff and major commands. In contrast to the "West Coast" group's technical focus, the Executive Committee sought to meet specific operational objectives. Much of the effort centered on a "requirements panel" of full colonels that directed Air Force Systems Command's Space Systems Division to prepare a program that conformed to specific strategic, reconnaissance, defense, command and control, and support "capability requirements." In early September, Space Systems Division presented an ambitious program of sixteen projects with a five-year cost of $9.8 billion. Yet by 9 November, when Secretary Zuckert submitted the Air Force fiscal year 1964 space budget request, the total figure had been progressively reduced to $2.85 billion. Even so, "in view of the magnitude of these amounts," the Secretary explained, he elected to request major funding increases totaling $200 million beyond currently approved Defense Department funding only for four of the programs—the Military Orbital Development System space station, the Blue Gemini manned spaceflight project to experiment with Gemini capsules, the MIDAS missile detection system, and Saint, the satellite inspector. Beyond the four on the Secretary's list, only Dyna-Soar and the large solid-fuel booster program could even expect to receive substantial funding.[62]

As the Five-Year Space Program study neared completion, the Air Staff already had finished its revisions to the 1961 Space Plan.[63] In its detailed review of space technology, the plan relied heavily on the "West Coast" study by projecting "state-of-the-art" in each of the twelve technical areas. It also defined objectives for each "capability requirement," and provided employment concepts and performance capabilities. Like the basic 1961 plan, the revised Air Force Space Plan emphasized the operational importance of manned military systems. "[M]an has certain qualitative capabilities which cannot be ignored," argued the planners, who proceeded to elaborate on potential roles for man-in-space described earlier in General Ferguson's congressional testimony and the previous year's space plan. They also noted that "requirements for manned military space systems seem inevitable despite present uncertainties concerning man's exact military role in space."[64] On 29 August 1962 planners circulated the revised draft for comment. Although most responses proved favorable, the Air Force never officially issued an approved version of the plan.

None of the three initiatives received formal acceptance from the Air Force or the Defense Department. Launor F. Carter, the Chief Scientist for the Air Force, pointedly remarked that the Air Force could hardly expect to formulate an effective space program without an approved space plan. Lacking initial agreement between the Defense Department and the Air Force on concepts and objectives, he argued, neither plan nor program would see the light of day. Like its September 1961 predecessor, the August 1962 Space Plan remained a draft study only, unapproved.

In early 1963 Carter subjected the entire 1962 planning and programming process to a scathing critique. He asserted that much of the Estes initiative proved ineffectual due to the absence of long-term plans approved by the Defense Department and the Secretary of the Air Force. Without these, operational commands could insist on unreasonable operational capability requirements which made an orderly research and development program impossible. Moreover, in preparation of the Five-Year Space Program, top-level decision makers envisioned a modest five year program, while the action panels established requirements calling for funding increases upwards of $5 billion. Realistic programming proved impossible under these circumstances. The chief scientist also criticized the practice of requesting from scientists only their opinion of technical feasibility without the additional complexities involving cost, timing, and alternative systems. In this regard, he singled out the Air Force's misuse of its best technical resource, the Aerospace Corporation. Rather than play a vital role in the study process, the service's major support contractor for space seemed to provide significant inputs only when "they happened to coincide with those of their military employers." Above all, Carter explained the failure of the space program development effort as the result of "distant relations" between the Air Force and DDR&E, characterized by the Air Force's failure to involve the Defense Department agency continuously in the process.[65]

From the chief scientist's perspective, the Air Force would have to establish better relations with the Defense Department, and especially DDR&E, before it could hope to achieve its space objectives. The unilateral pursuit of space objectives in a planning vacuum had proven unrewarding. At the same time, while the Air Force's relationship with the Defense Department by late 1962 had altered substantially, the service also had become much more dependent on NASA for participation in manned spaceflight operations.

Developing a "Partnership" with NASA

The Defense Department directive of 6 March 1961 and subsequent guidance had been no more specific on the relationship of Air Force and NASA space programs than it had on requirements for Defense Department approval of Air Force initiatives. Although the 1958 Space Act designated NASA responsible for civilian space activity, it also required the agency to support military needs by "making available to agencies directly concerned with national defense…discoveries that have military value or significance."[66] In declaring itself for an integrated national space program, the Kennedy administration reinforced the need to emphasize cooperative efforts and interagency coordinating mechanisms to provide mutual support and avoid duplication. The Air Force relationship with NASA in the 1960s involved four major aspects: shared programs and technologies; NASA's overwhelming dependence on the Air Force for launch and ground support; NASA's continued support of Air Force aeronautical research; and "persistent attempts by the Air Force to investigate the military applications of space," especially of manned earth-orbital operations.[67] Characterized by support, coordination, and rivalry, the Air Force association with NASA would depend less on the actions of the Air Force itself than on the evolution of both the Defense Department's and NASA's assertiveness and their interrelationship on space policy and programs.

Throughout 1961 the pervasive nature of NASA's dependence on military support —especially from the Air Force—and continued high-level coordination between the Defense Department and NASA tended to conceal the fact that NASA was evolving into the dominant space organization. By the spring of 1962 it had grown in one year from 57,500 to 115,500 personnel, and a year later had 218,000 on its roster.[68] Meanwhile, NASA's budget also signaled its phenomenal growth. Its fiscal year 1961 budget of $926 million, or 51.2 percent of the total space budget, represented the first year the civilian agency received more funding than the Defense Department. By fiscal year 1963, the NASA budget comprised 66.7 percent of the total space budget, while the Defense Department's figures indicated a decline from 45 percent of the total budget in fiscal year 1961 to 28.5 percent in fiscal year 1963.[*]

NASA's increased size and budgets reflected its responsibility for all manned spaceflight and strengthened its bargaining power and willingness to take a more active part in coordinating programs with the Air Force. Disagreement over pro-

---

[*] See Appendix 3-2.

cedures and responsibilities worked out for Cape Canaveral operations represented one aspect of NASA's new assertiveness, while differences over funding arrangements indicated another. In March 1962, NASA took the additional step of establishing independent field offices at both the Cape Canaveral and Vandenberg missile ranges in order to assert its "own identity" and prerogatives. The following year, it concluded an agreement with the Air Force whereby it signed on to use the Agena upper stage. In doing so, NASA officials became involved early in the planning stage and joined the Air Force Configuration Control Board for the Atlas, Thor, and Agena space vehicles. It also participated in the production phase by establishing special coordination groups at Air Force Systems Command to monitor production development. NASA's extensive involvement in Defense Department activities led in December 1962 to the appointment of a Deputy Associate Administrator for Defense Affairs. Under retired Admiral W. F. Boone, this office became a central coordination and liaison element between NASA and both the Defense Department and the individual military services.[69] By contrast, the earlier Air Force initiative to establish the AFSC Office of the Deputy Commander for Space at NASA headquarters represented the need for closer coordination and establishment of a strong Air Force presence with the increasingly important space agency. With the Air Force's disappointment over its failed campaign for a larger military space role, it became increasingly interested in cooperative programs with NASA. When the Defense Department continued to question the requirement for an Air Force man-in-space role, the particular focus for Air Force-NASA relations became manned spaceflight.[70]

By early 1963 both the Defense Department and NASA had become more determined to establish their own prerogatives and responsibilities for man-in-space activities, with the Air Force often playing the role of spectator as well as participant. The Project Gemini agreement of 21 January 1963, signed by Defense Secretary McNamara and NASA Administrator Webb, represented a major watershed in the evolving relationship between the three parties.

The Air Force Pursues a Dyna-Soar and a Space Station

In 1963, action in space involved manned spaceflight, and NASA possessed all of it. The Air Force, however, had in various stages of study and development a number of projects involving manned spaceflight, with which it hoped to claim a role of its own. Dyna-Soar represented the only program approved by the Office of the Secretary of Defense and the one reflecting the Air Force's strongest institutional commitment and interest. The remaining manned projects centered on some form of space station or laboratory.

Although the modern idea of a space station dates back to Hermann Oberth's work in the 1920s, Air Force researchers began actively studying the concept in 1957 when a Wright Air Development Center report examined the requirement for possible space research stations. In the wake of Sputnik the Air Force received a variety of contractor proposals for orbiting space stations, including one calling for an Atlas-launched, four-man crew orbiting at an altitude of 400 miles. However, when NASA received the manned spaceflight and space exploration missions, the Air Force found itself confined largely to space development activities with recognized military requirements or likely military implications. Even so, the space station concept continued to receive attention from Air Force planners like Brigadier General Homer A. Boushey, Director of Advanced Technology, who believed it might serve as an effective observation post and patrol or bombardment platform. In June 1960, the Air Research and Development Command approved a study requirement calling for a military test space station (MTSS) to assess the potential of military men and equipment to function in space.[71] By 1961 Air Force leaders had deemed the space station essential to the Air Force space program. The September 1961 Space Plan justified its acquisition as necessary for evaluating "space command posts, permanent space surveillance stations, space resupply bases, permanent orbiting weapon delivery platforms, subsystems, and components."[72]

Defense Department officials became aware of the Air Force space station concept late in the fall of 1961 during presentations of the Space Plan and correspondence between Secretaries Zuckert and McNamara. While the Defense Department studied the matter, General Ferguson told congressional committees in early 1962 that in order to conduct testing in "the true space environment…we are convinced that a manned, military test space station should be undertaken as early as possible." He went on to refer to possible coordination with NASA for use of the Gemini as the ferry vehicle for the orbiting station. Underway since December 1961, planning for Gemini, NASA's successor to Project Mercury, had always assumed substantial Air Force involvement.[73]

In a letter to Secretary Zuckert on 22 February 1962, Secretary McNamara encouraged the Air Force to pursue the concept by using Dyna-Soar and Gemini technology in the initial development phase. By late March Air Staff and AFSC planners had confirmed the technical feasibility of the project, now designated the military orbital development system (MODS). When submitted to the Pentagon for approval in early June, MODS consisted of a permanent station test module, a Gemini spacecraft, and the Titan III "building block" launcher. In August the Air Force had added a separate program for the spacecraft termed Blue Gemini, which focused specifically on rendezvous, docking, and personnel transfer functions. Air Force pilots would fly on six Gemini missions to gain astronaut experience for the MODS missions. But the Blue Gemini project did not elicit universal support within the Air Force. Some, like Chief of Staff General Curtis E. LeMay, worried that it might endanger the troubled Dyna-Soar program. Others argued that its use of available technology and equipment would make it operational before the X-20. NASA, on the other hand, saw in Blue Gemini a means of adding more defense funding to the entire Gemini project. By December 1962, however, Secretary McNamara had canceled Blue

Gemini, declined to support MODS in the fiscal year 1963 budget, and limited the Air Force to conducting a series of "piggy-back" experiments as part of NASA's Gemini mission.[74]

Although actions by the Office of the Secretary of Defense reflected Secretary McNamara's strong reservations about Air Force manned spaceflight projects, he remained unwilling to close the door entirely on determining a military role for man in space and leave the field of manned spaceflight entirely to NASA. Indeed, his view of an integrated national space program envisioned a continued major Defense Department voice in space decision-making, and he proved determined to assert the prerogatives of his office with Administrator Webb and his colleagues. In fact, during the week and a half before the signing of the Gemini agreement, Secretary McNamara attempted to take complete control of the Gemini project. Stressing the Defense Department's experience and the integrated nature of the national space program, he first informally proposed that all Defense Department and NASA manned spaceflight programs be centralized under Defense Department management. When Webb declined, the Defense Secretary countered by suggesting that Gemini be managed jointly by the Defense Department and NASA. Once again, to preserve its freedom of action, NASA refused the Secretary's advances. Nevertheless, in the agreement NASA concluded with the Defense Department on 21 January 1963, it went far to accommodate Defense Department concerns.[75]

Although managed by NASA, the project would involve Defense Department participation in every phase. The agreement created a joint Gemini Program Planning Board cochaired by NASA's Associate Administrator, Robert Seamans, and the Assistant Secretary of the Air Force for Research and Development, Brockway McMillan. Its charter called for it to plan and conduct operations to "avoid duplication of effort in manned spaceflight and to insure maximum attainment of both DoD and NASA objectives." Ultimately sixteen of the forty-nine Gemini experiments represented Defense Department projects that proved important for NASA, too. They focused on determining the military usefulness of manned spaceflight by testing extravehicular maneuvers with chest units and propulsion equipment designed for the Gemini space suit and the effects of weightlessness over extended periods of time in space. Additional projects included radiometric, radiation, and navigation experiments, and a variety of photographic and visual tests to determine the capability of acquiring, tracking, and photographing space objects and terrestrial features from the Gemini capsule. Because the Air Force considered many of these experiments classified, NASA officials worried about compromising their "peaceful" image. Despite considerable internal opposition, top agency officials agreed with the argument of NASA's Defense Affairs chief, Admiral Boone, that the national interest and NASA's charter warranted their inclusion.[76]

Above all, NASA submitted to McNamara's insistence that "NASA and the DoD would initiate major new programs or projects in the field of manned spaceflight in near-earth orbit only by mutual agreement." NASA officials worried that this provision might provide the Defense Department with veto authority over the civilian agency's scientific proposals on the basis of an unfavorable cost-benefit ratio while compelling the agency to agree to the Defense Department's manned spaceflight projects in the name of national security. Although NASA's fears did not materialize, this concession helped provide the Defense Department and the Air Force the leverage to secure future military inputs in national space decisions.[77] Yet the Air Force could not be entirely pleased with the Gemini decision. Despite retaining strong involvement with experiments and operational support, it did not represent the separate military manned spaceflight program it desired. Nor did it ease fears that NASA's Project Gemini competed with Air Force programs and might convince the Defense Department to cancel Dyna-Soar and other Air Force man-in-space projects. In fact, Gemini seemed to imply that there could be no Air Force manned space program independent of NASA.

By 1963 both the Defense Department and NASA confronted difficult questions about the nation's post-Apollo space future. For NASA, the main focus of what it called its Apollo Applications Program proved to be some form of space station, for which it had already initiated preliminary studies. Despite the already impressive performance of automated spacecraft, Air Force leaders continued to view the space future largely in terms of manned spaceflight and pressured a reluctant Defense Department accordingly. The task proved difficult. Following the Gemini agreement Defense Secretary McNamara established more stringent criteria for approving military space projects. As he explained to the Senate Appropriations Subcommittee in the spring of 1963, the space program must satisfy two basic criteria. "First, it must mesh with the efforts of...NASA...in all vital areas.... Second, projects supported by the Defense Department must promise, insofar as possible, to enhance our military power and effectiveness." He went on to defend the importance of cooperative efforts between the two agencies for the success of an integrated national program.[78]

For the Air Force, the new criteria seemed to mean that NASA came first, and space proposals would continue to suffer from the "requirements merry-go-round." By 1963, a cost-conscious Defense Department confronted crucial decisions on a number of major Air Force space programs for which research and development had reached important milestones. Consuming an ever larger share of the $1.5 billion space budget, now these projects faced more demanding Defense Department approval criteria.[79] Should the Defense Department support advanced development, proceed with development at scaled-back levels, or cancel the projects entirely? Programs under this kind of scrutiny included Bambi, MIDAS, Saint, and— especially—Dyna-Soar.

Armed with its new approval criteria, the Defense Department chose to "reorient" MIDAS with reduced funding and an extended development schedule in spite of its five successful flights in 1963. As Secretary McNamara explained, there still remained "unanswered questions regarding the technical feasibility, complexity, and cost-effectiveness of a space-borne [early warning] ballistic missile alarm system."[80] Determining that Bambi and Saint unfavorably competed with NASA programs and alternative Defense Department systems, he canceled Bambi entirely and reduced Saint to a "definition" study. Although the Air Force had argued that NASA's projects did not involve "non-cooperative" targets, the Defense Secretary had decided to turn from antimissile and antisatellite defense to more "reliable" and "cost effective" ground-based radar and missile systems. Above all, only ground-based systems qualified in terms of national policy of space for peaceful purposes.[81]

The one-man piloted Dyna-Soar faced the most intense scrutiny because it represented the costliest space project in the budget, and Defense officials continued to question what it would be used for since it could not be used for its original purpose of orbital bombing. As the Defense Secretary commented to the House Armed Services Committee in January 1963, "some very difficult technical problems still remain to be solved in this program, particularly in connection with the mode of reentry."[82] That same month he charged his DDR&E chief, Harold Brown, to assess the advantages and disadvantages of Dyna-Soar compared to expected benefits from NASA's two-man Gemini program.[83] Yet the technical challenges seemed to worry Secretary McNamara less than the high costs and especially the military purpose served. In March 1963 he consulted with NASA's Administrator, James Webb, on possible alternatives to spending $600 million for the Dyna-Soar program, with its "ill defined military requirement."[84] Later, in October, he visited the Martin-Marietta plant in Denver to review progress on the X-20 and Titan III. His concerns remained the same ones he had expressed in the spring. The Air Force focused primarily on getting into and out of orbit rather on the basic question: "what does the Air Force really want to do in space and why?" The Secretary left dissatisfied with the answers he had received.[85]

By the fall of 1963, while the door was closing on the Dyna-Soar program, it had opened for the concept of developing a military space station. Although the MODS project had been eliminated from the fiscal year 1963 budget, Secretary McNamara authorized the Air Force in the spring of 1963 to examine a similar concept known as the national orbital space station (NOSS). Apparently, McNamara approved the Air Force study in response to indications that NASA was ready to sign a $3.5-million contract study for a Manned Orbital Research Laboratory. At this point, both the Defense Department and the Air Force believed that a military version could be selected as the national space station in competition with NASA for post-Apollo space applications.[86]

During the spring and summer of 1963 senior Defense Department and NASA officials discussed the possibility of developing new manned earth orbital research and development projects. Secretary McNamara lobbied forcefully for the Defense Department's involvement from the start in any exploratory study effort. For him the Gemini agreement of 21 January 1963 did not go far enough to guarantee initial Defense Department participation to ensure its requirements would be incorporated into the design. He believed that the recommendation of the AACB's Manned Space Flight panel for coordination and exchange of information did not go far enough. He proposed a joint "sign off" clause for "initiation of any contractor study program or project in the field of manned orbital test stations of a magnitude equal to or greater than a $1,000 per year level of effort."[87]

The Secretary's tactic consisted of submitting to NASA officials signed draft Defense Department-NASA agreements for Administrator Webb's signature without preliminary staffing by both parties. McNamara's position and tactics alarmed Webb and his colleagues, who refused to allow the Defense Secretary veto power over initial studies NASA officials considered necessary to make effective planning and programming decisions.[88] With the two sides deadlocked, in late July Vice President Johnson asked for their views on space stations. The Defense Secretary took the opportunity to forcefully commit his agency to a space project that promised "immediate utility as a laboratory and development facility" that could evolve into an effective military vehicle. The Vice President's interest helped provide momentum for agreement. After declining to sign several proposed arrangements, officials from both agencies met informally and worked through the AACB to reach a compromise.[89]

On 14 September 1963 the Defense Department and NASA signed an agreement covering a "Possible New Manned Earth Orbital Research and Development Project." By terms of the accord, the two sides agreed on a "common approach" to projects involving new manned orbital research and development vehicles, particularly manned orbital systems larger and more complex than Gemini and Apollo. The goal would be a single project capable of meeting the requirements of both agencies. The Aeronautics and Astronautics Coordinating Board would coordinate the studies with the intention of submitting a joint recommendation for presidential approval. Management responsibility and funding apportionment would be determined jointly. Although Defense Secretary McNamara had reservations about NASA's head start and the method for handling disagreements, Administrator Webb reassured him with promises of full cooperation from the outset on all manned spaceflight projects. The Defense Secretary's concerns notwithstanding, the new agreement superseded the Gemini accord and ensured Defense Department an equal voice in post-Apollo national space decisions.[90]

Following the NASA-Defense Department space station agreement, Defense Secretary McNamara proceeded with his own plans for a military manned spaceflight research project to replace the Dyna-Soar manned orbital glider. By November DDR&E had completed the evaluation of Dyna-Soar's future that had engaged its attention since January. On the 14th Director Brown recommended that the Air

Force program be ended and replaced by a military space station and expansion of the Air Force's ASSET (aerothermodynamic/elastic structural systems environmental tests) project, previously a part of the Advanced Reentry and Precision Recovery Program begun in June 1960.[91]

Interestingly, of the six alternative Gemini-based space station proposals considered, Brown favored one far more ambitious than the Manned Orbiting Laboratory, the project announced by Secretary McNamara in December 1963. DDR&E's initial proposal called for a large, 2,140 cubic foot, four-room station with a crew of four astronauts on a thirty-day rotation, and launched by a Titan III. The ambitious plan included extensive ferrying, docking, and resupply operations. When the Director submitted the proposal to NASA as required by terms of the 13 September agreement, however, he encountered opposition from agency officials who believed the project conflicted with the civilian agency's mandate for such experiments. NASA countered with a more restricted alternative, an orbiting military laboratory. By considering the system a laboratory and not a space station, NASA could effectively argue that the military should leave ferry, docking, and resupply experiments to future NASA programs. Similar to the original Air Force MODS proposal, the NASA-proposed laboratory consisted of a Gemini capsule linked to a test module and launched by a Titan IIIC. The modest project seemed based more on the interagency Gemini agreement of January 1963 than on the September accord. As such, it would serve to postpone a formal decision on management responsibility for a national space station and, thereby, allow the Air Force to retain a man-in-space mission. Although Director Brown continued to advocate his original proposal, he agreed that the NASA alternative represented a credible "near-term" manned military space program. There the matter stood in December 1963 when the Defense Secretary made a major decision on the future of Air Force manned spaceflight.[92]

Setting Course on a Manned Orbiting Laboratory

Although the DDR&E recommendations precipitated a last-ditch effort by Air Staff officers to save Dyna-Soar, their arguments proved futile. At a 10 December 1963 press conference, Secretary McNamara announced the cancellation of Dyna-Soar and the approval of a Manned Orbiting Laboratory (MOL). The Secretary justified his decision to end the Dyna-Soar (X-20) program by citing imposing technical challenges to achieving an overly ambitious set of objectives that included maneuverable capability and precise reentry and landing techniques. Furthermore, the vehicle could carry only one man and had already moved beyond the Titan I and II to the Titan III. As the booster sequence suggests, budgetary concerns seemed uppermost in the Secretary's thinking. Already accounting for over half the budget for space research and development at $400 million, planners estimated a final program cost of $1 billion. Under existing constraints, the Air Force budget clearly could not accommodate both Dyna-Soar and the MOL.[93]

Two days after the press conference the Air Force began dismantling the program with the purpose of salvaging as much as possible for other projects. Although canceled nearly two years before its first scheduled orbital flight, Dyna-Soar left important legacies. Secretary Zuckert approved continuation of thirty-six specific activities in areas of advanced technology, hardware, and technical data. Improvements with high-temperature materials and fabrication processing contributed to development of other spacecraft and large rocket boosters. Data from over 2,000 hours of wind tunnel testing provided significant knowledge on aerodynamic stability and control and structural design problems. Engineers expected to adapt the X-20's environmental control system for future use, while the four guidance subsystems found immediate application in space activities.[94] The Dyna-Soar represented the first approved military spacefaring system, and the only one that initially included an offensive role. It kept the focus on manned military spaceflight and, most importantly, helped lead to the development of the Titan III, the "DC-3 of the space age." Its aerodynamic approach to space operations would reappear in the future in the form of the Space Shuttle. Meanwhile, Air Force space interests would now focus on examining man's capability to operate in the controlled environment of a space laboratory. This laboratory would have no offensive capability but, rather, would conduct passive defense functions in keeping with national space policy.

In the Manned Orbiting Laboratory, the Air Force at long last believed it would attain its man-in-space objectives, whatever they might be. The proposed laboratory, which closely resembled NASA's alternative to DDR&E's space station proposal, would rely on existing components from both Defense Department and NASA programs. Launched by a Titan III, a modified Gemini capsule would act as the transport vehicle for an attached laboratory canister "approximately the size of a house trailer." In the laboratory a two-man crew would conduct "shirt-sleeve" experiments, such as pointing cameras, for a three-day period.[95]

In one sense, the MOL represented a significant departure from the Defense Department's stringent requirements criteria. To this point the Air Force had faced a requirements paradox for military manned spaceflight projects. Because the Defense Department saw no specific requirements for military man-in-space, it had continued to oppose development of Air Force programs and authorized only participation in NASA-managed projects like Gemini. From the Air Force viewpoint, such projects did not provide necessary data on potential military capabilities on the frontier of space. Secretary McNamara's comments on the MOL reveal both his skepticism about manned spaceflight and his concession to the Air Force:

> This is an experimental program, not related to a specific military mission. I have said many times in the past that the potential requirements for manned operations in space for military purposes are not clear. But that, despite the fact they are not clear, we will undertake a carefully controlled program of developing the techniques which would

be required were we to ever suddenly be confronted with…[a]…military mission in space.[96]

In effect, the Manned Orbiting Laboratory would become the new military manned spaceflight "building block."

As for his established criterion requiring compatibility with NASA's projects, the Secretary stated that MOL did not duplicate NASA programs because, unlike Apollo and other current NASA projects, it filled a gap in the national space program by providing long-duration "near-earth orbit" manned spaceflight experiments under conditions of weightlessness. Furthermore, the Defense Department's laboratory would pursue military objectives like reconnaissance and satellite detection and inspection when possible. NASA had been invited to participate, although McNamara pointedly declared that "this entire program will be Air Force managed."[97] Later, NASA and Defense Department officials reaffirmed that the MOL did not violate the September 1963 space station agreement. The MOL, they said, was not a space station as defined by the agreement because it did not represent a future spacecraft "larger and more sophisticated than Gemini and Apollo." Therefore, it did not require a joint recommendation as a "national" project submitted for presidential approval. It would be a military program directed by the Air Force.[98]

The fact that the Defense Secretary had forcefully stressed the MOL as an Air Force-directed project suggests that he remained sensitive to the service's continued pressure for a military manned space role and to its concerns after the series of program cancellations and "reorientations" during the past year. From the Secretary's point of view, an Air Force MOL made good sense because, unlike Apollo, it would be based on Gemini, which offered the advantage of proven technology and use of the Titan III rather than NASA's Saturn IB. It also would keep the Defense Department active in the exploratory stage for the national space station. Air Force leaders clearly considered the MOL the first step to a permanent place for military man-in-space activities.[99] On the other hand, the Defense Secretary in December 1963 only authorized feasibility studies for the laboratory. The Air Force would have to establish convincing mission requirements before receiving approval for system production. Over the next twenty months, the Air Staff and Air Force Systems Command responded with organizational initiatives and intense study of system capabilities and potential mission functions.

The Air Force found itself reasonably well prepared when McNamara awarded it the Manned Orbiting Laboratory. Hoping for approval of its national orbital space station proposal, the Air Staff had been assessing organizational options since August 1963. That August General Ferguson urged Vice Chief of Staff General William F. McKee to provide a space station focal point in response to new organizational actions by both the Defense Department and NASA. The Defense Department had established a Deputy Director for Space, and NASA had under consideration a special management structure for its space station program. Impressed with General

Ferguson's argument, the Air Staff on 15 August created the Office of the Deputy Director of Development Planning, Space, headed by Colonel Kenneth W. Schultz. Colonel Schultz would support both Under Secretary of the Air Force McMillan and Alexander Flax, who succeeded McMillan as Assistant Air Force Secretary for Research and Development, on the Air Force side, and Albert C. Hall, the Defense Department's new Deputy Director for Space.[100]

A month after Secretary McNamara's decision, General Schriever proposed that he head a new MOL office at Air Force Systems Command headquarters to serve as "management agency" between the west coast Space System Division program office and the Office of the Secretary of the Air Force. Although Under Secretary McMillan found favor with Schriever's proposal, initially he pursued other options. First, he moved to upgrade and redesignate Colonel Schultz's position to that of the Office of the Assistant to the Deputy Chief of Staff for Research and Development for the MOL Program in order to accommodate the expected high degree of inter-agency and interservice coordination. Later, on 18 January 1965, he and Air Force Secretary Zuckert created the new office of Special Assistant for MOL under the Secretary's direct supervision and supported by a MOL Policy Committee. Finally, General Schriever received more responsibility than he first requested when he became head of a new MOL program office established at the Secretarial level under special security directives. The organizational evolution of the MOL's management structure reflected increasing high-level interest in the laboratory's mission. By mid-1965 it had become part of the sensitive national space reconnaissance effort.[101]

The long project definition phase, from December 1963 to August 1965, suggests the difficulty the Air Force faced in establishing convincing military missions for its astronauts to perform in space. It called on seventeen contractors to assess subsystems and experiments for possible incorporation in the MOL's mission. Areas examined included navigation, communication, observation, and biomedicine. Yet proposed mental and physical health studies, as well as experiments to determine if man could enhance the results produced by automated and semi-automated equipment, failed to convince the Defense Secretary of MOL's cost effectiveness, especially compared with automated spacecraft performing the same functions.[102]

During 1964, however, the Defense Department added two reconnaissance tasks involving radar and camera assembly and operation in space. The MOL launch site shifted from Cape Kennedy to Vandenberg Air Force Base, California, in order to conduct high-inclination launches needed for intelligence collection over Soviet territory. With the additional requirements for inspecting non-U.S. satellites when they passed in view and for ocean surveillance to meet naval concerns, the Defense Secretary found the MOL sufficiently important. Eventually, the requirements called for fifteen primary and ten secondary experiments.[103]

On 25 August 1965 President Johnson announced approval of the MOL for full-scale development with an initial budget of $150 million. The project involved three main contractors: Douglas would be responsible for the laboratory canister; McDonnell the Gemini capsule; and General Electric all space experiments. By this time the project's configuration differed somewhat from McNamara's description in December 1963. The laboratory canister now measured 41 feet long by 10 feet wide and weighed 14,000 pounds, with the reconnaissance payload comprising 5,000 pounds of the total 25,000-pound system. Once in orbit, the two astronauts would move through a specially constructed hatch into the laboratory, where one section housed pressurized living quarters and the other the experiments section with the reconnaissance telescope. The camera's lens would measure six feet in width, with a resolution between six and nine inches depending on atmospheric conditions. After completing their 30-day mission, the astronauts would close the laboratory, move back into the Gemini B capsule, and separate from the canister for the flight to earth and an ocean recovery. The laboratory would be left to burn up on reentering the atmosphere. The Air Force expected to launch the first of five MOLs in early 1968.[104]

At the decade's midpoint, Air Force leaders had renewed cause for optimism. It seemed that the service at last had a manned spaceflight project that would reach operational status. They confidently predicted that the laboratory's test of man's usefulness in space would ensure a permanent role for manned military spacefaring. By mid-decade the Air Force had also established a more effective working relationship with both the Defense Department and NASA.

Following criticism of Air Force space planning and programming by its chief scientist, both the Air Staff and the Office of the Secretary of the Air Force moved to develop closer rapport with the Office of the Secretary of Defense. Lieutenant General James L. Ferguson, Deputy Chief of Staff for Research and Development, Under Secretary McMillan and Assistant Secretary for Research and Development Alexander Flax led the way through many informal meetings with DDR&E's Harold Brown and his staff. As a result, Air Force space planning became more practical and realistic—and more modest. In late September 1963, when the Air Staff's Director of Plans proposed revising the 1962 Five-Year Space Program, General Ferguson recommended the Air Force forego another tedious official effort to define space goals and programs. He argued that the work involved in preparing the 1961 Space Plan and the 1962 studies had not been worth the effort and the acrimony that resulted. He also noted that the current proposed draft revision to the 1961 plan, now termed "USAF Space Objectives," offered no new space goals, thereby suggesting the soundness of past Air Force thinking on space. He reminded the Air Staff of the major headway achieved, largely through his office, in creating a more favorable attitude toward Air Force space issues in the Defense Department. Why take unnecessary action that might derail improving Air Force-OSD relations? The Air Staff persisted, however, and in the spring of 1964 General LeMay approved the "Space Objectives" paper. Yet, as a sign that relations between DDR&E and the Air

Force indeed had improved, Brown's office raised no objection, even though the list of Air Force space objectives included antimissile and antisatellite proposals already disapproved by the Defense Department.[105]

General Ferguson also referred to *Project Forecast* as offering nothing new on space. If so, this long-range projection of the Air Force's research and development requirements, which took place under General Schriever's direction from March 1963 to February 1964, provided what an official Air Force history termed "the most credible Air Force planning document on space yet."[106] It proposed "a balanced military space program" of systems necessary to support earth-based operations, studies of space-based "offensive" proposals, and advanced technical programs to improve launch vehicles and spacecraft subsystems. Taking into account the existing funding constraints, *Project Forecast* projected a "realistic" annual budget of just over $2 billion during the next five years. The more modest proposal also reflected the new reality of Air Force-Defense Department approaches to the military space program.[107]

Following the August 1965 decision to proceed with development of the Manned Orbiting Laboratory, Defense Department pronouncements remained encouraging, funding support continued, and NASA provided impressive assistance largely through the joint Manned Space Flight Policy Committee (MSFPC). In an agreement signed by Secretary McNamara and Administrator Webb on 14 January 1966, the MSFPC superseded the Gemini Policy Planning Board as the central joint planning and monitoring mechanism for Projects Gemini, Apollo, and the Apollo Applications program. Under its auspices, NASA furnished the Air Force a wealth of data, material, and experience for use in MOL development. This included three Gemini spacecraft, test capsules, a simulator, ground equipment, and subsystem hardware, as well as training aids, Apollo ships and tracking stations, and NASA engineers and technicians.[108]

The Air Force could point to significant progress in the MOL development program. In November 1966, the Air Force conducted successful tests with a smaller, simulated Gemini capsule that included nine on-board experiments, launched by a Titan IIIC. By this time, the experiments had increased the total weight to 30,000 pounds, which called for developing a more powerful Titan booster, the Titan IIIM. With its seven strap-on solid-fuel boosters producing a total thrust of 3.2 million pounds, the booster could launch the heavier spacecraft into polar orbit. By 1967 planners had completed design work on the basic Gemini-Titan MOL configuration, as well as the new west coast launch complex, and had selected for training twelve astronauts from the Air Force, Navy, and Marine Corps. Although Air Force Secretary Harold Brown doubted the Air Force could achieve its new projected initial launch date at the end of 1969, expectations remained high that the Air Force would receive its $600 million fiscal year 1969 budget request to complete the Vandenberg complex and final necessary MOL components.[109]

By 1968 more than technical challenges threatened the future of the MOL. In the latter half of the decade, the escalating financial burden of Vietnam and the domestic "Great Society" social agenda diminished support for the national space program across the board. Both Defense Department space programs and Project Apollo suffered reduced budgets.[*] In the competition for scarce resources, space generally and the MOL particularly became convenient targets for the budget cutters. Space represented a sizable twenty percent of the Defense Department's research and development budget. Of the Air Force budget, astronautics programs comprised one-third of the total, and half of this involved the MOL, the costliest project unrelated to the war in the Air Force budget for research and development.[110]

Cost-conscious critics also claimed that unmanned space systems could perform the MOL's experiments just as effectively at lower cost. Others raised the old cry of duplication with NASA's space exploration programs. Indeed, back in 1964, prior to President Johnson's announcement, the MOL had encountered considerable opposition during reviews by the President's Science Advisory Council and the Bureau of Budget. They concluded that NASA already had a major interest in orbiting a space station, while the military proposal seemed too small for the stated operational mission, and unmanned instrumented satellites could perform the functions identified more inexpensively. Charges of duplication became more persistent by the late 1960s, when NASA embarked on a large space station project as the centerpiece of its post-Apollo applications program. Although NASA and Defense Department officials argued that both the MOL and a civilian station would conduct necessary experiments that would not duplicate each others' efforts, critics remained unconvinced. A national poll taken in mid-July 1968 indicated that the majority of Americans thought the space program not worth the annual $4 billion price tag.[111]

Lower funding levels resulted in schedule "stretch outs," delayed milestone target dates and, ultimately, increased costs. Congress cut $85 million from the Air Force fiscal year 1969 request, which meant that final expected costs now totaled $2.2 billion rather than the fiscal year 1969 prediction of $1.5 billion. The Johnson administration's fiscal year 1970 defense budget that the Nixon administration inherited contained $576 million for the MOL, but the new Secretary of Defense, Melvin R. Laird, faced with continued high Vietnam war costs, targeted the MOL for reduction following a major review of the project. He chose to eliminate the fifth scheduled flight at a savings of $22 million and then cut an additional $31 million. This decision would delay until mid-1972 the first manned flight, leaving a total cost of $3 billion, twice the initial estimate. By June 1969, the administration determined additional defense cuts, and chose to cancel the MOL rather than eliminate competing satellite projects.[112] In his announcement on 10 June 1969, Deputy Defense

---

[*] See Appendix 3-4.

Secretary David Packard justified the decision as imperative in order to "reduce the defense research and development budget significantly." Moreover, "since the MOL program was initiated, the Department of Defense has accumulated much experience in unmanned satellite systems for purposes of research, communications, navigation, meteorology."[113] As Secretary Laird reaffirmed shortly thereafter, "these experiences as far as unmanned satellites are concerned have given us confidence that the most essential Department of Defense space missions can be accomplished with lower cost unmanned spacecraft."[114] The field of manned spaceflight now was left for NASA to exploit.

Immediately following the decision, the Air Force began closing down the project that by mid-1969 had cost $1.4 billion. Like its experience with the Dyna-Soar's termination a half-decade earlier, the Air Force salvaged a number of important elements for future use. One proved to be designation of the Vandenberg launch complex for future west coast Space Shuttle launches, while another involved the transfer of data and equipment to NASA for use in what became its Skylab space station operation. Most importantly the research experience gained from work on the Dyna-Soar and the MOL would prove instrumental in development of the new recoverable booster system—the Space Shuttle.[115]

An End and a Beginning

Termination of the Manned Orbiting Laboratory signaled the death knell of Air Force efforts to make manned spaceflight the center of a space-oriented military service. Although NASA's Gemini and Apollo programs included a number of military astronauts and experiments, the utility of military man-in-space activities remained untested.

Critics like retired Air Force Lieutenant General Ira C. Eaker declared that "cancellation…concedes to the Russians control of space."[116] Yet for other Air Force leaders, space represented abstract goals and assets that drained scarce operational funding from terrestrial needs. In the MOL's aftermath, former NASA Associate Administrator and now Air Force Secretary Robert Seamans knew spaceflight operations and requirements intimately. He nonetheless pointed to the shortcomings of conventional forces and the important requirement for F-15 fighters, C-5 transports, and an upgraded air defense posture. "The cost of a manned [space] system," he said, "is too great to be borne at this time." The Air Force, he said, must focus on modernizing its tactical and strategic forces rather than exploit the potential of space for future capabilities.[117] In effect, by decade's end, budgetary pressures and the impact of Vietnam compelled the Air Force to return to more traditional institutional interests. However desirable improved communications and navigation might be, space projects seemed more a luxury than a necessity.

On one level, Air Force manned spaceflight enthusiasts could look back on the decade of the 1960s as a graveyard of false optimism. High expectations at the onset of the Kennedy era for an expanded, "independent" Air Force space program proved unfounded. In the contest over manned flight projects between the Defense Department and the National Aeronautics and Space Administration, the Air Force emerged second best. Its campaign for more responsibility in the national space program diminished in the wake of NASA's Mercury—and later Gemini—successes and the growing détente between the United States and the Soviet Union. At the same time, elaborate, thoughtful efforts to formulate an acceptable Air Force Space Plan and a long-range development program received no blessing from a Defense Department determined to prohibit offensive systems in outer space and to put the brakes on spiraling space research and development costs by enforcing rigid mission requirements. The Air Force's man-in-space pretensions suffered most of all from skeptical defense officials increasingly who were obliged to rely on cooperative efforts with NASA.

An integrated national space program implied a mutually supportive relationship between civilian and military space agencies. Air Force leaders had hoped to make permanent NASA's early dependence on the "executive agent" for NASA support. Yet the lunar landing mission precipitated rapid growth in the civilian agency's responsibilities, independence, and funding. As a result, the Air Force's military manned spaceflight proposals became imperiled, and the service could never remove itself from NASA's shadow. Sensitive to public criticism of military encroachment on NASA's space exploration prerogatives, the administration reigned in aggressive Air Force space advocates and publicly questioned the usefulness of military manned space activities compared with automated satellites. By the end of 1962, the Air Force campaign for an ambitious, Air Force-led space program lay in shambles.

Air Force leaders responded by establishing closer, more effective working relationships with both the Defense Department and NASA. The price proved to be acceptance of a more modest space program without the schemes for antisatellite and antimissile orbiting space systems. Because the latter did not conform to U.S. space policy, the Pentagon elected to develop earth-based weapons instead. The Air Force, nevertheless, retained its man-in-space "mission" throughout the 1960s. Although compelled to forego Dyna-Soar and implement experiments only as part of NASA's Gemini and Apollo projects, approval of the Manned Orbiting Laboratory in 1965 seemed to promise an operational system by the end of the decade. Although President Johnson consistently supported the development effort, spiraling costs, schedule slips, and cost-effective satellites ultimately doomed the space laboratory.

At this point, the Air Force's space posture reflected changes within the service. Gone from the scene was General Schriever, long the service's most aggressive campaigner for Air Force space interests. In a sense, his retirement in 1966 confirmed the transition to the more modest and "practical" approach to military space. His able successor as commander of Air Force Systems Command, General Ferguson, proved more accommodating as an advocate of space interests within the framework of Defense Department and NASA relations. He also implemented a major reorganization within his command to respond to lower expectations and the changing state of space and missile development. By late 1966 General Ferguson and his staff decided that their west coast space and missile organizations should be reconsolidated. The Ballistic Missile Division's responsibilities had declined considerably with completion of most site activation work. As for the Space Systems Division, it never realized the potential General Schriever envisioned for it in the spring of 1961. NASA had garnered the bulk of the manpower and funding, while Secretary McNamara maintained severe limitations on defense research and development projects. On 1 July 1967, the Air Force created the Space and Missile Systems Organization (SAMSO) in place of the separate divisions.[118]

Yet, if the Air Force's space fortunes appeared to have plummeted at the end of the decade, the reality of space achievements proved very different. In 1969, President Richard Nixon established a Space Task Group to assess the nation's post-Apollo space requirements. Of the various options examined, it recommended development of a Space Transportation System (STS) based on a reusable launch capability. Earlier agreements between NASA and the Defense Department had ensured a joint military-civilian effort as part of the integrated national space program. Soon referred to as the Space Shuttle, its final configuration would reflect Air Force requirements. The development of the Space Shuttle also would precipitate a contest for operational responsibility among Air Force major commands, which would become a factor in quickening the pace for creation of an operational space command.[119]

Unmanned defense-support space systems represented another element in the evolution of a separate space command. Throughout the 1960s, the Air Force focus on its high profile man-in-space objectives overshadowed the growing importance of unmanned, instrumented satellites and the elaborate space infrastructure that had emerged to support them. In defending termination of the Manned Orbiting Laboratory program, Secretary Laird stressed the progress made in unmanned systems.[120] In effect, the end of the Air Force program for a manned space presence cleared the path for the dominance of unmanned military spacecraft with their important operational applications. By the late 1960s space programs increasingly moved from the realm of research and development to the operational arena where space could provide important support to traditional tactical as well as strategic mission areas. Although the dreams of a military man-in-space presence seemed over automated spacecraft proved to be making the "new ocean" an arena for military support applications and force enhancement.

CHAPTER 4

From the Ground Up:

The Path from Experiment to Operations

In the decade of the 1960s the Air Force turned its plans for military spacefaring into functioning systems. While manned spaceflight remained the centerpiece of the Air Force space agenda, the plans and programs for unmanned, automated satellites developed in the last Eisenhower administration now became reality. Communications, navigation, weather, and surveillance spacecraft came of age during the era that spanned the Kennedy, Johnson, and Nixon administrations.

Although the Defense Department proved unwilling to support the broad-based, Air Force-led military space program advocated by its leaders, the Air Force nonetheless forged ahead with development of spacecraft and the infrastructure to support them. The rapid pace of technological development over the course of the decade made possible more sophisticated instrumentation for these spacecraft. Equally important, developments in rocket boosters and the Air Force's efforts to achieve a more powerful, standardized launcher fleet produced reliable space boosters with greater lifting capacity capable of placing upper-stage vehicles like the Agena D and its satellites into geosynchronous orbit. Increasingly complex and larger satellites carried multiple payloads and performed a wide range of operational functions in space. At the same time, engineers succeeded in extending the lifetimes of satellites in orbit, thereby reducing the number of spacecraft needed. To support expanding satellite and booster capabilities, the Air Force created an elaborate space infrastructure of launch facilities, tracking and control networks, and research and development offices and laboratories. Taken together, the enormous growth in space capabilities by the early 1970s increasingly propelled space systems from the realm of research and development to the broader arena of operational applications.

Even though it could never achieve sole responsibility for military space, the Air Force found itself at the center of this fundamental transition. Critics rightly bemoaned the fragmented nature of military space responsibilities and organization that developed in the late 1950s and 1960s and that produced unnecessary delays, confusion, and severe security restrictions. General Bernard Schriever and other Air Force leaders valiantly attempted to have the Air Force assume ARPA's potential role as the sole military space agency equivalent to NASA on the civilian side. The Defense Department disagreed, and pursued a policy of tri-service management of space development in the name of cost-effectiveness and service cooperation rather than contention. Although the Air Force had achieved the dominant military space role through its authority to develop and launch military space systems and provide support to NASA, its more ambitious agenda would remain unrealized. As a result, by default, Air Research and Development Command and its successor, Air Force Systems Command, the Air Force's research and development organization, retained operational responsibility for the majority of space programs and systems for the Air Force and other space agencies. This set the stage for the intra- and interservice conflict over space roles and missions that would occur in the 1970s. Nevertheless, a fragmented military space program did not prevent the military space community—led by the Air Force—from compiling an enviable record of accomplishment. By the early 1970s military space dividends had become increasingly apparent at least to commanders who benefited from space-based systems in the Vietnam conflict and elsewhere. Instrumented earth satellites now offered the promise of providing the revolutionary applications predicted by space visionaries many years earlier.

Artificial Earth Satellites Become Operational

Seeking Global Communications—From Courier and Advent to the Defense Satellite Communications System (DSCS). The Second World War demonstrated the essential military need for electronic communications of longer ranges, greater security, higher capacities, and improved reliability. Orbiting earth satellites first proposed in 1945 by Arthur C. Clarke offered a revolutionary means of meeting these requirements. The British science fiction writer had suggested placing three satellites in geosynchronous orbit around the earth's equator. Equally spaced, they would put nearly every area on the earth within line-of-sight of one of the satellites; the spacecraft would receive signals from Earth and retransmit them back to Earth by means of solar power. Clarke's concept of synchronous repeater communication satellites attracted serious military interest, but remained only a theoretical possibility until technology could provide effective spacecraft and the boosters to place them in orbit.[1]

Immediately after the war, the Army experimented with passive relay space communications by using the moon and the planet Venus as signal reflectors. In the early 1950s the Navy also successfully bounced voice messages off the moon, and by the end of the decade had created two-way voice transmission between Washington and San Diego, then Washington and Hawaii using the earth's natural satellite. The Navy's project Communications by Moon Relay represented the nation's first operational space communications system and, except for navigation, the initial military application employing a satellite, in this case a natural one.[2]

After considering a number of proposals, the Defense Department's Advanced Research Projects Agency (ARPA) in July 1958 assigned the Army Project SCORE (Signal Communications by Orbiting Relay Equipment). On 18 December of that year an Air Force Atlas B booster launched the active (rebroadcasting) satellite into low-earth orbit, where on very high frequency (VHF) it broadcast President Eisenhower's recorded Christmas message. The Army followed this achievement in October 1960 with the successful launch of its Courier delayed-repeater communications satellite, which operated at ultra high frequency (UHF) in low-altitude (90-450 nautical miles) orbit. Meanwhile, the Air Force contracted with MIT's Lincoln Laboratory to produce 480 million hair-like copper dipoles, which, under Project West Ford, were launched on 9 May 1963 and reflected radio signals from an orbit nearly 2,000 nautical miles above the earth. Although initially scientists worried about potential interference with their radio telescopic observations, the dipoles ultimately proved benign, degraded rapidly, and three years later had completely disappeared.[3]

In the late 1950s military planners took another step on the road to translate into reality Arthur Clarke's dream of a global satellite communications system. In 1958 ARPA directed the Army and Air Force to plan for an equatorial synchronous (strategic) satellite communications system, with the Air Force responsible for booster and spacecraft, and the Army for actual communications elements aboard the satellite as well as on the ground. The program initially consisted of three projects: two, Steer and Tackle, involved medium-altitude repeater satellites; the third, Decree, called for a synchronous repeater satellite using microwave frequencies. In September 1959, the Secretary of Defense transferred communications satellite management responsibility from ARPA to the Army. Six months later, in February 1960, the Defense Department combined the three projects into a single program, Project Advent, and that September assigned it to the Army. In the meantime, it became apparent that neither the Army nor any other single service would have overall management responsibility for an operational military satellite communications (MILSATCOM) capability, because in May 1960 the Defense Department combined the strategic communication systems of the three services under a Defense Communications System (DCS) run by the newly created Defense Communications Agency (DCA).[4]

In the words of one observer of communications satellite developments, Advent proved to be "a not quite possible dream."[5] The ambitious program called for 1250-pound solar array-powered satellites, stabilized on all three axes, with the first group to be placed in a 5600-mile inclined orbit by Atlas-Agena B vehicles. The second set would achieve synchronous equatorial orbits when launched by the Atlas-Centaur booster combination. But Advent suffered from cost overruns, inadequate payload capability, and excessive satellite-to-booster weight ratios. At the same time, technology had advanced to the point where smaller satellites of 500 pounds or less could perform the same mission effectively. Advent's problems compelled Secretary of Defense Robert McNamara to cancel the program on 23 May 1962.[6]

With Advent's demise, Defense Department officials turned their attention to two alternatives that the Aerospace Corporation had been studying for the Air Force. In the summer of 1962 Secretary McNamara sanctioned the first Air Force proposal, which proposed randomly placed, medium-altitude (approximately 5,000 miles), nonstabilized satellites weighing 100 pounds each. He assigned the Air Force Systems Command's Space Systems Division responsibility for developing the spacecraft and communications payload and satellite operations. Unlike Advent, responsibility for orbiting elements would not be divided; the Air Force would handle spacecraft development and launch, while the Army's Satellite Communications Agency received authority to handle only the ground communications segment. Now termed the Initial Defense Communication Satellite Program (IDCSP), this would be another interservice project in which the Defense Communications Agency would coordinate Air Force and Army efforts to ensure compatibility. At the same time, the Air Force received permission to continue studies on a second alternative, which called for a future, stabilized synchronous system, later designated the Advanced Defense Communications Satellite Program (ADCSP).[7]

Progress toward full development of IDCSP proved difficult. In the spring of 1963, the Air Force received industry proposals for program definition studies based on using the Atlas-Agena D as the launch booster combination. Characteristically, the McNamara Defense Department required numerous studies and evaluations before funding an expensive new program, but the main reason for delay involved the new Communications Satellite (COMSAT) Corporation, established by Congress in early 1963. Before authorizing a more realistic MILSATCOM project to replace Advent, Secretary McNamara opened discussions with the COMSAT Corporation. McNamara questioned why the Pentagon should fund a separate, costly, medium-altitude MILSATCOM system if the Defense Department could lease links from COMSAT Corporation to satisfy military requirements at lesser cost. The Defense Department and COMSAT Corporation, however, could not agree on costs or the need for separate military repeaters aboard the commercial satellites.[8] Furthermore, the addition of military applications to a civilian system designed for use by other countries created international concerns. On 15 July 1964, after months of fruitless effort, Secretary McNamara ended the negotiations and opted for full-scale development of a dedicated military system, long favored by the Air Force to ensure security and reliability.[9]

By August 1964, when President Johnson announced immediate development of a military communications satellite system, the project had undergone a major change. The Defense Department decided to forego the medium-altitude system for the near-synchronous equatorial satellite configuration. The major incentive for the change proved to be the new launch vehicle under development, the Titan III, whose greater payload and altitude capabilities offered the prospect of launching a number of small satellites simultaneously into synchronous orbits. Defense officials elected to proceed with the more ambitious program despite concerns about solar heating at higher altitudes, the need to modify the original Philco-Ford satellites, and reliance on a booster yet to be launched. Taking a deliberate approach to reach synchronous orbit, the plan's first phase called for launching eight satellites into a near-synchronous equatorial configuration at nearly 21,000 miles in altitude rather than a more challenging synchronous orbit over 1,000 miles higher. Planners worried that, without "station keeping" capability, the satellites orbiting at the higher geosynchronous altitude might drift out of the desired position.[10]

Originally expected to function as an experimental system, IDCSP rapidly proved its operational worth and became the first in a three-phase evolutionary program to provide long-haul, survivable communications for both strategic and tactical users. The first seven IDCSP satellites, relatively simple in design to avoid the problems that had hampered Courier and prevented Advent from even getting off the ground, went aloft on 16 June 1966. Operating in the super high frequency (SHF) bandwidth, weighing about 100 pounds each, and measuring only three feet in diameter and nearly three feet in height, these spin-stabilized, solar-powered satellites contained no movable parts, no batteries for electrical power, and only a basic telemetry capability for monitoring purposes. The configuration of each IDCSP platform provided two-way circuit capacity for either eleven tactical-quality voice or five commercial-quality circuits capable of transmitting one million digital or 1,550 teletype data bits per second. The IDCSP satellite's 24-face polyhedral surface accommodated 8,000 solar cells that provided sufficient energy to power a single-channel receiver operating near 8,000 megahertz, a three-watt traveling wave tube (TWT) power amplifier transmitting in the 7,000 megahertz range, and one 20-megahertz double-conversion repeater. To launch the satellites, engineers placed them in a lattice framework mounted above the final booster, from where they would be released one at a time.[11]

On 16 June 1966, the Titan IIIC's fourth development flight successfully launched the first seven IDCSP satellites, along with an eighth experimental satellite designed to perform tests of gravity-gradient stabilization at high altitudes. Placing the satellites in almost exactly equatorial and nearly circular orbits involved the most complex series of orbital operations heretofore conducted in space. Buoyed by their initial success, IDCSP officials on 26 August 1966 launched a second set of eight satellites. But the fairing covering the satellites and its dispenser failed, and the Titan booster had to be destroyed eighty seconds after launch. With a redesigned fairing in place, a third launch on 18 January 1967 placed eight satellites into nearly the same orbits, while three additional IDCSP satellites joined their predecessors with a launch on 1 July. The latter flight also included three experimental satellites: a Navy gravity gradient spacecraft (*DODGE 1*); a despun antenna test satellite (*DATS 1*); and the fifth in the series of important Lincoln Laboratory experimental tactical satellites (*LES-5*). The final IDCSP Phase I group of eight satellites achieved orbit on 13 June 1968. With the last of the twenty-six satellites placed into proper orbit, the Defense Communications Agency declared the system operational and changed its name to Initial Defense Satellite Communications System (IDSCS).[12]

In mid-1968, thirty-six fixed and mobile ground terminals completed the satellite communications system. Originally designed for project Advent and later used in NASA's commercially targeted Synchronous Communication (Syncom) satellite program, two fixed AN/FSC-9 terminals with 60-foot diameter antennas, one located at Camp Roberts, California, and the other sited at Fort Dix, New Jersey, underwent modifications and began relaying IDSCS satellite data as early as mid-1968. Mobile terminals consisted of seven AN/TSC-54 terminals with 18-foot antennas and thirteen AN/MSC-46 terminals with 40-foot antennas. Additionally, the system included six 6-foot ship-based antennas. By the end of the decade officials were hard at work improving reliability and increasing terminal channel capacity. Additional ground terminal locations included Colorado in the United States, West Germany in Europe, Ethiopia in Africa, and Hawaii, Guam, Australia, Korea, Okinawa, the Philippines, South Vietnam, and Thailand in Asia.[13]

Already, by 1968, the new military satellite communications system had proved its value. A year earlier, the Air Force had established a link to Vietnam and publicly demonstrated its capability that summer at the 21st Annual Armed Forces Communications and Electronics Association convention in Washington, D.C. During the festivities Air Force Secretary Harold Brown spoke directly with the Deputy Commander for Air and Seventh Air Force Commander, General William Momyer, in Saigon, South Vietnam, about that day's air operations.[14]

The global IDSCS later became known as the Defense Satellite Communications System, Phase I, or DSCS I. Its exceptional reliability proved a very pleasant surprise to all involved in the project. By late 1971, fifteen of the twenty-six first-phase satellites remained operational. While several turned off after six years, as programmed, in mid-1976 three continued to function. The initial satellite system provided the Defense Communications Agency good service for nearly ten years. The IDSCS design, moreover, furnished the basic configuration for the communications satellites in the British Skynet and North Atlantic Treaty Organization (NATO) satellite programs that Air Force Thor-Delta boosters launched successfully in 1969 and 1970, respectively.[15]

Although the initial military communications satellites proved superior to available radio or cable communications, they remained limited in terms of channel capacity, user access, and coverage. Furthermore, military planners worried about the vulnerability of a command and control system that involved a central terminus connected to a number of remote terminals. The DSCS II design sought to overcome these deficiencies. Representing what planners had envisioned for Advent ten years earlier, DSCS II would encompass secure data and command circuits, greater channel capacity, and radiation protection features. In 1964 Secretary McNamara authorized preliminary work on the concept for a synchronous system offered by the Air Force after Advent's cancellation. In 1965 the Defense Communications Agency awarded six study contracts for concept definition. After numerous changes, in June 1968 the Defense Department approved the concept for procurement, and in March 1969 TRW Systems received the contract from Air Force Systems Command's Space Systems Division (formerly SAMSO) to develop and produce a qualification model and six flightworthy satellites that would be launched in pairs aboard a Titan III. Plans called for a constellation of four active satellites in geosynchronous orbit, supported by two orbiting spares. One satellite would be positioned over the Indian Ocean, one each over the eastern and western Pacific Ocean, and one over the Atlantic Ocean. Again emphasizing interservice development, the Defense Communications Agency would retain overall system management, with the Army responsible for ground terminals, and the Air Force responsible for the space segment, which included satellite acquisition, launch, and on-orbit operational control through the Sunnyvale, California, control facility's S-band space-ground link system.[16]

DSCS II represented a "giant step" in technical development over its smaller, lighter, and less capable predecessor. Each satellite measured nine feet in diameter, thirteen feet in height with its antennas extended, weighed 1,300 pounds, and was dual-spun for stability. An outer portion, consisting of an equipment platform, much of the satellite's structure, and cylindrical solar arrays, was spun to achieve stabilization. The inner section, housing X-band communications equipment and antennas, used a motor to despin, or remain stationary while the outer portion revolved around it, in order to keep the four communications antennas always pointed to the earth. Two horn antennas provided broad-area earth coverage, while two parabolic reflectors supplied narrow-beam coverage. The flexible, four-channel configuration provided a variety of communication links for achieving compatibility with various-size terminals. It possessed capacity for 1,300 two-way voice channels or 100 million bits of digital data per second, and rechargeable on-board batteries generated 520 watts of power to complement the satellite's eight solar panels. The five-year design life nearly doubled that of DSCS I, and the new system's redundancy, multichannel and multiple-access features and increased capability to communicate with smaller, more mobile ground stations especially pleased the Air Force and other users. While program officials readied the satellites for an initial late 1971 launch date, they proceeded to modify twenty-nine IDSCS ground terminals and build additional medium and heavy mobile and shipboard terminals for use with DSCS II.[17]

The orbital history of DSCS II satellites in the 1970s, beginning with launch of the first pair on 2 November 1971, revealed a somewhat spotty performance record. A Titan IIIC placed the first two DSCS II satellites into synchronous orbit, one positioned over the Atlantic Ocean and one over the Pacific Ocean. Problems occurred almost at once, when the first satellite's on-board receiver failed to respond to command signals and the absence of any telemetry signals from the second rendered it temporarily lost in space. Although Air Force technicians and engineers eventually succeeded in controlling both satellites, the Pacific satellite failed after ten months and the Atlantic satellite after nine months of operation. As a result, the Defense Department elected to continue using the IDSCS satellites until engineers could redesign the next two satellites. After balancing the despun platform and modifying the power-distribution system, the second pair of satellites was successfully launched and deployed on 13 December 1973. By February 1974, their performance convinced officials to declare DSCS II operational. Yet the launch of the final two satellites on 20 May 1975 proved disastrous. When the Titan IIIC's inertial guidance failed, the satellites deployed into low orbit and vaporized six days later during reentry. With only two satellites now operational, the Air Force responded by contracting with TRW for an additional six satellites of the original design and, later, four more with 40-watt TWT amplifiers in place of the 20-watt amplifiers. Despite another launch failure in March 1978 and continued high-voltage arcing in the power amplifiers, by the early 1980s the DSCS II constellation would not only fulfill global, strategic communications requirements through 46 DSCS ground terminals, but would also link the Diplomatic Telecommunications System's 52 terminals and the Ground Mobile Forces' 31 tactical terminals. Perhaps the best example of the satellite's durability is that DSCS II B4, launched on 13 December 1973, would last four times longer than its design life; and the Air Force would not turn it off until 13 December 1993. Meanwhile, in 1974 the Air Force began designing an improved DSCS III satellite to meet the military's need for increased communications capacity, especially for mobile terminal users, and for greater survivability.[18]

The Defense Satellite Communications System represented a global, strategic communications system. While the Air Force, in DSCS II, developed an operational strategic communications system, it also joined other agencies to produce an operational tactical satellite communications network. The road to tactical satellite communications took two paths. One involved Air Force activity that began in earnest in 1959 as part of ARPA's effort to develop a synchronous communication satellite. Although the Air Force supported the modest ARPA program, it focused on the Strategic Air Command's requirement for communications with its aircraft fleet in the polar region. As noted earlier, the ARPA concept could not meet this need, and the program became "reoriented" into three separate functions. Of the three, Steer proved most important for the Air Force because it envisioned satellites in polar orbit at 5,600 miles altitude capable of providing a single channel between aircraft and ground stations. But in the May 1962 reorientation that resulted in a single Advent program, Steer was canceled and, soon thereafter, officials terminated the program elements that had supported tactical communications.[19]

Air Force interest in tactical communications by satellite, however, did not diminish. With the conclusion of the passive West Ford dipole program in 1963, the Lincoln Laboratory turned its attention to active systems and began what would become a long history of tactical experimental satellite development. In short order MIT's laboratory produced a series of six Lincoln Experimental Satellites (LES) to test the technology for satellite-based communications with small mobile ground terminals. Over the life of the program, the Army and Navy participated by establishing UHF terminals on ships, submarines, jeeps and other small vehicles. The Lincoln satellites normally hitched a ride "piggyback" as a secondary payload on space launches. The first Lincoln satellite, *LES-1*, for example, entered orbit as part of a multiple-satellite payload aboard a Titan IIIA on 11 February 1965. Of the six satellites placed in orbit during the decade, the final two proved most significant for future tactical operational development. By 1 July 1967, when *LES-5* joined the three IDCSP satellites in subsynchronous orbit, technology had progressed to the point where the scientists could produce a 230-pound satellite with solid-state equipment capable of evaluating electronic despin logic. This proved important in developing DSCS II stabilization technology. *LES-5* remained operational until May 1971. Meanwhile, *LES-6*, the last experimental satellite, had been lofted into synchronous orbit by a Titan IIIC on 26 September 1968 along with three Office of Aerospace Research experimental satellites. *LES-6* represented a major technological advance over the *LES-5* that had been launched the previous year. Weighing nearly 400 pounds, *LES-6* housed a more powerful, all-solid-state UHF communications repeater and possessed electronic antenna despin capability. By connecting its amplifier directly to the satellite's solar array, scientists ensured that it would not compete for power with other equipment. Like its predecessor, *LES-6* also conducted experiments to measure the electromagnetic environment in space using a UHF radiometer.[20]

A second path on the way to operational tactical satellite communications involved a tri-service effort that, by late 1965, had agreed on producing a large satellite that would orbit at geosynchronous altitude. It would be designed with high-powered communications repeaters dedicated to the military UHF and super high frequency (SHF) wavelengths, with cross connections to other orbiting satellites, and the capability of switching bandwidths as desired. In December 1966, the Defense Department awarded Hughes Aircraft Company the satellite contract.[21]

TACSAT, as this satellite came to be called, represented "state of the art" communications technology. Measuring nine feet in diameter and twenty-five feet in height with antennas extended, and weighing 1,690 pounds, the cylinder-shaped spacecraft emerged as the largest communications satellite of its time and the first to be dual-spun for stability. Significant electronic, structural, and mechanical advances characterized its design and development. Generating one kilowatt of solar power, it possessed a 40-voice channel UHF capacity and an X-band capability of 40 voice circuits directed to a terminal on Earth with an antenna as small as three-feet. TACSAT's solid-state components provided 350 watts of power for UHF transmissions and 40 watts for SHF requirements using two traveling-wave tube amplifiers.

Unfortunately, funding limitations restricted the program to a single satellite. As a result, engineers and program managers conducted exceptionally thorough and challenging tests before declaring the satellite ready. On 9 February 1969, a Titan IIIC launched from Cape Kennedy and placed TACSAT into a near-synchronous 19,300 nautical mile orbit above the equator. Its performance exceeded all expectations. In March, twenty ship and land stations from Bermuda to Hawaii conducted a tri-service roll call, in which Air Force representatives successfully participated from Los Angeles using a battery-powered 22-pound portable transmitter and a six-pound receiver. On orbit for thirty-four months before an attitude control failure ended its operational capability, TACSAT well served the military by supporting a number of operations, including the recovery of *Apollo 9* in the Atlantic Ocean on 13 March 1969 by linking the carrier *USS Guadalcanal* recovery ship directly with the White House.[22]

The success of TACSAT also intensified interest in developing a tactical communications system for the Navy that could link ships, shore installations, and aircraft. Concept development, with Air Force participation, commenced in 1971 on a four-satellite, near-synchronous equatorial configuration that would become known as the Fleet Satellite Communications (FLTSATCOM) Program. Although the Navy provided funding and ground terminals, the Air Force served as the Navy's agent in all spacecraft areas and received use of a portion of the system's capacity. The Air Force realized that participation in the Navy's program could satisfy its long-term need for global tactical communications for its strategic aircraft. Meanwhile, the success of TACSAT could not prevent the program's termination on the basis of high cost. After several reconfigurations, the project reemerged as the Air Force portion of the FLTSATCOM. In effect, the Air Force Satellite Communications (AFSATCOM) System would use current and planned FLTSATCOM spacecraft to provide global communications for strategic Single Integrated Operations Plan (SIOP) forces. The planners expected to launch the first of the four satellites in the late 1970s.[23]

In the decade ahead, planners faced daunting challenges in their efforts to master the new communications technology in order to provide operational service to an ever-increasing number of users. Along the way, they would have to fend off more attempts from cost-cutters to combine the nation's civilian and military communications systems. Nevertheless, despite the rocky course, communications satellites had proven their value, and a global network of tactical and strategic space-based communications appeared on the horizon. By the early 1970s, the Air Force had begun to fulfill the dream of Arthur Clarke and the designers of Advent for point-to-point worldwide communications by placing sophisticated communications satellites in synchronous equatorial orbits.

Watch on the Weather—From TIROS to the Defense Meteorological Satellite Program (DMSP). Meteorological satellites comprise a second functional category of artificial earth satellite support to tactical and strategic military operations. Like Arthur Clarke and other space visionaries, dreamers and scientists had also envisioned orbiting satellites that could observe and report on weather phenomena from space. Although military officials had long recognized the importance of weather data for military operations, they often remained unable to gather needed information with conventional weather equipment over land and sea controlled by the enemy during conflicts and normally inaccessible during peacetime. Moreover, significant weather conditions frequently originated over water, where total coverage proved lacking, and spotty reports from ships or aircraft remained inadequate. In the aftermath of the Second World War, military authorities recognized the potential for weather reporting offered by "earth-circling" artificial satellites. As the 1946 Rand report predicted, "the observation of weather conditions over enemy territory" represented a most important kind of satellite observation.[24]

By 1961, the Air Force, largely through the Aerospace Corporation, studied the requirements for military weather satellites. Such satellites could provide photographs of cloud characteristics and their distribution for flight planning. Yet what appeared below the cloud cover, along with atmospheric temperatures, pressures, and wind velocities and directions, usually remained less susceptible to satellite measurements. At the same time, too much photographic information might saturate data processessing capability. Satellites might provide the perfect solution to these challenges, but only if scientists and engineers could develop capable sensors and supporting equipment.[25]

If technical problems presented military planners with one dilemma, civilian satellite operations already underway created another. NASA had received authority to develop weather satellites for all government users. It led the way with the low-altitude Television and Infra-Red Observing Satellite (TIROS) for the Weather Bureau. The successful launch of the 273-pound *TIROS I* by a Thor-Able II booster on 1 April 1960 from Cape Canaveral, Florida, opened a new era in meteorology.

Operating for only three months, it completed 1,302 orbits and transmitted nearly 23,000 photographs of global cloud cover from its position 450 miles in space. While *TIROS I* was establishing the feasibility of satellites for global weather observations, officials from the Departments of Defense and Commerce and from NASA met to consider development of a single weather satellite system that could satisfy the needs of both the military and civilian communities. Such a program would require civilian management to accord with the national policy of the peaceful use of space. After agreeing in principle, the Panel on Operational Meteorological Satellites by April 1961 had developed a plan for a low-altitude spacecraft termed the National Operational Meteorological Satellite System (NOMSS). But the NOMSS did not satisfy military requirements for coverage, readout locations, timeliness, operational flexibility, and security. Specifically, TIROS did not provide coverage of high-latitude and polar regions, while the satellites' cameras pointed to the earth little more than twenty-five percent of the time and observed specific earth areas at different times every day. Moreover, data transmission and processing weaknesses could not allow rapid operational data use. Reminiscent of the communications satellite issue, the Defense Department worried that political leaders, who viewed the weather satellite program as an example of the nation's peaceful space and foreign policies, might not allow such satellites to be available for military use in times of international tension. Although Nimbus, the second-generation weather satellite developed by NASA and the National Oceanic and Atmospheric Administration, improved upon many TIROS deficiencies, it proved to be a large, three-axis stabilized spacecraft that the space agency relegated to research use.[26]

As a result of these problems, early in 1963 the Aerospace Corporation recommended that the Air Force develop a dedicated military system, and the Defense Department agreed. The main emphasis would be on cloud-cover photography, but planners expected to add more sophisticated equipment when it became available. Later, when civilian weather satellites improved their capabilities and could satisfy most military requirements, the Defense Department continued to prefer a separate system responsive to the "dynamic" needs of the military. As a result, the Air Force embarked on the first segment of what became known initially as the Defense Satellite Applications Program (DSAP), or Program 417. Because the Air Force weather satellite program began with the mission of providing specific weather data to support Strategic Air Command and National Reconnaissance Office (NRO) requirements, the project remained classified until 17 April 1973, when Secretary of the Air Force Dr. John L. McLucas decided that the Defense Department's decision to use satellite weather data in the Vietnam conflict and to provide it to both the Commerce Department and the general scientific community warranted declassification of the DSAP mission and release of some of its performance data. In December 1973 the Defense Department changed the name to the Defense Meteorological Satellite Program (DMSP).[27]

The initial DSAP military weather satellites of the 1960s were relatively inexpensive and unsophisticated. Weighing 430 pounds and measuring approximately five feet in height and five feet in diameter, the twelve-sided spacecraft produced daytime visual and nighttime infrared weather photographs with resolution of one-third and two nautical miles, respectively. From polar, sun-synchronous orbit, two satellites transmitted weather data both early in the morning and at noontime to readout stations in Washington state and Maine, and from there to the Air Force's Global Weather Central facility at Offutt Air Force Base, Nebraska. Along with furnishing tactical weather information to Vietnam mission planners, the satellites passed auroral data to the Air Force's Cambridge Research Laboratory and the National Oceanic and Atmospheric Administration.[28]

By the early 1970s, the Air Force had launched four series of military weather satellites, each more capable than its predecessor. By the middle of the decade the Air Force prepared to launch the first of its fifth block of DMSP satellites with the Thor-Burner II booster pairing. This new generation of polar orbiting satellites, known as Block 5D, represented a major technological leap over previous models. Weighing 1,140 pounds and measuring four feet in diameter and twenty feet long, they tripled the size of the earlier satellites. Designed to provide both day and nighttime very high quality weather pictures, they also contained three times the number of special sensors. Most important among the latter proved to be an upgraded Operational Linescan System (OLS) to provide cloud-pattern images. A new integrated design that combined the satellite and upper stage of the booster created substantial weight savings. This made possible the use of redundant components which would increase the operational lifetime of the satellite from nine to as many as twenty-four months.[29]

Along with major advances in satellite capability came increased complexity, and problems with satellite stabilization and other technical difficulties led to questions about the system's reliability, its higher costs, and predictable scheduling delays. During the 1970s, DMSP also would face critics who sought to cut costs by combining the two parallel low-altitude weather satellite systems. As with the communications satellite issue, Defense Department officials relied on much the same argument to successfully withstand the pressure. To better retain a dedicated weather satellite system, the Air Force in 1973 sought, and achieved, active Navy and Army participation. After four years of discussion among the services, DMSP became a triservice program.[30]

Despite developmental problems, no one could doubt DMSP's important contributions to military operations. The success of the initial program convinced officials to broaden the satellites' SAC-oriented mission to include tactical weather support. Their confidence proved justified when, in the late 1960s, planners in Vietnam relied extensively on DMSP data for conducting combat operations, while others counted on weather satellite data to provide accurate hurricane warnings and perform Apollo recovery operations in the Pacific Ocean. By the early 1970s, weather satellites reached the level of performance predicted for them a quarter of a century earlier.

The Quest for Precise Location—From Transit to Navstar/Global Positioning System (GPS). Navigation satellites represent a third major functional area of space applications that came of age by the early 1970s. Throughout history, one's location on the earth and the ability to navigate from one point to another have remained essential requirements. For the military commander, whether in the air, at sea, or on land, there can be no more important questions requiring answers than: where am I and where am I going on the route selected? Early navigation involved simple pilotage, the process of determining position using observable landmarks to move from point to point. Ancient mariners also used the positions of the sun, moon, planets, and stars as reference points and developed instruments such as the astrolabe and quadrant to provide basic measurements in latitude. Celestial navigation became increasingly accurate once the chronometer appeared in the eighteenth century to calculate longitude. Since that time improvements in sextants, compasses and other instruments have enabled aircraft and ships to determine their positions on the globe within a mile through celestial navigation. But mariners needed an answer to the dilemma of cloud cover and dense fog that often made celestial navigation impossible.[31]

The answer came in the 1920s with the development of radio, which led to various techniques of radio navigation, first based on using a radio receiver with a simple loop antenna to calculate the radio signal's direction and relative bearing to the transmitter. Later, experimenters relied on the difference in time of arrival of a signal from two correlated stations. Position is determined by the intersection of two hyperbolas produced by the time differences in arrival of the signal at the receiver. One of the most effective hyperbolic systems appeared early in the Second World War, when MIT's Radiation Laboratory developed LORAN (long range navigation), which used synchronized pairs of transmitters at different locations to produce measurable time differences for aircraft at great distances from the transmitter. Accuracy could reach approximately a fifth of a mile at a range of 1,000 miles. But LORAN and similar radio navigation techniques were two-dimensional systems, designed to determine latitude and longitude only, not altitude or velocity of the aircraft. Moreover, weather disturbances and ionospheric conditions made low-frequency radio waves subject to errors, while high frequency transmissions depended on line-of-sight capabilities and the synchronization of ground stations. Artificial earth-circling satellites, on the other hand, could provide ideal platforms for radio navigation transmitters.

On 13 April 1960, an Air Force Thor-Ablestar launched from Cape Canaveral, placed a 120-pound Navy *Transit IB* satellite into a 700-mile altitude, circular polar orbit, thereby making the Transit navigation system the first to use radio transmission from satellites. It proved to be a simple, reliable two-dimensional system based on Doppler measurements. Scientists from the Johns Hopkins Applied Physics Laboratory (APL) had discovered that measuring the Doppler shift in frequency of Sputnik's continuous-wave transmitter provided sufficient data for determining the complete orbit of the satellite. Conversely, knowing such satellite information, termed its ephemeris or almanac, one could establish precise positions on Earth using the same Doppler calculations. Transit satellites provided position accuracy to about 600 feet, which met the Navy's need for accurate location of slow-moving ships and ballistic missile submarines. But the Transit system proved too slow and intermittent—and two-dimensional—to satisfy the more demanding requirements for precise positioning of high-speed aircraft and ground-launched cruise and ballistic missiles.[32]

The answer would prove to be the Global Positioning System (GPS), which would improve the Transit approach and supply a three-dimensional system to provide position, velocity, and altitude by a process closely related to the LORAN technique for measuring time differentials. The initial concept for a modified LORAN-type system involving altitude together with latitude and longitude appeared in a 1960 study prepared by Raytheon Company scientists to support a mobile version of the Minuteman intercontinental ballistic missile (ICBM) force. As described by one of its creators, Ivan A. Getting, the new system, called MOSAIC (Mobile System for Accurate ICBM Control), "used four 3000-MHz (S-band) continuous-wave transmitters at somewhat different frequencies, with their modulation all locked to atomic clocks and synchronized through communications links."[33] When Getting left Raytheon to become the first president of the Air Force's nonprofit Aerospace Corporation, he supported further research on this concept and the challenges associated with a satellite navigation system applicable for tactical aircraft and other vehicles moving rapidly in three dimensions.

By 1963 the Aerospace Corporation's engineers and scientists convinced the Air Force that the path to accurate measurement lay in calculating distances to satellites with known positions. That October the Air Force charged the corporation to pursue its satellite ranging study, now termed Project 621B (Satellite System for Precise Navigation), with support from Air Force Systems Command's Space Systems Division in nearby Inglewood. From the start such a system would include the capability of supplying accurate, all-weather position data to an unlimited number of users anywhere on or near the surface of the earth. Planners believed they could achieve position accuracies within fifty feet in three dimensions (latitude, longitude, and altitude). At the same time, the system had to be cost-effective. By mid-1966 successful studies of this satellite navigation concept led the Air Force to award study contracts for system hardware design to Hughes Aircraft Company and TRW Systems. From 1967 to 1969 additional studies envisioned a global network of twenty satellites in synchronous, inclined orbits using atomic clocks synchronized with a master system clock. The ground tracks of the satellites would comprise four oval-shaped clusters extending thirty degrees on either side of the equator. Because the satellites would be placed in orbit periodically during the development phase, the system could achieve a limited operational capability well before the entire system deployed.[34]

Meanwhile, the Air Force work stimulated the Navy to continue its own advanced navigation research. In the mid-1960s Roger Easton of the Naval Research Laboratory developed a system he called Timation, for Time Navigation, based on using precise atomic clocks. In 1967 and 1969 the Air Force launched Navy Timation satellites carrying sophisticated crystal oscillators and rubidium atomic clocks, which transmitted UHF signals for ranging and time transfer. By 1971 the Navy and RCA, its main contractor, proposed a system of 21 to 27 satellites in inclined eight-hour orbits. Earlier the Army had proposed its own system called SECOR (Sequential Correlation of Range). In 1968 the Defense Department organized a tri-service committee, later called NAVSEG (Navigation Satellite Executive Committee), to coordinate the various projects.[35]

Tests of operator equipment at White Sands Proving Ground in 1971 and 1972 using ground and balloon-carried transmitters achieved accuracies within fifty feet. Yet the Defense Department proved reluctant to approve full development of the expensive, technically ambitions Air Force system. In late 1972 the satellite navigation program received a new leader, Colonel Bradford W. Parkinson, who opened talks with the Navy to combine the Air Force's Program 621B and the Navy's Timation. On 17 April 1973 William P. Clements, Deputy Secretary of Defense, called for a joint development program, termed Defense Navigation Satellite Development Program, with the Air Force acting as executive agent. By September 1973 a unified program adopted the Air Force signal structure and frequencies and the Navy's satellite orbits. The satellite orbits would be raised to 7,500 miles altitude to produce twelve rather than eight-hour periods. The system would also use atomic clocks, which the Navy had already successfully tested in its Timation program. By December, Secretary Clements had authorized the first in a three-phase development effort. The initial four-year period comprised a four-satellite configuration in 10,500 NM twelve-hour orbits to validate the concept. On 2 May 1974 the Air Force renamed the planned system the Navstar Global Positioning System (GPS).[36]

In the coming years, GPS development would be beset by critics who worried about the vulnerability of the satellites, the susceptibility of the receivers to jamming, or the possibility of an enemy using the system to its own advantage. The global economic recession of the 1970s also made it difficult to obtain Defense Department funding that seemed more available for weapon systems than for defense support systems. Even so, by the early 1970s the Air Force found itself

playing the key role in the creation of a space applications program that, if successful, promised to revolutionize the tactical battlefield.

Surveillance from Space—From Vela Hotel and the Missile Defense Alarm System
(MIDAS) to the Defense Support Program (DSP). In many ways, surveillance from
space for missile detection and early warning represents the most important space-
application function in the military space program. By the early 1960s the outlook
for space reconnaissance proved immensely successful. On the international level,
all nations came to accept the principle of "open skies" with right of overflight
through space, while negotiations between the United States and Soviet Union
produced international agreements prohibiting weapons of mass destruction in
outer space. On the technical level, engineers and scientists demonstrated that
satellites in synchronous equatorial orbit would remain above the same point of
land, because the earth rotates beneath the satellite at the same rate as the satellite
travels in its orbit. With the advent of solid-state microelectronics, satellites could
collect vast amounts of data by means of increasingly powerful sensors, and rapidly
transmit to ground receiving stations information that could be made available to a
global network of users in near-real time. By the early 1970s such operations would
become increasingly routine.[37]

Space reconnaissance involved the so-called "black world" of highly classified
national space programs that, since 1961, were outside the purview of direct Air
Force management responsibility. When the Defense Department terminated the
Air Force Samos reconnaissance satellite program in 1962, it left the space reconnaissance field to the increasingly successful, CIA-Air Force CORONA project under
the National Reconnaissance Office (NRO). Although the Air Force furnished
personnel, boosters, and infrastructure support for the CORONA effort, the highly
classified project continued as a national reconnaissance effort, outside mainstream
Air Force space activities.[38] The Air Force, however, directed and managed two other
important space surveillance satellite programs. MIDAS, or Program 461, involved
developing an effective early warning satellite to detect the launch of ballistic rockets
using infrared radiometers. The other, Vela Hotel, comprised a space-based system
to detect nuclear/thermonuclear detonations in the atmosphere and outer space. It
provided the "space watch" necessary to ensure compliance with the limited nuclear
test ban treaty of 1963 and supported a variety of disarmament initiatives. In effect,
it became a crucial element of the "national technical means" for verifying compliance with nuclear weapons agreements.[39]

The perceptive Project Rand report of 1946, which considered the military applications for surveillance satellites, called attention to "the spotting of points of impact of bombs launched by us [the United States] as one major type of observations
provided by satellites."[40] The Vela program altered this prediction by directing
sensor attention to nuclear detonations in space in all locations. Serious efforts

to develop a satellite capability to monitor high-altitude nuclear tests date from international conferences at Geneva in 1958 and 1959, followed by congressional hearings in April 1960. These discussions prompted ARPA to develop the Vela program to detect all types of nuclear testing. The Atomic Energy Commission (AEC) and the Defense Department, through ARPA, managed the program jointly. Vela comprised three segments. Vela Uniform focused on underground or surface nuclear detonations using seismic techniques, while Vela Sierra involved ground-based detection of nuclear explosions above the earth's surface. Vela Hotel, the third element, served as the "watchman" for space-based detection of nuclear bursts in the atmosphere and outer space. Vela Hotel's challenge from the start was to be able to discriminate between nuclear detonations and natural background solar or cosmic radiation.[41]

Under ARPA's direction the Air Force became responsible for providing Vela Hotel the boosters and spacecraft. The Atomic Energy Commission laboratories furnished the instrumentation and Lawrence Radiation Laboratory the sensors. On 22 June 1961 ARPA authorized a test program of five Discoverer/Project CORONA Atlas/Agena launches of two Vela spacecraft each at three-month intervals. Planners scheduled the initial launch for April 1963. Vela proved to be an exceptionally well-managed program, and only twenty-eight months elapsed between program approval and data received from Vela Hotel sensors. Built by TRW, the spacecraft themselves measured 58 inches in diameter and weighed about 500 pounds. The intriguing Vela shape, an icosahedron, consisted of a solid with twenty equilateral-triangle faces and twelve vertices to allow X-ray detectors to view more than half a hemisphere. Other detectors appeared at the vertices and inside the spin-stabilized satellite. Orbiting at 60,000 miles altitude and spaced 140 degrees apart, the spacecraft's powerful four-watt transmitter sent 256 data bits per second to sixty-foot ground antennas by means of dipole antennas. In addition to its main function, additional instruments determined background radiation levels and fluorescence produced by nuclear blasts.[42]

Not surprisingly, the Vela "treaty monitors" made their initial appearance shortly after the United States ratified the Limited Nuclear Test Ban Treaty. On 16 October 1963, an Atlas-Agena B lifted the first two Vela satellites into a 67,000-mile circular orbit at thirty-eight-degrees inclination. A second pair followed on 17 July 1964 and a third the following year. The launch success and on-orbit reliability of the first six satellites convinced planners to cancel the final two launches and modify the fourth and fifth pairs for atmospheric surveillance. Expected only to remain operational for six months, the initial Vela satellites operated for a period of five years. The fourth and fifth earth-oriented pairs became operational in 1969 and 1970 and functioned superbly, well beyond their predicted eighteen-month lifespans. To the relief of all involved in the program, the sun's X-ray bursts did not produce an excessive number of unrecognizable false alarms in the Vela sensors.[43]

So successful was the program that the Air Force in 1965 turned Vela over to TRW, which became responsible for all future work. The contractor developed larger and more sophisticated satellites, with the last pair, the eleventh and twelfth in the series, launched in 1970. The Vela program demonstrated that a complex system could be developed and successfully deployed in a period of only five years, then turned over to contractor management for an additional five years of "routine" operations. In the 1970s Vela satellites gave way to nuclear detectors placed on other Air Force satellites. As part of a defense policy "of launching fewer but larger spacecraft and using them for multiple functions," officials redesignated the nuclear detection system the Integrated Operational Nuclear Detonation Detection System (IONDS), and in the 1970s sent detectors into space aboard Defense Support Program early warning and GPS navigation satellites.[44]

The Defense Support Program (DSP) succeeded the MIDAS missile detection and early warning space watch program in the late 1960s. Unlike Vela Hotel, MIDAS experienced problems from its inception. MIDAS relied on advanced electronic and cryogenic technology to move beyond the visual spectrum to the spectrum of much longer infrared wavelengths. By recording heat emissions from objects on Earth, infrared radiometers in aircraft could produce thermal pictures during darkness and identify camouflaged targets. MIDAS envisioned using polar-orbiting Agena satellites with infrared scanners mounted on a rotating turret that scanned the earth continuously to detect ICBM exhaust flames within moments of their launch. Initially, planners expected to launch MIDAS satellites into polar orbits at 300 miles altitude, but the high-intensity background radiation from sunlit clouds and other phenomena convinced officials to raise the altitude to 2,000 miles. Even so, the challenges remained formidable.[45]

The MIDAS story illustrates a number of complexities faced by Air Force space planners determined to develop a much-needed but technologically challenging system during the McNamara era. Where, for example, was the balance between keeping a program in the study phase, or the development phase, before deploying it as an operational system? Where was the point beyond which the value of the data produced by the system failed to justify the high cost of its development, deployment, and operation? How did the Air Force achieve the proper booster-payload combination during the advance of technology and changing satellite mission requirements? The approaches to these questions normally found the Air Force operational commands favoring early operational capability for MIDAS and, as a result of early failure, the Defense Department preferring a more deliberate, research-oriented focus. As a result, MIDAS experienced a rocky development road, often appearing to end in premature cancellation of the project. Nevertheless, MIDAS established the groundwork for its incredibly successful successor, DSP, which would become the central component in the nation's global missile warning network.

When the Kennedy administration took office in early 1961 MIDAS already faced major survival hurdles. During the reorganization of the satellite reconnaissance program in August 1960, MIDAS' technical difficulties convinced Defense Department officials to reemphasize technical development. Air Force leaders, concerned about the growing Soviet ICBM threat, lobbied hard for an early operational date for the infrared detection system. That fall General Laurence S. Kuter, the commander-in-chief of the North American Air Defense Command (NORAD), and the commander of the Air Force's Air Defense Command (ADC), Lieutenant General J. H. Atkinson, urged Chief of Staff General Thomas White to authorize an expedited and expanded MIDAS development program. If the Joint Chiefs of Staff were to approve the preliminary MIDAS operational proposal of February 1960, which had raised the ire of Army and Navy representatives, NORAD would receive operational control and ADC designation as the "using Air Force command." General White reminded the commanders that not only the operational plan awaited action by the Joint Chiefs of Staff, but the basic MIDAS development plan had yet to receive Defense Department approval. He convinced DDR&E chief Herbert York to authorize two radiometric tests aboard upcoming Discoverer/Project CORONA flights. The planners hoped that these experiments could answer the basic question surrounding the future of MIDAS: could the infrared detectors distinguish between missile radiation in the boost phase and high-intensity natural background radiation? Meanwhile, in September 1960, Dr. W. K. H. Panofsky of Stanford University headed a panel of the President's Scientific Advisory Committee, which concluded that the MIDAS concept remained sound and that every effort should be pursued to overcome engineering problems and produce an operational system by 1963.[46]

Early in 1961, following a considerable number of program revisions, planners at Air Force Systems Command's Ballistic Missile Division continued work on a "final" development plan that excluded any reference to operational funding or capabilities in favor of concentration on research and development. This dichotomy pitted Air Force commands, including Air Defense Command, that favored accelerated satellite development and early deployment against the Air Force research and development community and an Office of the Secretary of Defense whose worries about technical feasibility and high costs led them to favor a more cautious approach. It would characterize the course of MIDAS development throughout the 1960s. The "final" MIDAS development plan appeared on 31 March 1961. It scheduled twenty-seven development launches rather than the twenty-four proposed earlier, with initial operational capability set for January 1964. Meanwhile, the Joint Chiefs of Staff and Secretary of Defense on 16 January 1961 approved the operational plan that assigned MIDAS responsibilities to NORAD and the Air Defense Command. In mid-March Air Defense Comand authored a proposed operational plan calling for a constellation of eight satellites spaced in two orbital rings to ensure continual coverage of the Soviet landmass. Data from the sensors would be transmitted to Ballistic Missile Early Warning System (BMEWS) radar sites, then re-layed to the NORAD command post. Planners hoped to achieve a twenty-four month satellite lifespan, but by mid-June 1961 Under Secretary of the Air Force Joseph Charyk balked at authorizing an operational configuration without additional infrared sensor data from forthcoming flights. Operational and logistic planning priorities gave way to emphasis on demonstrating acceptable early warning techniques.[47]

During the summer of 1961, Harold Brown, the Kennedy administration's new DDR&E chief, conducted an extensive review of the MIDAS program for Secretary of Defense McNamara. He predicted that, ultimately, engineers could solve severe problems associated with system reliability and the detection of both low- and high-radiance missile emissions, but he raised doubts about the system's ability to detect small, Soviet Minuteman- and Polaris-type solid-fuel missiles. He estimated the warning time for a potential high-radiance liquid-propellant ICBM attack at five to twenty minutes. Was the additional warning time worth the effort required to solve the technical problems and the estimated $1 billion price tag for operational capability, not to mention the $200 million needed for annual operations? At this time Secretary McNamara was reassessing the broader concept of how the country should respond to an enemy attack. If the nation's leaders chose not to retaliate on warning of a missile assault, but to rely primarily on the ICBM second-strike capability, the additional strike aircraft made available by a MIDAS alert would prove of little value. Not surprisingly, the Air Force vigorously countered by showing that ten minutes of additional warning time would guarantee that fourteen percent of the Strategic Air Command bomber force could become airborne, while fourteen minutes would raise this figure to sixty-six percent.[48]

The technical and political uncertainties, along with Air Force criticism, compelled Brown that summer to appoint a study group headed by John Ruina, to examine the issues of MIDAS technical capabilities and mission importance. Although the Air Force considered the Ruina study just one more in a long line of investigations that had delayed MIDAS development, General Schriever went to the heart of the matter when in the fall of 1961 he wrote Air Force Chief of Staff General Curtis E. LeMay that "complete satisfaction can only be achieved by a conclusive demonstration of system feasibility through an orbital flight test that detects and reports the launch of ballistic missiles and has a reasonable orbital life." Such capability appeared far in the future. MIDAS experimental flights occurred as part of Project Discoverer/CORONA. The first two flights, on 26 February and 24 May 1960, produced little significant data. The first launch failed after an explosion occurred upon separation of the second-stage Agena from the Atlas booster, while *MIDAS 2's* sensors operated successfully for two days from its 300-mile-high orbit before its communications link failed. The third MIDAS spacecraft, launched on 12 July 1961, returned data from its experimental infrared telescope for only five orbits before failure of the solar array auxiliary power. Although *MIDAS 4* successfully achieved a near circular polar orbit at a 2,200 nautical mile altitude, on 21 October 1961, it operated for only seven days without meeting any of the flight's objectives.[49]

Even before the Ruina group issued its report, the Office of the Secretary of Defense deleted all fiscal year 1963 MIDAS nondevelopmental funds and refused to sanction an operational system. The Ruina report deepened a mood of doom and gloom. Issued on 30 November 1961, it faulted the current MIDAS design as too complex for reliable use, expressed skepticism regarding the system's ability to detect solid-propellant missiles, and criticized the Air Force for focusing on immediate operational capability to the detriment of essential research and development. The report recommended a major reassessment to produce a simplified MIDAS with more attention directed to research and development. In December Brown directed the Air Force to implement the group's findings. Referring to the report's "serious allegations," General LeMay reacted sharply by requesting several alternate development proposals and by working to defer the DDR&E directive. Air Force Systems Command's Space Systems Division moved quickly to form an advisory group under Clark Millikan of Cal Tech to assess the Ruina report. The Millikan group faulted the Ruina panel for being unaware of the scope of available test data, and for erroneously analyzing the cloud-background-clutter data in assessing the infrared sensor's capability. A simplified system, the group asserted, could be operational before 1966.[50]

Of the various plans Air Force Systems Command prepared, the most convincing one stressed research and development and more test flights. During February and March 1962 Air Staff members repeatedly met with DDR&E officials to convince them to accept the Air Force proposal, which Space Systems Division completed on 29 March 1962. It called for as many flights as possible leading to an initial operational capability between mid-1965 and mid-1966. It also projected a fiscal year 1962-1963 budget increase from the programmed $290 million to $334 million. Then, on 9 April 1962, the Air Force finally found itself in a position to break the logjam on MIDAS development. On that date a fifth MIDAS flight achieved polar orbit and began transmitting data which demonstrated that it could discriminate between cloud background and rocket exhaust plumes. The very next day Air Force Assistant Secretary Brockway McMillan requested that DDR&E approve the 29 March plan. In response, Brown released funds to sustain the program through the fiscal year, but he declined to authorize development.

In fact, the DDR&E chief sponsored another review of MIDAS. This time Stanford's Panofsky reappeared to chair another panel. Unlike his favorable 1960 conclusion, this time he agreed with the Ruina panel's findings and criticized the Air Force for proposing operational prototype flights when basic missile detection capability remained in doubt. Harold Brown notified the Air Force that he would not release further funds until MIDAS proved capable of detecting low-radiance missiles. While unhappy Air Force officials prepared yet another plan—one involving an accelerated research schedule—to accommodate DDR&E concerns Secretary McNamara told Air Force Secretary Zuckert that the Defense Department would conduct a "full-scale" analysis of MIDAS in light of the importance of early warning and the seriousness of the Soviet ICBM threat. At the same time, Brown criticized the Air Force for focusing on an early operational capability without first solving basic questions about low-radiance, noise background, and system reliability. By the summer of 1962, MIDAS supporters had little reason for optimism, and in early August, Secretary McNamara announced reduction of MIDAS to a limited research and development program because of its expected slow development, high costs, available early warning alternatives, and the decreased value of early warning occasioned by the growing importance of hardened missile sites compared to the strategic bomber force. The Defense Department subsequently cut funding for fiscal year 1963 to $75 million and for fiscal year 1964 to $35 million.[51]

By the spring of 1963 it appeared that MIDAS might be doomed to extinction as another system too ambitious technologically to warrant operational development. Then, in May 1963, the fortunes of MIDAS seemed to make an abrupt recovery along the lines forecast by General Schriever two years earlier. On 9 May an Atlas-Agena launched Flight Test Vehicle 1206 from Vandenberg Air Force Base into a near-perfect 2,000-mile-altitude circular orbit. Over the next six weeks, the satellite vindicated its supporters by detecting nine launches of solid-propellant Minuteman and Polaris as well as liquid-propellant Atlas and Titan missiles. A subsequent flight on 18 July confirmed "real time" detection of an Atlas E launch as well as the ability to monitor Soviet missile activity. Above all, the flights convinced officials that MIDAS could provide real-time data on missile launches without interference from earth background "noise." The successful flights prompted Secretary McNamara to reevaluate the possibilities for tactical warning and the future of MIDAS.[52]

Although at this time Air Force Systems Command responded with four alternative proposals designed to achieve an operational system, the Air Staff adopted a more flexible response that called for a prototype approach on the assumption that neither current technology nor funding constraints warranted an entirely operational system. The Air Staff Board recommended that Air Force Systems Command improve system tracking and launch site identification techniques as well as the real-time detection of low-radiance, short-burning solid-fuel missiles, and that it consider additional defense applications. Most interesting, the Air Staff, in the name of cost-effectiveness, favored the development of more simplified, more reliable satellites with longer orbital lifespans; such satellites also would orbit at higher altitudes to provide greater coverage of the earth with fewer spacecraft. On 1 October 1963 the Chief of Staff approved a three-phase flight test program extending throughout the remainder of the decade with initial fiscal year 1965 funding set at $100 million.[53]

In early November 1963, Brown suggested that Program 461 be reoriented to include detection capability of submarine-launched ballistic missiles (SLBMs) and medium-range ballistic missiles (MRBMs), while later in the month the Office of the Secretary of Defense cut the Air Force proposed fiscal year 1965 figure of $100 million to $10 million. In early 1964 Brown agreed to release only half of the fiscal year 1964 MIDAS budget allocation, explaining that the "drastic reduction" resulted from alternative early warning systems and anticipated high deployment costs for MIDAS. Nevertheless, he agreed the recent flight successes warranted continuing the program, but with four objectives beyond its initial strategic warning function. His list included reliability, global coverage, launch point determination, and real-time detection of nuclear detonations as well as SLBM and MRBM launches. If the Air Force reoriented the program according to his guidelines, MIDAS could expect increased funding support in future. The latest modification of the MIDAS effort, the DDR&E chief admitted, envisioned a major deviation from a system originally designed to detect a mass raid of Soviet missiles.[54]

Given the budget cutback, the Air Force remained concerned about the program's future. Already scheduled flights would have to be canceled resulting in termination of contracts, substantial investment losses, and a four-year hiatus between the series of radiometric and system detection flights. Throughout the spring of 1964, Air Force officials negotiated with DDR&E to reach an acceptable compromise. By late spring the Air Staff proposed a minimal program designed to preserve both near- and long-term objectives by the increasingly prevalent method of slipping the flight schedule and accepting greater technical risk.[55]

Budget cuts and skepticism within Defense Department circles continued to plague the infrared-detection satellite early warning program. In late 1964 and throughout 1965 the Defense Department's proposed fiscal year 1966 through 1969 budget reductions prompted major efforts by the Air Staff and Air Force Systems Command's Space Systems Division to keep MIDAS afloat without having it revert to development status. Their dilemma did not benefit from delays caused by Lockheed's difficulties with sensor components, a labor walkout at payload producer Aerojet General Corporationand launch site availability problems at Vandenberg Air Force Base. As revised, the MIDAS program in the latter half of the decade called for two phases of tests. Between 1966 and 1968, flights would conduct a variety of experiments in three stages at altitudes from 2,000 to 6,000 miles; in 1969 and 1970 more tests and a final operational assessment would occur with satellites launched by Titan IIICs to a 6,500-mile orbit.[56]

MIDAS remained a test program. Although Program 461 had shown conclusively that satellites could provide early warning of a missile attack by detecting and tracking missiles of all sizes, in the late 1960s mounting costs, low budgets, technical problems—and ambitious expectations—outpaced the original MIDAS program.

Moreover, with the advent of the Titan III booster, it became increasingly possible to contemplate launching larger, more capable infrared-detection satellites into geosynchronous orbits, where fewer satellites could cover more ocean and earth areas. As a result, DDR&E in August 1966 approved Program 949. Originally designed to monitor the Soviet Fractional Orbital Bombardment (FOB) threat, it soon came to be regarded as the replacement for ground-based warning systems such as BMEWS. As the MIDAS successor, it could ensure simultaneous warning of all three potential space and missile threats—ICBMs, FOBs, and SLBMs. In the spring of 1969 a breach in security eventually led officials to rename Program 949 the Defense Support Program (DSP).[57]

In November 1970, a Titan IIIC launched the first TRW-built DSP satellite into an elliptical, rather than the intended synchronous-altitude, orbit over the Indian Ocean. Referred to as "the best mistake we ever made," the spacecraft successfully transmitted data on American and Soviet launches to tracking stations as it circled the globe every five days. In April 1971 the newly completed overseas ground station at Woomera, Australia, took control of the satellite when in view, while at other times the Satellite Control Facility at Sunnyvale, California, assumed control. An additional ground tracking station at the Buckley Air National Guard Base near Denver, Colorado, joined the system in late 1972. By this time a second DSP satellite had been successfully lofted into synchronous orbit. In early 1973 a third early warning infrared satellite joined the constellation in synchronous equatorial orbit over the western hemisphere, where it helped monitor the SLBM threat from the Atlantic Ocean.[58]

DSP satellites represented a major technological leap over their MIDAS predecessors. Weighing 5,200 pounds, the DSP spacecraft measured thirty-three feet in length and nine feet across. Four solar panels covered one end of the cylinder-shaped satellite which housed the electronics. The other end housed a twelve-foot telescope with an array of 2,000 infrared detectors. In contrast to MIDAS, the DSP satellite itself rotated, at six revolutions per minute. Planners expected it to far exceed MIDAS in both coverage and reliability.[59]

In the months and years ahead, Air Force planners would worry about the challenge of developing DSP computer software which could receive and process an incredible amount of data then transmit the results almost instantaneously to users worldwide. They also remained apprehensive about loss of coverage due to adverse sensor angles over the pole and uncovered nadir holes, and they lobbied for additional satellites to provide redundant capability. Citing budget constraints in the 1970s, Defense Department officials proved unresponsive. They effectively noted the unexpected, outstanding performance achieved by the three-satellite network that immediately came to serve as the bedrock of early warning protection against the Soviet and "nth" country missile threats. With DSP performing space watch, a surprise attack became next to impossible.[60]

Ground-Based Space Surveillance Comes of Age

The Space Detection and Tracking System (SPADATS). SPADATS furnished additional protection against surprise attack by providing a space watch for detecting, tracking, and monitoring satellites and debris from low earth orbit to deep space through its network of ground-based sensors and control and data processing facilities. Operational responsibility for SPADATS had been a contentious issue between the Air Force and the Navy in the closing years of the Eisenhower era. On 7 November 1960 the Joint Chiefs of Staff acceded to Air Force wishes by assigning NORAD operational control and the Continental Air Defense Command (CONAD) operational command of SPADATS. In January 1961 the Secretary of Defense confirmed the decision, and the Air Force followed by making the Air Defense Command responsible for technical control of Spacetrack, the Air Force segment of the surveillance system. In mid-February 1961, the 1st Aerospace Surveillance and Control Squadron was activated to operate the new SPADATS data collection and catalog center as part of NORAD's Combined Operations Center at Ent Air Force Base, Colorado. The latter assumed the responsibilities previously handled by the Interim National Space Surveillance and Control Center (INSSCC) data filtering and cataloguing center at the Air Force's Cambridge Research Center in Massachusetts.[61]

The Air Force Spacetrack sensor network at that time included the Millstone Hill radar in Massachusetts, Baker-Nunn satellite tracking cameras, available observatory telescopes, and a variety of research and development and missile early warning radars. None of the equipment belonged to the Air Defense Command, and the performance of the detection, tracking, and cataloguing network suffered severely from accuracy, timeliness, and range weaknesses. On 1 February 1961 NORAD had assumed operational command of the Navy's Space Surveillance (Spasur) east-west minitrack radar fence and its data processing facility in Dahlgren, Virginia. During the acceptance ceremonies, NORAD Commander-in-Chief General Laurence J. Kuter observed that with thirty-two American and three Soviet spacecraft in orbit, the need for a constant, accurate space watch had arrived.[62]

Over the next few years NORAD, CONAD, and ADC worked diligently to expand system capabilities through computerized "volumetric" track-and-scan sensors to provide immediate detection and identification of multiple space objects. At the same time, they also labored to expand their area of operational and ownership responsibility. By 1965 the ADC's Spacetrack Operations Center at Ent Air Force Base received data from assigned detection fan and tracking radars at Shemya, Alaska, and Diyarbakir, Turkey. The new Shemya FPS-80 tracking radar, for example, proved capable of detecting within minutes Soviet satellites launched from Kapustin Yar and providing highly accurate satellite positioning data to a distance of 2,500 miles. The previous year, ADC had accepted AFSC's FPS-49 long-range pulse Doppler tracking radar at Moorestown, New Jersey, and a scanning radar and 84-foot dish tracking radar at Trinidad, British West Indies. Although hampered by azimuth and elevation restrictions, the Trinidad tracker could follow one-meter-square targets at 2,000 miles, and the site proved especially important in detecting and tracking Soviet satellites in low-inclination orbits.[63]

Spacetrack also relied on a network of Baker-Nunn cameras at Oslo, Norway; Edwards Air Force Base, California; and Sand Island in the Pacific for deep-space surveillance. Far superior to electronic sensors of the time, the optical cameras could record one-meter-square targets to a range of 50,000 nautical miles, provided the camera operated in twilight or darkness, free of clouds, against illuminated targets. Data readout delays put the Baker-Nunn cameras in the category of contributory systems. In addition, the Ballistic Missile Early Warning System (BMEWS) radars at Thule, Greenland; Clear, Alaska; and Fylingdales, England, represented a supplementary Spacetrack component. The Air Force also had under development an FSR-2 prototype electro-optical satellite sensor at Cloudcroft, New Mexico, and a new FPS-85 phased array radar at Eglin Air Force Base, Florida. Both sensors, however, could not meet their programmed 1965 initial operational dates. The Cloudcroft radar experienced technical problems that prevented it from joining Spacetrack immediately, while a fire in January destroyed seventy-five percent of the Eglin radar and delayed its completion until 1967.[64]

Spasur and Spacetrack underwent continual improvements in their capabilities to detect, track, and monitor a space population that by 1969 found NORAD's Space Defense Center in Cheyenne Mountain observing 20,000 objects daily. By the early 1970s the Air Force and NORAD exercised responsibility for nearly the entire SPADATS sensor and control system, and they planned to monitor objects in deep space by replacing the Baker-Nunn network with the Ground-based Electro-Optical Deep Space Surveillance (GEODSS) System. The latter would include three optical telescopes supported by sophisticated computer software and processing and display equipment. Like other space systems, by the 1970s the growth in sophisticated sensors and supporting equipment served to make Spacetrack and SPADATS increasingly responsive to operational requirements.[65]

A Fleet of Space Vehicles Sets the Course

While orbiting satellites increasingly provided important space-based information for larger numbers of operational commanders and other users, their performance remained dependent upon boosters and upper-stage vehicles capable of placing them in the desired orbits. Available space boosters had enabled the Air Force to achieve early space supremacy among the services, and the responsibility it received as the "booster service" guaranteed its central space role throughout the 1960s.

Before 1960, ARPA, NASA, and the Army and Navy carried out all but two of the American space launches. In 1960, the Air Force began its dominance of the space launch business with fourteen of the twenty-nine service-sponsored flights that year, and the trend would continue. When the Air Force initiated its space program it had a ready-made advantage in the liquid-propellant ballistic missile force designed and built in the 1950s. The Thor IRBM and Atlas ICBM could not compete favorably with their heavier Soviet counterparts and soon were superseded by the more capable solid-fuel Minuteman ICBM and Polaris SLBM. Nevertheless, the Thor and Atlas would continue to provide effective, reliable space boosters for a wide variety of unmanned space flights well into the era of the Space Shuttle.[66]

The Douglas Aircraft Company's Thor, measuring 65 feet in length and 8 feet in diameter, relied on liquid oxygen and kerosene to produce 150,000 pounds of thrust from its single main engine. It began its impressive flight history with the initial December 1959 Discoverer/Project CORONA launch and continued to operate primarily from the Western Test Range, at Vandenberg Air Force Base, where it launched in a southerly direction to achieve polar orbits. The more capable 71-foot-long, 10-foot-wide one-and-a-half-stage Atlas ICBM, built by General Dynamics-Astronautics, could produce at lift-off 387,000 pounds of thrust from its three main and two vernier engines. Atlas began its booster career by launching heavier payloads from the Eastern Test Range at Cape Canaveral. Both Thor and Atlas would remain workhorses for the Air Force for the remainder of the century. In the 1960s, the Air Force augmented its booster fleet with Martin-Marietta's Titan IIIC, consisting of a two-stage liquid-propellant rocket core with two enormous solid-propellant strap-on motors and a "Transtage." Measuring 108 feet long by 10 feet in diameter, this launch combination generated nearly 3,000,000 pounds of thrust and could place up to 33,400 pounds into low-earth orbit and nearly 4,000 pounds into a synchronous equatorial orbit. With the Titan, the Air Force possessed a booster of vastly increased size, capable of launching a wide range of satellites into higher-altitude orbits.[67]

For both Thor and Atlas and their heavy-lift successor, the Titan, upper-stage vehicles immediately became fundamental for mission success. Thor initially used Space Technology Laboratories' Able, a modified Vanguard vehicle, consisting of an solid upper stage and two liquid lower stages, and the improved two-stage Aerojet Able Star, whose liquid-propellant engine was restartable in space. The Thor booster became a favorite NASA launch vehicle for its own and foreign satellites when, in 1959, the civilian agency developed two more sophisticated solid-propellant Vanguard upper stages, and renamed the three-stage spacecraft Delta. The Air Force preferred to use the Agena, which Lockheed had begun developing in 1956. More than any other booster-satellite craft, the Agena "put the Air Force in space." Serving as a satellite once placed in orbit, the Agena went through several models, with the Agena B in use by the Air Force and NASA until 1966. Seeking a basic Agena upper-stage vehicle, Lockheed responded to an Air Force request by developing the standard, thirty-seven-foot-long and five-foot-wide Agena D. First launched atop of a Thor from the Western Test Range on 27 June 1962, the Agena D would continue to serve the Air Force into the early 1980s before the rocket-powered space glider, the Space Shuttle, became operational. The Agena D's common configuration included four usable modules containing the major guidance, beacon, power, and telemetry equipment, a standard payload console, and a rear rack above the engine for plug-in installation of optional gear-like solar panels, "piggyback" subsatellites, and an optional Bell Aerosystems engine that could be restarted in space as many as sixteen times.[68]

The Air Force's efforts to achieve standardization also embraced the stable of launch vehicles. It sought to emphasize similarities for the various missions while keeping deviations to a minimum. With a more powerful booster on the drawing board by 1961, Thor and Atlas became known as medium launch vehicles, with Thor designated SLV-2 (standard launch vehicle) and Atlas SLV-3. Although the small Scout booster received the designation SLV-1, it normally served NASA mission interests.[69]

The original single-stage Thor booster could support a variety of upper stages, such as Aerojet-General's Able and AbleStar, Lockheed's Agena A and B, and McDonnell Douglas' Delta. The standardized version included a Thor, modified with additional tankage and an upper-stage Agena D. It proved capable of launching 1700 pounds into a 115-mile circular orbit from Vandenberg Air Force Base, or just over 3,000 pounds from the Eastern Test Range. Later in the decade, a standard Thor with the addition of three strap-on Castor solid rockets, became the SLV-2A Thrust Augmented Thor with a 30-percent increase in payload capability. In 1966 a further modification took place by lengthening the Thor's propellant tanks. Known as the SLV-2H, this version demonstrated sixty-five seconds additional burning time and a 35-percent payload capability increase over the SLV-2A. Other versions included the Delta and Boeing Burner II upper stages. The Thor achieved a remarkable performance record that included only three failures among 154 launches from 1962 to 1972. In the 1970s the Air Force designated the Thor to launch the Defense Meteorological Satellite Program (DMSP) satellites.

The Atlas booster attained an equally enviable record of accomplishment. Initially, the Air Force used Atlas D missiles with slight modifications to accommodate the Agena upper-stage vehicle. In 1963 additional changes produced the Atlas SLV-3, while two years later the SLV-3A appeared with propellant tanks lengthened by twelve feet. Normally launched from the Eastern Test Range, it could place an Agena D payload of 8,600 pounds into a 115-mile low-earth orbit. The Atlas SLV-3D version used a Centaur upper stage for NASA launches, and in the early 1970s officials selected it to launch the Navy's Fleet Satellite Communications (FLTSATCOM) satellites. When the Minuteman intercontinental ballistic missile replaced the Atlas ICBM in the mid-1960s, the Air Force determined that savings could result by refurbishing Atlas missiles from silos in the midwest rather than purchasing new SLV-3s. Redesignated the Atlas E and F, these "wheatfield" boosters proved highly reliable from 1967 to 1979 in support of Air Force research experiments as well as

TIROS and Global Positioning System launches. Atlas also served as the basic booster for NASA's Mercury program.[70]

With the first successful launch of a Titan IIIC on 18 June 1965, the Air Force had a booster sufficiently powerful to launch satellites into geosynchronous orbit. The following year a Titan IIIC successfully launched the first series of IDCSP military communications satellites into a near synchronous orbit, 21,000 miles above the equator. In the 1960s and early 1970s, the Air Force developed two other versions of the Titan III. Originally designed for the manned orbiting laboratory, the Titan IIIB used a lengthened core to enable it to place an 8,200-pound payload into a 115-mile polar orbit when launched from the Western Test Range. By 1971 the Air Force had developed the Titan IIID, which added two five-segment solid rockets to the core but used a third-stage Agena in place of the Transtage. Intended as a transition to the Shuttle, it operated exclusively from the Western Test Range to place payloads weighing as much as 24,600 pounds into a 115-mile polar orbit. On 2 November 1971, it successfully launched the first pair of 1,200-pound DSCS II satellites into synchronous orbit.[71]

The success of the Titan III in the 1960s vindicated its proponents who sought to create a "DC-3 of the space age." During the period from 1964 through 1979, 111 of the 119 launches proved successful. Of these, six failures occurred with the Titan IIIC, primarily with the Transtage portion. While Air Force leaders like General Schriever bemoaned the micromanagement approach taken by the McNamara Defense Department, officials in the Office of the Seccretary of Defense could proudly point to the Titan's record of both launch and budget success. The total development cost of $1.06 billion proved well within estimates, considering inflation, two significant program changes, and a scheduled "stretch-out."[72]

By the 1970s, space launch vehicles had matured to the point where Air Force planners could consistently count on available standard Air Force boosters for launching substantial payloads, placing them into complex orbits, and demonstrating reliable performance. Nevertheless, with the advent of the reusable Space Shuttle, the future appeared uncertain for the Air Force's fleet of expendable space boosters.

Space Infrastructure Provides the Support

The integration of Air Force space systems also depended on the supporting infrastructure of booster launch centers, a tracking network and control center, and processing centers to evaluate and transmit data to users. In the late 1950s, the Air Force's Weapon System 117L and the many-faceted Project Discoverer/CORONA precipitated a major expansion of space infrastructure that continued unabated with NASA's rise in the 1960s.

Two major Air Force launch centers supported the nation's satellite program from its inception. One, the Eastern Test Range at Cape Canaveral, Florida, and renamed Cape Kennedy following the President's death, began in the late 1940s as a joint long-range proving ground run by the Air Force. Comprising the northern-most wedge of a barrier island fifty miles east of Orlando on the Florida coast, the Cape remained separated from the mainland to the west by the Banana River, Merritt Island, and the Indian River, which comprised a portion of the Intercoastal Waterway. The area saw little activity until the Second World War, when the Navy established the Banana River Naval Air Station fifteen miles south of the Cape. After the war, activity declined until the Air Force selected Cape Canaveral as the western end of its new Long Range Proving Ground and supported it by constructing Patrick Air Force Base on the site of the naval air station.[73]

The location proved ideal for testing cruise missiles and, later, launching ballistic missiles and space flights. Launches in a southeasterly direction avoided major population centers by passing over islands that served as tracking stations along a 7,500-mile course from the Bahamas to Ascension Island in the South Atlantic to the coast of Africa. As a result, burned-out missile stages and expendable boosters avoided densely populated land areas. Moreover, with an easterly launch the earth's rotation added greater velocity, which enabled boosters to orbit heavier loads. In the 1960s the Cape underwent an enormous buildup resulting especially from NASA's rapid expansion and the Air Force's development of the Titan III. The Eastern Test Range became the center for Vela and communications satellite launches as well as NASA's Mercury, Gemini, and Apollo manned flights and all American spacecraft launched eastward into low-inclination equatorial orbits.[74]

The Western Test Site at Camp Cooke, later Vandenberg Air Force Base, California, also began operating as a missile test base. In 1956, the Air Force selected the Army's old Camp Cooke, which extended over twenty-five miles along the coastline some sixty miles west and north of Santa Barbara. Used for testing ICBMs and IRBMs, it became part of the Pacific Missile Range, which also encompassed the Navy's Point Mugu between Santa Barbara and Los Angeles. In 1958, when the Air Force renamed Camp Cooke Vandenberg Air Force Base, Lockheed already had started work on the Agena upper-stage spacecraft. Officials selected the California site for launching satellites into near-polar orbit. Missiles and reentry vehicles launched westerly over the South Pacific and space boosters launched in a southerly direction avoided population centers on their path into high-inclination polar orbits, which proved essential for effective satellite coverage of the Asian landmass. As a result, the Western Test Site became the location for the nation's high-inclina-tion sun-synchronous surveillance missions and, from 1971, the designated location for proposed near-polar-orbit Space Shuttle operations. Like its eastern counter-part, the western range depended on a long line of space tracking stations stretched across the Pacific from California to the South and Southeast Asian coasts. The Western Test Range served as the launch site for the important Samos, CORONA, MIDAS, and DMSP satellites that required near-polar orbits.[75]

A second group of space facilities comprised the tracking network and its control center that made possible the crucial integration of satellites, launch sites, and processing centers. Designated the Satellite Control Facility (SCF), it included a global system of remote tracking, telemetry, and command stations, a central control center sited in California, and the communications links that bound together all the equipment and software needed to track and control spacecraft during launch, orbit, descent, and recovery.[76]

Tracking stations functioned effectively only when satellites remained in range of the ground antennas. Because this time was brief for satellites in low orbits, ground stations were scattered widely but tied to the control center. By the end of the 1960s, the Air Force relied for its space operations on six key radio tracking and command stations, located in 1959 at Vandenberg Air Force Base and New Boston, New Hampshire; Thule, Greenland, in 1961; Mahe Island in the Indian Ocean in 1963; Guam in 1965; and the oldest, Kaena Point on the island of Oahu, Hawaii, in 1958. The last proved especially valuable in recovery of CORONA reconnaissance film capsules. During the 1960s, the Air Force worked to standardize and upgrade tracking site operations and equipment. This included use of a standard Defense Department telemetry, command and control system designated the Space-Ground Link Subsystem, and adoption of two uniform dish antennas measuring between forty-six and sixty feet in diameter that could pivot rapidly to monitor low-earth satellites.[77]

The control center, the second element in the SCF network, received the designation Satellite Test Center. The Air Force's early and close relationship with Lockheed led to locating the command center in Sunnyvale, near Palo Alto, California. The first center in 1959 amounted to little more than several rooms with plotting boards next door to Lockheed's computer complex in Palo Alto, where members of the 6594th Test Wing (Satellite) successfully controlled the first Discoverer flight in 1959. By June 1960 the Air Force had constructed a permanent facility eleven miles away in Sunnyvale. After another year of equipment improvements, the Satellite Test Center could support three satellite missions at once with its two Control Data Corporation (CDC) 1604 computers. Improvements continued throughout the 1960s. In 1965 the Air Force replaced the single control room with separate rooms for each flight. Over the next three years, the center upgraded its computer capability with five CDC 3600 and seven CDC 3800 computers to handle the increasingly complex software programs and growing satellite population. Early advances culminated in 1968 with completion of the so-called "Blue Cube," a new ten-story windowless "Advanced Satellite Test Center" scheduled to handle Manned Orbiting Laboratory flights. With cancellation of the Air Force's manned mission in 1969, however, the Blue Cube provided controllers vastly increased capabilities to support, twenty-four hours per day, seven days per week, real-time operations for instrumented satellite missions. Statistics help explain the phenomenal develop-

167

ment in command and control capability. In 1960 the Satellite Test Center recorded 300 satellite contacts and 400 hours of flight operations. Fifteen years later, Satellite Test Center ground stations logged 52,445 hours and made contact with more than 30 satellites a total of 60,536 times.[78]

The communications network represented the third element of the Satellite Control Facility. For the initial Discoverer flight in 1959, it consisted only of landlines, radio links, and submarine cables that connected the Satellite Test Center with tracking stations confined to the continental United States, Alaska, and Hawaii. In 1962 the Air Force extended to its overseas stations secure circuits capable of 100 words per minute. During the next two years, a "multiple satellite augmentation program" provided the Sunnyvale Satellite Test Center with high-frequency radio capability through four independent voice channels and the addition of twenty-eight teletype machines, with transmission links to the remote tracking stations.[79]

The Satellite Control Facility's communications network underwent dramatic improvement with the launch of the first seven military communications satellites in June 1966. Each of the satellites in the Initial Defense Communications Satellite Program could transmit 600 voice or 6,000 teletype channels. With the addition of eight more satellites in January 1967, the Air Force could implement its "advanced data system" communications net designed to support the more challenging near-real-time command and control operations from the Blue Cube. A new sixty-foot dish antenna located at Camp Parks Communications Annex near Oakland, California, served as the network terminus. By 1970 the Camp Parks facility passed all data it received directly to the Sunnyvale control center over land lines and microwave relay links. In the future, the larger, more powerful satellites of the DSCS II program promised a wideband satellite relay communications system capable of transmitting 1.5 million bits of data per second. In less than a decade, the Satellite Control Facility had proven capable of expanding to meet the challenging demands of a burgeoning space community.[80]

Organization Provides the Focus for Space

The rapid growth of Air Force space infrastructure during the 1960s compelled planners to provide new, more effective organizational structures for range management, launch and on-orbit authority, payload recovery, and operational command and control of satellite systems. With the establishment in 1961 of Air Force Systems Command as an independent management headquarters for space and all Air Force research and development, it came as no surprise to find organizational responsibility for Air Force space resources increasingly associated with AFSC and its focal point for space, the Space Systems Division, and its successor organizations.

First came reorganization of the Atlantic and Pacific ranges. In January 1964, AFSC created the National Range Division (NRD), with provisional headquarters at Patrick Air Force Base, Florida. This followed agreement with NASA on range responsibilities in early 1963 and, later in the year, Secretary McNamara's decision to transfer Pacific Missile Range responsibility from the Navy to the Air Force and to assign the worldwide satellite tracking network to the Air Force. The National Range Division assumed responsibility for coordinating Defense Department and NASA activities at both the eastern and the western launch sites, and it established a provisional Air Force Space Test Center (AFSTC) at Vandenberg Air Force Base to manage Pacific Range activities. In January 1964, the National Range Division also gained the Air Force Satellite Control Facility at Sunnyvale, California. In May, the Air Force relocated the division to the site of Air Force Systems Command headquarters at Andrews Air Force Base, Maryland, near Washington, D.C., and redesignated the two ranges as the Eastern Test Range and the Western Test Range. Operations at Sunnyvale, however, proved awkward, with AFSC exercising direct control of the range, while Space Systems Division retained on-site responsibility for launch operations. A reorganization in July 1965 reassigned the Satellite Control Facility to Space Systems Division at Los Angeles.[81]

Following the establishment of the Space and Missile Systems Organization (SAMSO) on 1 July 1967, which recombined Air Force missile and space functions in a single entity, additional organizational changes served to enhance the space role of the Los Angeles headquarters. On 1 April 1970, by forming the Space and Missile Test Center (SAMTEC) at Vandenberg Air Force Base, California, the Air Force centralized all launch operations at the Pacific site for the first time. By assigning SAMTEC to SAMSO, the Los Angeles headquarters became responsible for nearly all military space program facilities. The consolidation became complete seven years later when, on 1 February 1977, the assignment to SAMSO of the Eastern Test Range at long last brought all space and missile launch facilities under one organization.[82]

The organizational changes of the 1960s helped lay the groundwork for Space Systems Division and later SAMSO to direct the development of the unmanned communications, weather, navigation, and early warning satellite programs that made the military community increasingly aware of, and dependent upon, space systems. At the same time the organizational developments enhanced the control of a research and development command over space systems that were becoming increasingly operational.

Vietnam Offers the First Military Space Test

Satellites first demonstrated their tactical battlefield defense support capability in Vietnam. There, meteorological and communications satellites provided vital near-real-time data essential for mission planning and execution. During a nationally-televised CBS interview in May 1967, General William Momyer, the Seventh Air Force Commander declared:

> As far as I am concerned, this weather picture is probably the greatest innovation of the war. I depend on it in conjunction with the traditional

forecast as a basic means of making my decisions as to whether to launch
or not launch the strike. And it gives me a little bit better feel for what
the actual weather conditions are. The satellite is something no commander has ever had before in a war.[83]

Indeed, weather satellites proved to be an invaluable feature and key innovation of
the war. Air missions in Southeast Asia often depended for success on the availability of a cloud-free environment for low-level fighter, tanker, and gunship operations. Few in number and limited by the dangers of operating in or over hostile
territory, conventional weather sources proved inadequate to the challenge. Satellite
imagery, relayed throughout the region, provided the answer.

The Air Force did not furnish the only satellite weather data for Allied forces in
Southeast Asia. In the mid-1960s Nimbus satellites developed by NASA for the
Weather Bureau used their Automatic Picture Transmission capability to transmit
imagery from their sun-synchronous orbits daily between 0700 to 0900 and 1100 to
1300 hours. Beginning in 1965 DMSP imagery proved more useful to Air Force and
Navy meteorologists and mission planners. From an altitude of 450 NM, the sun-
synchronous satellites furnished day and night, visual and infrared imagery consistently at 0700, 1200, 1900, and 2400 hours local time. DMSP data, however, did not
become available to the Navy until 1970, when the *USS Constellation* acquired the
necessary readout equipment. With timely, accurate satellite weather data available,
planners knew when the weather would break over a target area, and used night-
sensor imagery to determine the extent of burning rice paddies to forewarn pilots of
likely smoke coverage. Weather information proved especially useful in the Navy's
lengthy effort to destroy the important Thanh Hoa Bridge in North Vietnam. The
Son Tay raid to rescue American POWs in 1970 also depended on satellite imagery.
In this case DMSP data provided extremely accurate three-to-five-day forecasts that
allowed the planners to schedule the raid to coincide with a break in two tropical
storms moving across the South China Sea onto the mainland.[84]

Communications satellites also proved their worth in Vietnam, where for the first
time satellite transmissions provided communications from a real-world theater of
operations. In June 1966, a satellite communications terminal operated from Ton
Son Nhut Air Base using the limited one-voice and one-record circuit capability of
the NASA-developed Synchronous Communications Satellite. It operated between
Saigon and Hawaii until its demise in 1967 owing to satellite drift. Improvements
arrived with the installation of two ground terminals at Saigon and Nha Trang to
support the Initial Defense Communications Satellite Program (IDCSP). Operational by July 1967, each terminal had expanded from five to eleven circuits by
January 1968. Under Project Compass Link, IDCSP provided circuits for transmission of high-resolution photography between Saigon and Washington, D.C. As a
result of this revolutionary development, analysts could assess near-real-time battlefield intelligence far from the battlefield. On the other hand, this raised questions

of major import for command and control of operational forces. Although IDCSP satellites made possible more centralized operational control, at this time they also comprised part of a vulnerable system connecting a number of remote terminals with a single central terminus.[85]

Commercial systems also supplied satellite circuits to support area communications requirements. The Communications Satellite Corporation (COMSAT) leased ten circuits between its Bangkok facilities and Hawaii, while the Southeast Asia Coastal Cable system furnished part of the network for satellite terminal access between Bangkok and Saigon. Satellite usage during the Vietnam conflict established the military practice of relying on commercial space systems for routine administrative and logistical needs while trusting more sensitive command and control communications to the dedicated military system.[86]

Communications and weather satellites brought space into the realm of combat operations. They provided much needed real-time weather information and communications support to battlefield commanders and planners in Vietnam, and they linked them regionally and globally. Weather and communications satellites established their operational value for future defense support combat—as well as peacetime—operations.

The Military Space Community in Transition

For the burgeoning Air Force space program, the decade of the 1960s represented a transitional period in which experimental programs became effective operational systems. By the end of the decade communications and weather satellites operated by the Air Force provided crucial information to commanders in Vietnam. Air Force-led engineers found themselves on the brink of developing a three-dimensional satellite navigation system that promised to revolutionize battlefield command and control capabilities. In the area of surveillance, the early 1970s witnessed the operation of Air Force infrared early warning satellites that immediately became the central element in the nation's missile warning network. Already Air Force-developed Vela nuclear detection satellites helped make possible verifiable nuclear test ban treaties and potential arms limitation agreements. In this, as in more sensitive areas of strategic intelligence, automated satellites made an invaluable contribution to strategic reconnaissance and thereby considerably diminished the ability of any nation to launch a surprise attack. Unmanned, instrumented satellites had largely met the military requirements that President Eisenhower had set in the 1950s for a major missile and satellite program.

By the early 1970s military space had come of age. Both within the Air Force and among the other services and defense agencies, the contribution of space-based systems to the Vietnam war, as well as a growing range of peacetime defense support requirements, led to increased acknowledgment, if not acceptance, of space operations. This, however, set the stage for a return to the intense service competition of the Eisenhower era. The Defense Department traditionally sought to avert service rivalries through joint funding and management ventures and by designating the Air Force the service for military space research and development. Policy promulgated by the Office of the Secretary of Defense effectively stymied Air Force efforts to gain sole responsibility for military space activities, while joint management did not always prove successful—or diminish the voice of congressional critics of separate civilian and military systems. Although interservice competition remained muted for much of the 1960s, it certainly did not disappear.

This became clear at the end of the decade when the Navy reopened the issue of space management responsibility by challenging the Air Force monopoly on space development. In the Navy's view, the 1961 directive had become outdated and only served to prevent wider exploitation of space for important military requirements. Back in 1960, Air Force Chief of Staff General Thomas D. White fended off the Navy's Admiral Arleigh Burke by arguing that the Air Force would provide effective leadership for the nation's space program and be responsive to the needs of the other services. In 1970, General White's successors could point to a decade of successful, responsible management for the benefit of all the services. But with space programs providing support to an increasing number of users throughout the military community, the dominant Air Force position came under fire from the other services and their allies in Congress and the Defense Department.

Unconvinced by Air Force arguments, Secretary of Defense Melvin Laird, on 8 September 1970, issued Directive 5160.32, which declared that space systems would be acquired and assigned according to the guidelines pertaining to other defense weapon systems. Ongoing programs, however, would remain unaffected. As a result, the Air Force would retain responsibility for developing and deploying "space systems for warning and surveillance of enemy nuclear delivery capabilities and all launch vehicles, including launch and orbital support operations." On the other hand, all three services could now compete "equally" for future programs, including "communications, navigation, unique surveillance (i.e., ocean or battlefield), meteorology, defense/offense, mapping/charting/geodesy, and major technology programs." While this provision reinforced traditional naval interest in ocean surveillance and navigation and the Army's preeminence in geodesy, it left open the question of future management responsibility and operational relationships for communications (DSCS), battlefield command and control (GPS), weather (DMSP), and the crucial area of "major technology programs."[87]

Rather than attempt to overturn the directive and fight to reclaim the Air Force space monopoly, Air Force Secretary John McLucas wisely chose to focus on its portent for unbridled competition in the other services and agencies. Such groups, he warned, would likely bypass the Air Force and its vast reservoir of space resources and experience, and the resulting duplication and wastefulness would not be in the nation's best interests. On this issue, Secretary McLucas could refer to the often unmentioned, yet impressive, range of research and development activities that had characterized Air Force space efforts during the 1960s in support of a wide range of NASA and Defense Department requirements. These would include rocket testing at Edwards Air Force Base, California, Agena target analysis at the Arnold Engineering Development Center in Tennessee, and the fundamental work done by associate contractors on environmental and system testing projects, normally in conjunction with Air Force Systems Command offices and laboratories. Highlighting the latter efforts would be the development of solid and liquid rocket technology for both NASA and the Defense Department.

Impressed by McLucas' argument, Secretary Laird in February 1971 modified the 1970 decision by directing that all space program development be coordinated with the Air Force beforehand. The Defense Secretary's revision and the work of Secretary McLucas and other Air Force leaders to cooperate on space defense matters helped to mute the impact of the original directive for the immediate future. Nevertheless, the door now stood open to fierce competition for scarce space resources and for control of space systems increasingly important to ever-larger numbers of users.

To retain its space supremacy, the Air Force now needed to get its own house in order. Space systems demanded less emphasis on lengthy control by research and development agencies and more focus on operational organizational and management decisions. The advent of the Space Shuttle in the 1970s would compel the Air Force to face this challenge. Development and operational control of the Shuttle involved intense competition among Air Force commands, while interservice rivalry and civil-military management issues remained unresolved. The Space Shuttle would serve to crystallize the thinking of the Air Force community on space issues in the decade ahead. Unmanned, instrumented space systems had brought space to the operational threshold. It remained for Air Force leaders to determine the proper place for space within the traditional Air Force.

CHAPTER 5
Organizing for Space:
The Air Force Commits to Space and an Operational Space Command

In the early 1970s the American space community found itself in disarray. The post-Apollo future for civilian and military space agencies brought indecisive space policy, uncertainty over roles and missions, and fragmented organization. As the primary military space service with responsibility for 80 percent of the Defense Department space budget, the Air Force reflected the weaknesses in the national space program. For most Air Force leaders, space seemed more an element within the research and development arena than an operational field. Many doubted whether space programs represented dedicated Air Force programs per se. Rather, the Air Force seemed to manage space activities for others, as part of larger tri-service or joint efforts that Defense Department officials favored to lower costs and minimize interservice rivalry. As a result, Air Force leaders and the wider Air Force community did not make space operations a genuine institutional commitment.

A decade later, however, military space had undergone a remarkable transformation. Gone was the disarray over policy, organization, and future roles and missions. By the early 1980s, the nation boasted a clear and decisive space policy supported by initiatives to improve and expand space programs and infrastructure to the end of the century. Above all, the Air Force created a centralized organization for space that committed the service to an operational rather than a research and development focus. Normalizing and integrating defense-support space missions throughout the service would become the major space objective for the future as leaders moved to take advantage of the growing importance of space for operational forces.

Pressures from without and within the Air Force account for the upturn in fortunes. At the national level, space became a central focus against the backdrop of the decade-long debate over the merits of détente and arms control. Both critics and opponents of arms control agreements increasingly came to rely on space systems to provide crucial national technical means of surveillance and verification. The deficiencies in the space arena alarmed leaders and convinced them to support major policy and organizational initiatives.

Air Force leaders felt compelled to reassess the importance of space for operational commanders and the service's institutional commitment to space operations. Beginning in the mid-1970s leaders began a long process of building consensus for some type of centralized space organization and integrating military space requirements into mission and system architecture planning. The process seemed incredibly slow and contentious to space enthusiasts, who found allies primarily within the middle strata of the officer corps rather than the senior leadership. As a result, it took several years of space studies and forums to create a better understanding of space and an appreciation of its importance for corporate Air Force interests. Space proponents received major help from the operational maturity of space systems themselves—the Defense Meteorological Satellite Program (DMSP), the Defense Satellite Communications System (DSCS), the Global Positioning System (GPS), and the early warning Defense Support Program (DSP)—which, over the course of the decade, became increasingly important throughout the defense community. Above all, the advent of the Space Shuttle crystallized the pressure within the Air Force for change. This, the most expensive and technologically complex space project in the nation's history, raised important questions about cooperation between the civilian and military space communities, the future role of military manned spaceflight, the feasibility of exclusive reliance on a reusable launch vehicle, and the most appropriate organizational structure for Shuttle operations. The Shuttle precipitated an intense competition for operational responsibility among four major Air Force commands, each of which considered itself the logical choice to become the operational space command. By the end of the decade, the Air Force found itself in the midst of a series of important organizational changes that set the stage for the creation of an operational space command to follow.

In this era of change and reassessment a number of space "missionaries" played vital roles in moving space to the forefront of Air Force interests. Dr. Hans Mark and Lieutenant General Jerome O'Malley led the charge for an operational space commitment, often in the face of reluctant or overly cautious senior leaders. With the arrival of the Reagan administration in early 1981, the pace of events threatened to outrun the ability of Air Force leaders to control it. The overwhelming momentum for change compelled senior leaders to act before outside elements imposed solutions that might not reflect institutional interests. The result proved to be a major victory for the operational Air Force with establishment of the Air Force's Space Command on 1 September 1982.

A Space Community in Disarray

From the vantage point of the early 1970s Air Force space enthusiasts would be hard pressed to envision an operational space command only a decade into the future. At the national level, the budget-conscious Nixon administration responded to the public's disinterest in major post-Apollo space initiatives by canceling the final two Apollo lunar missions along with the Air Force's central manned program, the Manned Orbiting Laboratory (MOL). Additionally, the President eliminated important advisory bodies for space issues, the President's Scientific Advisory Council (PSAC) and the Federal Council on Science and Technology, while Congress transferred space responsibilities from its standing space committees to the more widely focused House Science and Technology and Senate Commerce committees. The concerns of both Congress and the White House centered more urgently on budget priorities to deal with the legacy of the Great Society's social agenda and the incessant demands of the Vietnam conflict. Oil shortages following the 1973 Arab-Israeli conflict heightened the financial crisis and set a tone of lower expectations and malaise for the remainder of the decade. An ambitious military space agenda could hardly flourish in this atmosphere, and the declining space budgets during the early 1970s provide the best evidence of the military space program's woeful status.[1]*

Disarray and disinterest best characterized the condition of military space activities during the first half of the 1970s. Air Force commanders seemed reluctant to accept the importance of space and to support space program initiatives during crucial budget proceedings. Military space missions—communications, meteorology, early warning, and navigation—comprised defense support functions rather than the traditionally more prestigious and appealing offensive operations. Moreover, while the Air Force controlled the newly-operational Defense Support Program early warning satellite network, it shared all other military satellite programs with other civilian and military agencies or, in the case of the sensitive national reconnaissance program, played a significant supporting role.

At the same time, the dispersed nature of space systems within the Air Force, as well as throughout the military and the civilian space communities, created more immediate management problems. The Office of the Secretary of Defense, for example, through the Defense Communications Agency, often participated in day-today management of communications satellites, while the Air Staff monitored an increasing number of space programs and functions that normally would have been

---

* See Appendix 3-2.

assigned to a single major command. Army and Navy planners faced similar hazards in handling their terrestrial mapping and navigational satellite programs, respectively, while other government and civilian agencies often followed their own relatively independent courses of operation. The lack of central direction not only led to management inefficiency and duplication, it prevented the creation of constituencies to effectively advocate and support space systems during the budget process.[2]

The fragmentation nature of the military space program in the 1970s reflected the absence of both a comprehensive employment doctrine for space and any significant change in executive policy or military space strategy since the Eisenhower and early Kennedy years. Military activities in space received little open attention in the age of Apollo and the era of Soviet-American détente in the early 1970s. The Defense Department's directive of September 1970, which overturned the Air Force's decade-old exclusive responsibility for military space research and development programs, further fragmented operational space planning and control by allowing each service and Defense Department component to pursue its own course. In short, dispersed authority made it difficult to coordinate military space requirements and operational concepts from a broad, national security perspective.[3]

The space policy vacuum and organizational fragmentation did not go unnoticed by interested observers. In a widely quoted article in late 1974, retired Air Force general and NASA manager, Jacob E. Smart, accurately described the condition of the space community.

> Presently there are multiple agencies of the U.S. government engaged in space related activities, each pursuing programs to fulfill its own missions. This of course is proper but points up the question: Does the sum of the individual agency's perceived roles adequately fulfill the total national need? There is no central policy coming from the top, guiding and coordinating these efforts.[4]

Given these developments, space did not acquire the status of a dedicated "Air Force" mission or lead to a specific "user" space community. Moreover, without an Air Force major command for space, officer career progression and space program advocacy suffered. For many, space seemed to represent an additional level of abstraction, one in which commanders often felt insecure about relying on supporting elements beyond their direct sight or control. As one Air Staff planner observed, "space...requires first of all, a psychological adjustment to and philosophical acceptance of the use of space assets and warfare conducted in space." Air Force commanders needed to understand the operational importance of space activities for themselves. Despite the contribution of communications and weather satellites during the Vietnam war, an institutional commitment to space seemed far off in the early 1970s.[5]

Nevertheless, the space community was on the threshold of change. A number of developments already underway would lead to a major reassessment of the military's role in space. The most important forces for change involved, first, the many-sided national strategic debate over the policy of détente and the efficacy of arms control measures and, second, the advent of the Space Shuttle. Both would bring space issues to the forefront of the national agenda.

The National Debate over Détente and Arms Control

The debate over America's strategic nuclear policy in the 1970s took place against the backdrop of the continuing shift in national defense policy from deterrence of the 1960s to the countervailing nuclear warfighting strategy of the Carter administration. The evolution in nuclear strategy paved the way for the emergence of a reinvigorated, modernized strategic policy and force structure under the Reagan administration's Strategic Modernization Program and centerpiece Strategic Defense Initiative (SDI).

During the 1970s, it became clear both to strategic policy analysts and the public alike that America could no longer take comfort in its traditional position as the dominant nuclear superpower. By 1974 the Soviet Union had overtaken the United States in total number of ICBMs and SLBMs, achieving a figure of 2,195 ballistic missiles in contrast to the United States' 1,710, and appeared hard at work developing a multiple independently targeted reentry vehicle (MIRV) capability.[6] Given the alarming increase in Soviet offensive nuclear weapons, events during the decade increasingly centered on the vulnerability of the Minuteman retaliatory force and what should be done to protect it. Could the nation's traditional policy of mutual assured destruction—or second-strike retaliation—continue to reflect the nuclear warfighting realities of the 1970s?[7]

To President Richard M. Nixon and his advisors, the assured-destruction strategy seemed to offer the dilemma of the single alternative. One faced the choice of either massive nuclear retaliation or not launching missiles at all, which could very well amount to surrender in the nuclear age. Nixon wanted more options along the spectrum of deterrence, and his Secretary of Defense, James R. Schlesinger, responded by focusing on flexibility and increased targeting options. Under the so-called "Schlesinger Doctrine," he developed the concept of providing "selective, small scale options" or target packages for rapid use in a variety of nuclear contingencies along the "spectrum of deterrence." The Schlesinger Doctrine reflected the concerns of many, both inside and outside of government, that the United States should prepare more effective contingency plans for fighting a nuclear war.[8]

At the same time, many looked to détente and arms control agreements as offering the best hope for underpinning and establishing rough nuclear equivalence at lower force levels and, thereby, reducing the danger of nuclear war. By terms of the Strategic Arms Limitation Treaty (SALT) I Interim Agreement on Strategic Offensive Weapons, signed in May 1972, the Soviet Union and the United States agreed to a five-year freeze on missile launcher construction as a prelude to further, more sweeping arms control measures. However, the agreement capped the Soviet ICBM arsenal at 1,618 ICBMs, in contrast to the American figure of 1,054, and did not include MIRVs. During the same month, the two sides recognized the impossibility of protecting their countries from a large-scale missile attack by agreeing to limit further deployment of their anti-ballistic missile systems. The Anti-Ballistic Missile (ABM) Treaty restricted both sides to two limited ABM systems, one deployed around the national capital, and the other at an ICBM site. The two sides formally recognized the role of satellite surveillance by agreeing that verification would be conducted by "national technical means…consistent with generally recognized principles of international law." As John Newhouse, former Assistant Director of the U.S. Arms Control and Disarmament Agency pointed out, "each side surrendered any meaningful right to defend its society and territory against the other's nuclear weapons."[9] In short, the ABM treaty made credible the policy of mutual assured destruction. Yet, another provision of the treaty would prove contentious in future years. According to Article V, "each Party undertakes not to develop, test, or deploy ABM systems or components which are sea-based, air-based, space-based, or mobile land-based." This proviso would seriously challenge the legality of President Ronald Reagan's Strategic Defense Initiative. Critics of the arms control process, however, bemoaned what they considered the failure of détente to prevent the establishment of a permanent state of American strategic inferiority.[10]

In this arena, both critics and proponents of détente depended on space reconnaissance and the related surveillance systems and associated warning infrastructure to provide the so-called "national technical means" of arms control verification. Consequently, the nation's satellite systems and ground-based space surveillance network became increasingly important to verify arms control compliance or to support charges of a growing potential Soviet threat from space. Moreover, a policy calling for flexible response required more sophisticated strategic surveillance, warning and, possibly, active defensive systems. For administrations searching for greater options along the spectrum of deterrence, improvement of space capabilities would become a growing priority over the course of the decade.

Indeed, by the mid-1970s, the primary mission of NORAD, the binational U.S.-Canada command, had become surveillance and warning of impending attack rather than active defense, and the once elaborate air defense structure controlled by the Air Force's Aerospace Defense Command (ADCOM) continued to decline in terms of quality and quantity of forces as it underwent organizational restructuring. Interceptor aircraft could not respond to ICBMs. Henceforth, space systems assumed greater importance, and ADCOM and NORAD commanders looked to space to preserve command prerogatives—with wider implications for the future of the Air Force space community.[11]

The Air Force Commits to a Space Shuttle

A second development in the rise to prominence of space involved the national commitment to develop as successor to the Project Apollo lunar program a national space transportation system that would serve both civilian and military agencies. The Space Shuttle represented tremendous potential with its promise of routine access to space. At the same time, it presented enormous challenges because of its technical complexity, high cost, and promise, as a joint civil-military program, to satisfy both NASA and Defense Department requirements. For the Air Force, the Defense Department executive agent for the Shuttle, the advent of the Shuttle represented a new era of military manned spaceflight, an end to dependence on its fleet of costly, expendable launch vehicles, and the reassertion of Air Force dominance in the national space program. Along the way, the Air Force also found itself compelled to reassess its institutional commitment to space if it intended to realize its claim to space leadership. Over the course of the 1970s, the Shuttle prompted planners to increasingly reassess space policy, technological feasibility, and optimum organizational structures in preparation for what advocates confidently proclaimed to be the "age of the Shuttle."

Shortly after taking office in 1969, President Nixon, as part of his initial program review, formed a Space Task Group to determine the best direction for the nation's post-Apollo space program in a future beset by declining interest in space and budget constraints. In September, shortly after *Apollo 11*'s historic July lunar landing, the group's report outlined three long-range possibilities. The first two comprised variations on an expensive, ambitious program to launch in the 1980s a manned mission to Mars. This would occur after first establishing a lunar base and a fifty-person earth-orbiting space station supported by a fully reusable transportation system to "shuttle" between Earth and the space station. The third alternative, which involved only the space station and Shuttle, appealed to a cost-conscious Nixon administration determined to pursue a less challenging post-Apollo space future. Before giving formal approval, however, NASA and the Defense Department needed to assess the Shuttle's technical feasibility, projected cost, and civil and military requirements.[12]

For NASA, the Shuttle represented the centerpiece of its future manned space program in the wake of the administration's cancellation of the final two Apollo lunar flights and reduction of the Apollo Applications program to the Skylab mini-space station. For the Air Force, initial enthusiasm was tempered by NASA's central responsibility for Shuttle design and development and by questions about the system's long-term benefits. At first the Air Force focus centered on the project as a cost-effective replacement for launching future larger, heavier satellites that would require lifting capacity greater than the Atlas and Titan expendable boosters could provide. Very soon, however, Air Force leaders came to see in the Shuttle a multipurpose vehicle with the means of preserving the Air Force's traditional interest

Right: General Henry H. "Hap" Arnold; below (left to right): Dr. Theodore von Kármán, chairman of the Scientific Advisory Group; Brigadier General Donald L. Putt, Director of Research and Development, Office of the Deputy Chief of Staff for Materiel; and Dr. Albert E. Lombard, Jr., head of the Research Division under General Putt.

Above: Officials of the Army Ballistic
Missile Agency at Huntsville, Alabama:
(counterclockwise from top right) Major
General H. N. Toftoy, Commanding
General; Dr. Ernst Stuhlinger; Hermann
Oberth; Wernher Von Braun; and Dr.
Eberhard Rees; right: Major General
Curtis E. LeMay.

Above left: Dr. Robert H. Goddard beside his liquid-fuel rocket before launch, Auburn, Massachusetts, 16 March 1926; above right: Trevor Gardner, Assistant Secretary of the Air Force for Research and Development; below: German V-2 rocket.

Above: Atlas intercontinental ballistic missile assembly plant; below: General Bernard A. Schriever with models of the missiles he helped develop and build.

Dr. Simon Ramo, a founder of the Thompson-Ramo-Wooldridge (TRW) Corporation

Dr. John von Neumann

Above: Technicians prepare a Thor intermediate range ballistic missile for a test launch, *ca.* August 1957; left: Model of *Sputnuk I*, the first man-made satellite to orbit the earth; launched 4 October 1957.

Above left: Secretary of Defense Neil H. McElroy; above: Secretary of the Air Force Eugene M. Zuckert; lower left: Secretary of the Air Force Donald A. Quarles.

Above: President Dwight D. Eisenhower and Chief of Staff of the Air Force General Thomas D. White (center) view the *Discoverer 13* capsule, the first object recovered from space, at the White House; right: Ivan A. Getting, one of the creators of MOSAIC (Mobile System for ICBM Control), a precursor of the Global Positioning System.

Left: Air Force systems and facilities, from the Atlas booster to ground-based range systems, were critical to the success of the NASA manned spaceflight program. Pictured is the launch of *Friendship 7* with astronaut John H. Glenn, Jr., aboard; below: Air Force Thor launch, *ca.* 1963.

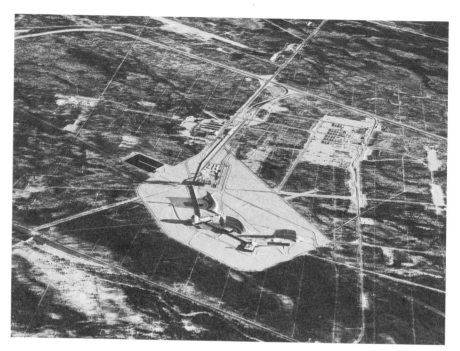

Above: Aerial view of Site II of the Ballistic Missile Early Warning System (BMEWS) at Clear Air Force Station in central Alaska; below: Ground-level view of a radar fan (left) and tracking radar radome at BMEWS Site II.

Above: Baker-Nunn satellite tracking camera, a workhorse in the Air Force's Spacetrack network for three decades; below: Space Detection and Tracking System (SPADATS)-Spacetrack Operations Center at Ent Air Force Base in Colorado Springs, Colorado, in the early 1960s.

Above: Space Defense Center at the Cheyenne Mountain Complex in Colorado Springs, Colorado, in 1973; right: Artist's conception of an Initial Defense Communication Satellite Program (IDCSP) satellite.

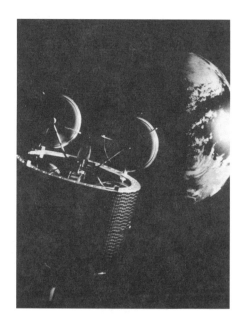

Above: Air Force Satellite Control
Network station at Anderson Air Force
Base, Guam; left: Secretary of Defense
Harold Brown.

Top right: Defense Satellite Communications System (DSCS) III satellite (artist's rendition); right: Defense Support Program (DSP) satellite (artist's rendition); below: Defense Support Program Overseas Ground Station at Woomera, Australia.

Above left: General James E. Hill; above right: General James V. Hartinger; below: Space Launch Complex (SLC)-6 at Vandenberg Air Force Base, California, in 1986, at the height of preparations for west coast Shuttle operations.

Above left: Global Positioning System (GPS) satellite; above right: An F-15 fighter carrying an air-launched antisatellite weapon, a system cancelled by Congress in the mid-1980s; left: An Air Force Delta II launch vehicle lifts off from Cape Canaveral Air Force Station carrying a Global Positioning System satellite into orbit.

in manned spaceflight following cancellation of its Manned Orbiting Laboratory program in 1969. The Shuttle would represent the Air Force's third attempt to achieve a man-in-space capability, a quest that began with the aerospace plane "lifting body" experiments of 1950s, proceeded with the ill-fated Dyna-Soar boost-glide space-plane, and culminated with the MOL.[13]

By 1972, Air Force leaders like Secretary Robert C. Seamans, Jr., chose to emphasize the variety of services they expected the Shuttle to provide:

> The shuttle offers the potential of improving mission flexibility and capability by on-orbit checkout of payloads, recovery of malfunctioning satellites for repair and reuse, or resupply of payloads on orbit thus extending their lifetime. Payloads would be retrieved and refurbished for reuse and improved sensors could be installed during refurbishment for added capability.[14]

The Secretary's rationale, which became the Air Force position in the years ahead, also encompassed the requirements of the surveillance and national reconnaissance "black world" space programs.

Moreover, Air Force leaders quickly realized the advantages of supporting a joint program that found NASA in the forefront. One legacy of the Kennedy-McNamara era continued to be the integrated nature of the nation's space program, which called for agreement between the civilian and military agencies on major national programs like the Shuttle. Although the Shuttle became a "NASA program," the civilian agency realized that Congress would not support the project unless military requirements could be satisfied. Tactically, the Air Force let NASA promote the Shuttle's man-in-space mission—and supply the bulk of project funding—while it stressed the economic advantage of saving up to 50 percent of projected launch expenses by adopting for the 1980s reusable boosters that, according to NASA projections, would average 60 flights annually. Characteristically, in the 1970s NASA would focus on its always uncomfortable budgetary battles with a parsimonious congress while the Air Force stayed in the background and remained uncompromising on military requirements. Evolving mission needs and technological challenges involving the most complex spacecraft yet attempted both added to the Shuttle's checkered course of development. Design changes would lead to cost increases, new launch-site requirements and, ultimately, schedule delays.[15]

Equal Air Force representation with NASA on the newly-formed Space Transportation Committee ensured that military requirements would be included in the various contractor design studies that assessed technology, scope, timing, and cost. From the start NASA and the Air Force differed over design and performance specifications—most notably those for payload weight and Shuttle size. NASA favored a cargo compartment 12 feet in diameter by 40 feet in length, but the Air Force insisted on dimensions of 15 feet by 60 feet. Likewise, the Air Force favored an expanded Shuttle design capable of launching a 65,000-pound payload into a low-inclination earth orbit (38.5 degrees), and a 40,000-pound spacecraft into low-earth (100nm) near-polar orbit (98 degrees). It estimated that fully half of its future launches would involve heavy payloads in higher or geosynchronous orbit. This meant that the Shuttle would need to accommodate these payloads as well as Lockheed's so-called Orbit-to-Orbit Shuttle (OOS), or Space Tug, that would "shuttle" the spacecraft to higher orbits and return to the orbiter. NASA, on the other hand, preferred reduced requirements to keep down steadily rising projected development costs that threatened to jeopardize congressional funding approval for both the Shuttle and the agency's unmanned programs.[16]

Although the Air Force, fully supported by Defense Department officials, remained inflexible on its weight and size requirements, during 1971 the two sides reached agreement in a number of important areas. NASA responsibilities would continue to encompass design, development, and fabrication, with the Air Force serving as Defense Department agent responsible for military requirements. The two agencies would act jointly to choose launch sites, with the Air Force funding a second site, if needed, and launch rates and costs would be apportioned according to the type of mission and amount of supporting equipment used. Meanwhile, congressional scrutiny continued to compel NASA to extend design study deadlines in order to consider ways of achieving lower development costs. By the end of the year, NASA had decided to forego its earlier plans for a reusable, manned, flyback booster and to accept, instead, a simplified booster design in conjunction with a smaller, more efficient orbiter using an external, liquid hydrogen-liquid oxygen tank. Final design specifications, however, remained unsettled when President Nixon, with one eye on the ailing aerospace industry, gave formal approval to the Space Shuttle on 5 January 1972.[17]

In announcing the $5.5 billion, six-year development program, the President declared the future Shuttle the "work-horse of our whole space effort." He said it would replace all expendable boosters except the smallest (Scout) and the largest (Saturn). By March of that year, NASA and the Air Force had reached agreement on the Shuttle's design. A delta-winged orbiter would be launched into low-earth orbit by the force of its three 470,000-pound-thrust liquid rocket motors in the orbiter, and two water-recoverable, solid-fuel rocket motors on the booster, each capable of four million pounds of thrust. An expendable, external, liquid-fuel tank completed the basic design. Following reentry, the orbiter would land on a conventional runway using a high-speed, unpowered approach. In effect, the Orbiter and Solid Rocket Boosters would be recovered, refurbished, and reused. Significantly, the 156-inch-diameter booster motors were the product of the Air Force's large-rocket development program that dated back to 1960. Although the new booster concept resulted in a drop in overall development cost from $5.5 billion to $5.1 billion, operational cost rose to $10.5 million per mission, more than twice the original estimate. In the future, cost-efficiency would be dependent on achieving the high launch rate projected for the 1980s. By this time, however, NASA had canceled plans for both the Space Tug and a fifth Shuttle orbiter, which contributed to a drastic reduction in annual flights and an increase in operational costs.[18]

In April 1972, NASA and the Air Force chose both the Kennedy Space Center and Vandenberg Air Force Base as sites for future Shuttle operations. Each would allow for water recovery of the booster motors. The Florida site would support research and development test flights and easterly launches, while Vandenberg would be used for payloads requiring high inclination polar orbits. The development schedule called for the first "horizontal" flight test in 1976, to be followed by manned and orbital flights in 1978, with full operations commencing by 1980.[19]

Yet, precise Shuttle objectives remained undetermined. As one author has noted, "the Space Shuttle emerged, but no decision on the goals of future spaceflight. Apollo was a matter of going to the moon and building whatever technology could get us there; the Space Shuttle was a matter of building a technology and going wherever it could take us."[20] Such uncertainty, however, applied more to the civilian side of the Shuttle than to the military. To establish military utility, specific missions, and coordinate with other military departments, the Defense Department created in November 1973 the Defense Department Space Shuttle User Committee chaired by the Air Staff's Director of Space. By the end of 1973, the Air Force and the Defense Department had agreed on a December 1982 operational date for Vandenberg based on refurbishing the old MOL Space Launch Complex 6 (SLC 6), and it had scheduled a phased replacement of the expendable boosters extending from 1980 to 1985. At the same time, an Upper Stage Committee appointed by the Space Transportation Committee examined Space Tug requirements and reaffirmed that a full-scale Space Tug with retrieval capability should be developed by NASA. In order to ease NASA's ever-present budget hurdles and provide the agency a more deliberate development schedule, the Upper Stage Committee suggested the Air Force demonstrate its commitment to the Shuttle by developing a less costly Inertial Upper Stage (IUS) based on modification of an existing vehicle. The Air Force agreed, and accepted as its responsibility the interim upper-stage vehicle along with the Shuttle launch site at Vandenberg.[21]

Although the basic elements of the Shuttle program had fallen into place by 1974, technical and political problems would continue to play havoc with developmental and operational milestones. Along with its responsibilities for constructing the Vandenberg launch site and producing an interim upper-stage vehicle in place of a Space Tug, Air Force concerns would focus on how best to protect and control classified military space missions from NASA's Johnson Space Center (JSC). Should they be handled by NASA's controllers alone, or by an Air Force element collocated at the JSC? Or should the Air Force develop a new organization to replace or augment its overworked Satellite Control Facility in Sunnyvale, California? This organizational issue became one of many that confronted the Air Force in the latter half of the decade. Already in 1974, however, the Shuttle had precipitated another, more contentious internal organizational dilemma. Because of the poorly defined line separating experimental from operational space systems, AFSC performed an operational role with on-orbit spacecraft. Understandably, AFSC received military development responsibility for the Shuttle through its SAMSO program office. But what of future responsibilities? Would SAMSO also serve as the military's "operational" organization for the Shuttle?[22]

In April 1974, NORAD and ADC commander General Lucius D. Clay, Jr., seized the initiative by submitting a ten-page position paper to the chief of staff calling for an immediate decision to award ADC operational responsibility for the Shuttle. He argued that his command possessed the requisite experience through its service as the operational command for the ground-based space surveillance system and the newly operational Defense Support Program. Less direct in General Clay's argument was his motivation to justify the importance of his command's space role through award of the Shuttle. With the waning of ADC's air defense mission, the Shuttle could perhaps serve to preserve the existence of the command itself. Shortly thereafter, the Military Airlift Command (MAC), as the Air Force "transportation" agent, along with the Strategic Air Command (SAC) and Air Force Systems Command entered the bidding, each staking out its claim to the Shuttle.[23]

The imminent operational status of new systems like the Shuttle compelled Defense Department and Air Force officials to begin reassessing whether systems should continue to be assigned to commands on an individual basis or, by contrast, be centralized in a single operational Air Force command. Traditionally, the Air Force and the Defense Department assigned space systems on a functional basis to the command or agency with the greatest need. As a result, Air Force Systems Command, for example, controlled military communications satellites, and Strategic Air Command managed meteorological satellite outputs, while the Aerospace Defense Command (ADC) operated the space surveillance and missile warning system. This arrangement proved workable as long as defense officials had to handle only a few satellites with modest capabilities. By the latter 1970s, programs such as the Global Positioning System (GPS), which possessed multiple capabilities serving a variety of defense users, promised to blur the functional lines enormously. If the Space Shuttle presented Air Force leaders with their greatest dilemma, by the end of the decade defense satellite systems provided an impressive array of potential applications for battlefield commanders. In a fragmented space community, however, many questioned their operational effectiveness.[24]

The Growing Prominence of Space Systems in the Late 1970s

Few would argue that space systems in the 1970s achieved important milestones and became more important to military commanders. Yet, they experienced a variety of problems that prevented them from reaching their full potential. Both the operational capabilities they achieved and the frustrating limits on realization of their full potential made the military simultaneously more dependent on, and more concerned with, space systems by decade's end.

For example, by February 1974 the second series of Defense Satellite Communications System (DSCS) satellites had reached its full orbital configuration of four operational and two spare satellites positioned in synchronous equatorial orbit to provide global coverage to 72 degrees latitude. The DSCS II satellite network provided super-high-frequency communications support, without the problems of orbital drift and limited channel capacity that its predecessor series, the Initial Defense Communications Satellite Program (IDCSP) experienced. In February 1977, the Defense Communications Agency authorized full-scale development of the next generation, DSCS III, which would provide greater flexibility and security over six channels instead of two, as well as better jam-resistant and nuclear-hardening capabilities. Planners estimated DSCS III satellites would have a ten-year design life. On the other hand, developmental problems and funding shortfalls had pushed the expected operational date from 1981 to 1984.[25]

The Navstar Global Positioning System also made good progress by decade's end. Rockwell International commenced full-scale engineering development in mid-1979, although four test satellites had been launched the previous year. Despite failures of atomic clocks which required the replacement of two satellites, by 1981 the full complement of five test satellites provided three-dimensional data one to two hours per day in support of a variety of Navy requirements. Originally, the three-phase project was to have a full complement of 24 satellites operational by 1984. By the early 1980s, however, budget shortfalls and technical problems led planners to conclude that it would be late in the decade before a fully operational system could be deployed to provide a 24-hour-a-day capability for global, three-dimensional positioning and weapons delivery. At the same time, budget reductions now had resulted in eliminating three satellites, which meant deploying an 18-satellite configuration with three spares. Program managers hoped that constant funding uncertainty would not produce "stretch outs" that would further delay operational milestones.[26]

The Defense Support Program early warning satellites had performed admirably since the first launch in May 1971. Three years later the three operating satellites had detected nearly 1,300 missile launches, including 966 Soviet and 16 Chinese test flights, and they had exceeded or were approaching their estimated 15-month design life. Fortunately, in the mid-1970s, the unexpectedly long life of the satellites allowed engineers to retrofit those in the inventory with improved infrared sensors that provided more accurate missile launch counts and launch point determination. By the end of the decade, upgrades included a more sensitive Mosaic Sensor System to offset scanning limitations by continuously "staring" at the earth's surface, and a Sensor Evolutionary Development program. The latter included developing mercury-cadmium-telluride sensor cells to give a larger number of infrared detectors greater sensitivity. Improved ground station computers and software completed the scheduled modifications underway by the early 1980s. At the same time, NORAD operators worried about coverage deficiencies and their inability to convince the Defense Department to provide backup satellites before deficiencies appeared with the operational spacecraft.[27]

While all three military satellite systems experienced technical, managerial, and budget challenges that tended to characterize highly complex and advanced technological projects, none approached the difficulties surrounding the Air Force-managed, joint service Defense Meteorological Satellite Program (DMSP). A new generation of polar-orbiting satellites, known as Block 5D-1, was to provide better quality and more reliable weather data from its Operational Linescan System and twelve new or improved secondary optical and infrared sensors. From the first, delayed launch on 11 September 1976, however, a variety of technical and management failures continued to limit operational effectiveness of the two-spacecraft system to the point where Defense Department users at times had to rely on low-altitude National Oceanic and Atmospheric Administration (NOAA) meteorological satellites for weather data. Like the communications satellite program, DMSP also underwent protracted "convergence" discussions in the late 1970s to determine the feasibility of combining civil and military polar-orbiting weather satellite programs in order to avoid duplication and cut costs. Once again, the dedicated military program survived, but the uncertainty about its future contributed to delays in development of an improved Block 5D-2 series satellite, which would produce 25 percent more power from it solar array and use larger on-board computers and eleven special advanced sensors. Moreover, program manning and funding had failed to keep pace with the increased complexity and risk of the program. Indeed, the DMSP system remained largely nonoperational from December 1979 until the first launch of the new series on 20 December 1982.[28]

The DMSP experience revealed the limitations in attempting too wide a technological leap between generations of satellites in an evolutionary system. At the same time, all satellite programs suffered from inconsistent funding, technical deficiencies, management weaknesses, and political interference. In the early 1980s all would require reconfiguration at great cost, first to allow launch by the Air Force Titan 34D booster and, then, to accommodate transition to the Space Shuttle. The advent of the Shuttle and the challenges involved in achieving reliable operational status of the satellite programs demanded greater attention in the last half of the decade.

The Space Detection and Tracking System's (SPADATS) ground-based space surveillance sensor network also improved its capabilities with the acquisition of three major new systems in the late 1970s and early 1980s. Operational since 1977, the Cobra Dane radar located on Shemya Island in the Aleutian chain, employed the new phased array technology which permitted the system to maintain tracks on multiple satellites simultaneously. Three years later, two additional phased array radars joined SPADATS as collateral sensors. PAVE PAWS radars at Beale Air Force Base, California, and Otis Air Force Base, Massachusetts, functioned primarily as missile warning detectors for SLBMs, but also provided precise detection and tracking of satellites. A third new system joined SPADATS in the early 1980s first as a supplement, then as a replacement, for the aging Baker-Nunn deep space optical telescopes. Earlier, in 1978, the Air Force added the Maui Optical Tracking and Identification Facility (MOTIF) to the network. Unlike the Baker-Nunn cameras, the Maui system provided near-real-time observations by means of linking an optical telescope to a computer and television camera. The major improvement in deep-space detection capability, however, proved to be the Ground-based Electro-optical Deep Space Surveillance (GEODSS) System. Beginning in 1982, the Air Force expected to inaugurate the first three of five sites, each of which would operate two deep-space tracking sensors and one wide-area search telescope for coverage of lower altitudes. Yet, despite these improvements, proliferation of satellites in deep space and persistent coverage gaps promised to challenge the capabilities of SPADATS in the years ahead.[29]

From its inception, the worldwide, ground-based space surveillance infrastructure had remained operationally focused, with assets owned by the Air Force and centralized under NORAD's operational control. By contrast, the satellite infrastructure, under the tutelage of Air Force Systems Command, always emphasized the research and development elements of its growth and operations. Only the Defense Support Program (DSP) early warning satellite system linked the two military space communities. Developed by AFSC, but operated by NORAD, DSP also represented the only operational satellite system wholly controlled by the Air Force. All others reflected tri-service or joint management and development. Little wonder that within the Air Force the perception developed that space systems perpetually remained in research and development rather than transition to the operational side of the service in the traditional manner, or that space systems represented Defense Department rather than Air Force programs and, consequently, did not deserve Air Force advocacy or funding. Any future organizational initiative would have to stress operational applications and more effectively combine the orbital and ground-based space communities.[30]

NORAD and ADCOM commanders hoped their experience with both orbital and ground-based systems would enable ADCOM to serve as the space command of the future. Clearly the growing maturity of, reliance on, and problems associated with space systems in the late 1970s increased the pressure on Air Force leaders to normalize space operations by means of a more centralized organizational focus. At the same time, without a stronger national space policy, centralized management and control, and more capable systems, political and military leaders could not rest assured that increasingly important space systems would survive in the face of what seemed a concerted Soviet effort to develop the capability to threaten Western satellites and their supporting facilities with antisatellite (ASAT) weapons. The fragmented military space program needed improvement across the board. For the Air Force, this challenge, if handled properly, could be a superb opportunity for the service to lead the effort to improve the military space program and, in so doing, perhaps regain its preeminent position as sole executive agent for Defense Department space activities.

Soviet ASAT Testing Prompts Space Initiatives

The ongoing national debate over détente, arms control, and the Soviet military buildup entered a new phase late in the Ford administration when the Soviets resumed antisatellite testing after a four-year moratorium. Renewed Soviet anti-satellite testing in early 1976 provided the impetus for political leaders, who already were alarmed about Soviet military expansion, to reassess all facets of the American space program. The resulting momentum for change produced major policy and organizational initiatives that had by decade's end put the nation firmly on the path to what advocates termed space "normalization," the integration of space assets in all phases of military planning and operations.

When the Soviets resumed co-orbital satellite interception testing in mid-February 1976, the United States had no antisatellite system operational or proto-typical. The previous year it had finally terminated its only operational antisatellite system, Program 437, an Air Force project which involved launching nuclear-equipped Thor boosters from Johnston Island in the Pacific. Hampered by reliabil-ity and cost problems, as well as diminishing interest in a nuclear capability in the era of détente, officials placed the program on standby status in 1970 when Air Force launch personnel transferred to Vandenberg, and they deactivated the system on 1 April 1975. Later, in the fall of 1976, President Ford authorized a ground-based replacement in National Security Decision Memorandum 333, which resulted in a program calling for launch of a Miniature Homing Device from an F-15 aircraft.[31]

Shortly after the initial Soviet antisatellite test on 12 February 1976, Dr. Malcolm Currie, Director of Defense Research and Engineering, testified before Congress that "satellite vulnerability has to be a major issue for us, a major topic of study and of planning over the next few years. The question is, can we maintain space as a sanc-tuary or not?"[32] That April, Dr. Currie requested from Air Force officials a thorough assessment of the Space Detection and Tracking System, the worldwide network of sensors linked to the NORAD Space Defense Center and responsible for locating and tracking all objects in space. Likewise, when his counterpart, Assistant Secretary of the Air Force for Research and Development John Martin, responded on 1 June with the first of three major studies, titled "Plan for the Evolution of Space Surveillance Capabilities," he approached the problem not by emphasizing the potential Soviet threat from weapons in space but from the "significantly enhanced military operations in other theaters made possible by the use of space systems." In the years ahead, Air Force and Defense Department officials would continue to stress this theme of normalization, the application of space activities to military forces, throughout their struggle to bring some degree of order to the space program. The Air Force study submitted to Dr. Currie in the spring of 1976 proved to be just one of many analyses of space systems prepared during the next several years. The long and difficult process of developing an integrated systems approach to space requirements would culminate in 1982, with publication of the first portion of the Air Staff's *Space Systems Architecture 2000*, which established plans and programs through the end of the century. Likewise, policy and organizational studies abounded during this period. If the many studies and analyses appeared symptomatic of problems in the space program, they also proved to be a means of promoting agreement on space issues throughout the defense community and, especially, within the Air Force. Without general consensus, necessary support for major organizational and doctrinal changes leading to an operational space focus would not be forthcoming.[33]

In August 1976, the Ford administration continued to demonstrate its serious interest in space defense improvements by directing a "significant" increase in the Air Force's space research and development funding and, in early November, establishing a Space Defense Working Group that consisted of representatives from the Defense Department, including the Defense Advanced Research Projects Agency (DARPA), the Joint Chiefs of Staff, and the services. Secretary of Defense Donald Rumsfeld recognized the divergent views present in the space community and called for a deliberate approach designed to "educate" people, create general agreement, develop a "broad-based understanding of DoD-wide Space Defense efforts…, facilitate the exchange of information…, and illuminate the important issues." Although the working group could accomplish little before Ford administration officials left office in January 1977, it nevertheless helped pave the way for the Carter administration's space initiatives.[34]

Despite President Jimmy Carter's reputation as a nuclear weapons "disarmer," his administration ultimately belied its critics and profoundly affected the direction of strategic aerospace defense.[35] Under considerable pressure and much against his basic view of the country's military requirements, President Carter found himself compelled to take increasingly bold measures to improve the nation's defensive posture. Responding to changing perceptions of the Soviet threat, efforts to make the country's command, control, and intelligence systems function effectively pointed toward a greater need for improved air, missile, and space defense. On the surface, one saw the stream of now predictable studies dealing with space policy and operations, which might suggest that officials achieved little of substance. Yet, by 1981, when the Carter team left office, the nation had received a revised, comprehensive national security space policy as part of what his Defense Secretary, Harold

Brown, termed the "countervailing strategy." According to the new national defense strategy, the nation must be capable of responding to any type of nuclear or conventional threat.[36] The expanded national defense commitment resulted in strategic defense improvements in all areas, including space. The Carter years would prove decisive in setting the direction of the country's space program, and Air Force leaders could cite major progress in organizational, doctrinal and system upgrade initiatives. Indeed, at decade's end, the Air Force appeared on the verge of consolidating its claim to sole-agent status for Defense Department space matters.

Normally one would expect a new administration to take a deliberate approach to reassess defense programs through the customary procedures for policy review. Although the Carter team's space policy did, in fact, emerge following a lengthy internal review, the new president's interest in arms control issues prompted immediate action in space matters. During a press briefing on 9 March 1977, President Carter announced that he intended to propose to the Soviets mutual restrictions on antisatellite weapons, which accorded with established policy on peaceful uses of space. On the eve of Secretary of State Cyrus Vance's visit to Moscow later that month, where he would raise the subject with the Soviets, President Carter signed Presidential Review Memorandum 23 which directed the newly established National Security Council Policy Review Committee to "thoroughly review existing policy and formulate overall principles which should guide our space activities." Chaired by Frank Press, Director of the Office of Science and Technology Policy (OSTP), the committee intended to pay special attention to the problem of ineffective coordination and friction among the four major space users: the intelligence community, the Defense Department, federal space agencies like NASA and the National Oceanic and Atmospheric Administration (NOAA), and the commercial sector.[37]

For over a year the OSTP assessed space policy issues before President Carter, on 20 June 1978, issued Presidential Directive 37, which described the "basic principles" of the nation's space program. His directive focused on defense priorities. A National Security Council policy review committee would provide "a forum to all federal agencies for review of space policy and...rapid referral of open issues to the President." NASA would continue to bear the overwhelming financial burden for the Shuttle, while the Defense Department, in the name of national security, had priority on all future Shuttle flights. Moreover, the Defense Department would emphasize satellite survivability and develop the antisatellite system that President Ford had approved. As the press release on the directive concluded, "the U.S. space defense program shall include an integrated attack warning, notification, verification and contingency reaction capability which can effectively detect and react to threats to U.S. Space Systems."[38]

Although the directive sounded like music to the ears of Air Force space enthusiasts, in the aftermath of its public unveiling the civilian space community found the policy too heavily weighted on the military side and pressed for an adjustment. As a result, the Policy Review Committee prepared what became Presidential Directive 42, "US Civil Space Policy," issued on 11 October 1978. In this directive, the president focused on the potential for new nonmilitary space applications and explorations after Shuttle development resources had diminished. The administration also addressed the old issue of fragmentation of resources and stressed the importance of cooperative efforts to eliminate duplication. This set the stage for the "convergence" efforts to combine military and civilian communications and weather satellite programs. With the two directives, the nation now possessed a forceful, declaratory military space policy to serve as a point of departure for developing an effective, long-term space program. Even so, Air Force officers later would criticize the directive as not going further toward explicitly establishing the need for a warfighting capability in space.[39]

Air Force leaders had not remained idle during the Carter administration's policy review. In fact, Air Force deliberations had begun late in the Ford administration with a reassessment of several tenets that over the years had developed as extensions of general national policy guidelines. Echoing General Thomas D. White, Air Force leaders affirmed that aerospace was a medium for performing missions rather than a mission in itself. Second, they said military space programs should be centered in one service, the Air Force, to promote maximum efficiency and economy. Third, the major factor in deploying space systems would be their potential effectiveness for space applications. Fourth, the Air Force would vigorously guard the principle of "space for peaceful purposes" while maintaining the military options to guarantee these purposes. Finally, Air Force policy called for strong, consistent support of, and cooperation with, NASA.[40]

To generate momentum on space policy within the Air Force, the Air Staff's office of the Deputy Director for Plans and Operations proposed several new space tenets and sought support by soliciting comments in mid-November 1976 from Air Staff agencies and field commands. By January 1977, it reported to Vice Chief of Staff General William V. McBride that the respondents generally supported the tenets dealing with space for "peaceful purposes," cooperation with NASA, and the importance of the "aerospace" medium for the Air Force. Many of the respondents seriously disagreed, however, with the remaining two tenets, because assertion of Air Force space prominence had strong interservice implications. A cautious Air Force Chief of Staff General David Jones chose to await the incoming Carter administration's initiative on space issues before taking further action.[41]

By that spring, with the administration's policy review underway, the Chief of Staff acted. In a major Air Force policy letter issued in May 1977, General Jones seized the opportunity to establish an updated set of space policy tenets describing Air Force responsibilities that, he asserted, could provide the framework for further Air Force "efforts to develop plans and capability objectives for space." At the same time, he hoped to influence the administration's deliberations on space. In response to "increasing reliance on space operations" and a "growing threat to the free use of space," Air Force space policy would comprise the following:

> 1. The Air Force affirms that among its prime responsibilities are military operations in space, conducted by the letter and spirit of existing treaties and in accordance with international law;
>
> 2. As DoD executive agent for liaison with NASA, the Air Force affirms its responsibility for close coordination on projects related to national security and for cooperation and support on projects of mutual benefit;
>
> 3. The Air Force affirms its responsibility for maintaining the freedom of space by providing needed space defense capabilities.

Space doctrine and employment discussions initiated at the time, however, would continue to founder for lack of consensus, and General Jones' policy letter would remain the official Air Force position on space for the next six years.[42]

Although Air Force leaders carefully omitted claims to preeminence in the space community, the series of space studies and analyses that appeared in the next few years were less restrained. One of the first, and most important, appeared that July when the Air Staff's plans and operations office issued *A Study of Future Air Force Space Policy and Objectives*. It provided a crucial point of departure for efforts to achieve consensus on space within the wider Air Force, and it asserted the service's pretensions to space leadership.

Taking the now traditional view of "aerospace," the study's authors asserted that space should be viewed as a continuation of the atmospheric arena, where activity was most efficiently performed under a single manager. The Air Force possessed the expertise and a history of exploiting technology that made it most qualified to be that manager. In other words, it should actively pursue a "sole agent" policy by seeking, as a minimum, recognition as de facto executive agent for the Defense Department in all space matters and, ultimately, formal designation as sole executive agent. This represented the most explicit declaration in favor of advancing a "sole agent" policy within the Defense Department. As a first step, the study argued, the Air Force must put its own house in order by updating important documents and establishing a set of "corporately-endorsed" goals and policies for space. Overcoming this major obstacle would clear the path to the best use of space systems, whose capability objectives should focus on providing greater operational support to ground forces and credible deterrence at all levels of conflict. Although the authors endorsed the policy promulgated in May by General Jones, they believed further action would be necessary to convince the skeptics who questioned the survivability and usefulness of space systems. The authors harbored no doubts about the situation at hand. "We are presently at a juncture which presents the Air Force with a unique opportunity to set unambiguous policy and objectives for the future to maintain US military leadership in space."[43]

In the wake of the July report, General Jones established a Space Operations Steering Group to provide a central Air Staff focus for space issues. Although the study's ringing call to action did not result in additional major initiatives during the remainder of 1977, subsequent comments and decisions suggest that the July 1977 Air Staff study established a benchmark for stimulating discussion and action on key space issues.[44] Unfortunately for advocates of rapid change in the space community, attention within the Air Force for the next two years too often focused on the slow death of the Aerospace Defense Command.

ADCOM's Demise and the Search for Space Consensus

Periodic threats to ADCOM's existence were nothing new in the first half of the 1970s. With its original anti-bomber defense mission in decline, it continually faced congressional and Defense Department pressure to streamline operations by cutting costs as well as eliminating personnel and subordinate headquarters functions.[45] In 1977, however, ADCOM's future also became enmeshed in the ongoing assessment of space organization and issues.

Early that year, an internal Air Staff evaluation of ADCOM led to a major review titled, *Proposal for a Reorganization of USAF Air Defense and Space Surveillance/ Warning Resources*, known informally as the "Green Book" study. It proposed eliminating ADCOM entirely and parceling out its air defense resources to the Tactical Air Command (TAC), its communications facilities to the Air Force Communications Command, and its space assets to the Strategic Air Command (SAC). ADCOM's flamboyant commander, General Daniel "Chappie" James, labored mightily to save the command but to no avail. Although in the end he failed, his often contentious counterattack against the Air Staff's agenda sharpened the focus of the debate on space and contributed to major changes following his departure in late 1977. James' successor, General James E. Hill, arrived in December 1977 to assume command of a NORAD that had shrunk to 25 percent of its original 1957 contingent. Although he had no significant air defense experience, he became a strong advocate of air defense. He sought to prevent reorganization of the command by stressing the importance of the Canadian role in NORAD and the "logic" that would have ADCOM remain as a major command with responsibility for space operations.[46]

General Hill found a key ally in Under Secretary of the Air Force Dr. Hans Mark, who objected to the element in the reorganization plan that called for combining strategic offensive and defensive forces in one command, SAC. Not only did the Canadians find this disquieting, said Mark, but advocacy for essential space modernization improvements would not receive sufficient attention from an offensive-oriented command. The energetic under secretary looked to ADCOM to establish a preeminent role for the Air Force in space. After a visit to NORAD/ADCOM headquarters in April 1978, for example, he expressed to General Hill his belief in the future importance of space operations, and of ADCOM taking a "leading role in developing the requirements for the kinds of operations that we will have to carry out in space." Mark actively—if unsuccessfully—pursued his argument for ADCOM's space role at the highest levels of the Pentagon.[47]

General Hill also vigorously lobbied the Joint Chiefs of Staff, General Jones and his successor, General Lew Allen, as well as his colleagues at the "Corona" conferences of four-star generals. His argument centered on the need for an Air Force space operations command. In the absence of an operational focus for space activities, he declared, key issues were surfacing which required immediate attention. Among these was the need to designate an operator for the Space Shuttle and Navstar GPS as well as a focal point for the man-in-space program and military use of the Shuttle. He also stressed the need for a responsive, dedicated military launch capability, and the integration of space systems into normal Air Force logistic-engineering channels. Overall, there had to be a focal point for operational requirements. Whereas the service had designated Air Force Systems Command to oversee the development of space systems and ADCOM to manage defense of space systems, no single command had been selected as the operator of space systems. General Hill proposed ADCOM as the logical choice.[48]

The ADCOM commander next took his case to the Corona Pine meeting of four-star commanders in October 1978. He commented afterward that space had received considerable discussion, but he was surprised and disappointed at the lack of understanding of the issues among leaders from the other major commands and the meager support they demonstrated for his position. He noted that Air Force Chief of Staff General Lew Allen had responded coolly to his presentation, advocating instead the development of a "Space Defense Operations Center" and declaring that space represented a "long term thing." He did not think a space command was not needed at that time. General Allen did, however, mention that he would form an Air Staff group to examine the feasibility of a future space command.[49]

In fact, in September Dr. Mark had convinced Air Force Secretary John Stetson to suggest such a study group to General Allen. It would prove to have a major impact on Air Force space thinking. In early November General Allen appointed a nine-member Space Mission Organization Planning Executive Committee to examine all facets of space mission management including organizational responsibilities, operator participation, and command and control of space mission resources. Although general officers comprised the committee's steering group, action officers under Colonel G. Wesley Clark of the Air Staff plans office bore the major research and analysis responsibilities. One of these, Lieutenant Colonel Thomas S. Moorman, Jr., future commander of Air Force Space Command and Vice Chief of Staff of the Air Force, recalled the crucial impact this committee had on later developments because of what he termed its "extraordinarily important socialization process," which took place in two ways. Over the course of several months, group members, who represented the entire spectrum of the space community, debated all aspects of the space program among themselves and with the generals who, in Moorman's words, became "socialized for the issues." The action officers benefited similarly, and many would play key roles in the subsequent establishment of the Air Force's Space Command. Colonel Moorman's experience also suggests that a recurring theme in the search for consensus on space involved bridging the gap between the larger group of committed middle-echelon officers and the more hesitant, skeptical senior leadership.[50]

The *Space Missions Organizational Planning Study* that appeared in February 1979 described five alternatives ranging from the status quo to establishment of an operational space command. Although it had found consensus for greater centralization of space operations and for seeking Air Force designation as Defense Department executive agent for space, timing and specific organizational structure remained unclear. The Air Force's four-star generals received a briefing on the study at their February 1979 Corona meeting, where General Hill's efforts to provoke wide-ranging discussion on space once again proved unsuccessful. Although he noted that most generals favored aggressive, centralized management of space activities in principle, they remained divided on the specifics. Perhaps most significantly, General Allen at this time did not seem to favor centralized organization for space at all.[51]

General Hill continued to argue that ADCOM's space resources should be left in place to serve as the core of a future space command. Nevertheless, Air Force leaders proceeded with the reorganization of the air defense and space missions, which resulted in disestablishment of ADCOM as an Air Force major command (leaving the specified ADCOM in place*) with its air defense systems and its missile and space defense systems parceled out to Tactical Air Command and Strategic Air Command, respectively, in 1979 and 1980. Following extensive discussions, General Hill succeeded in retaining operational control of aerospace defense forces and responsibility for systems advocacy with the Joint Chiefs of Staff and Air Staff. Reflecting later on the turbulent years of ADCOM's demise and his efforts to retain a space mission, General Hill strongly believed that had Air Force leaders agreed with his proposal, they could have successfully achieved an operational space command four years earlier than they eventually did. On the other hand, commenting in 1982, on the eve of the Air Force's new Space Command, the retired former NORAD/ADCOM commander admitted that there existed, then, "a crystallization, an understanding, of where we are in space and an appreciation for the requirement for a space

---

* ADCOM, the Air Force major command, was inactivated on 31 March 1980 and replaced by a direct reporting unit, the Aerospace Defense Center. ADCOM, the specified command serving as the United States component of NORAD, continued until 16 December 1986, when it was inactivated and replaced by US Element NORAD.

195

command now that didn't exist 4 years ago and that I was unable to persuade people of."[52]

As the ADCOM experience demonstrated, Air Force leaders like General Jones and his successor, General Lew Allen, preferred to compromise and proceed cautiously until they could be assured that parochial command interests had abated and that a sizable consensus within the Air Force would support new departures. The deliberately slow pace also reflected their own doubts, as well as their intention not to create unnecessary opposition from the Joint Chiefs of Staff, the other services, and Defense Department agencies. While the ADCOM reorganization plan postponed rather than precipitated final decisions on Air Force space arrangements, the process allowed time to build greater appreciation for space within the institution. The reorganization of ADCOM in 1979 represented only one of a number of important organizational changes for space that took place during the last two years of the Carter administration. These changes reflected pressure to accommodate the advent of the Space Shuttle, as well as the difficulty of achieving consensus on centralization of operational space activities.

The Organizational Prelude to an Air Force Space Command

The *Space Missions Organizational Planning Study* set the stage for a number of changes from 1979 to 1981 that consolidated space activities and emphasized the normalization of space operations. Understandably, most of the key changes involved the space research and development community and the imminent arrival of the Space Shuttle. On 1 October 1979, the day the Air Staff announced the inactivation of Aerospace Defense Command, Secretary Mark officially approved the Air Force decision to split out the functions of the Air Force Systems Command's Space and Missile Systems Organization (SAMSO), replacing it with two organizations—the Ballistic Missile Office (BMO) and the Space Division. Similar to the reorganization of 1961, this change reflected the strain placed on a single organization to manage ambitious missile and space programs, in this case the Missile-Experimental (MX) and the Space Shuttle. Unlike the earlier reorganization, however, the eagerly anticipated Shuttle had promised to produce the expanded "space age" that had failed to materialize in the 1960s.[53]

Already the Air Force had for the first time centralized all launch functions under a single management headquarters, having reassigned the Eastern Test Range to SAMSO on 1 February 1977. Later that year, SAMSO formed a Space Transportation System Group at Vandenberg to prepare for Shuttle launch operations from the Western Test Range. With the activation of Space Division on 1 October 1979, the Air Force renamed the Patrick and Vandenberg Air Force Base range sites the Eastern and Western Space and Missile Centers, respectively, and subordinated them to the newly-designated Space and Missile Test Organization (SAMTO) at Vandenberg, which replaced the earlier Space and Missile Test Center.[54]

The importance of impending Shuttle operations led planners to create two additional organizations. On 1 September 1980, a Space Division Deputy Commander for Space Operations was made responsible for all non-acquisition space functions, including coordination with NASA and the integration and operational support of all military Shuttle payloads. In effect, for the first time, the research and development community separated acquisition and non-acquisition activities, but it did little to clarify the line between experimental and operational systems. As a briefing paper produced by the Air Staff plans directorate stated, "We have recognized that space systems are different from other Air Force systems and have affirmed a much closer relationship between operator and developer for space." The old issue of development or operational priorities and responsibilities remained unresolved. Even for the plans and operations office, which invariably led the effort to have operational commands play a larger role in space activities, operational space systems still seemed to require special consideration from the research and development community.[55]

The new organization also reflected the Shuttle's operational impact and challenges. Since early 1975 the Air Force had realized that NASA possessed inadequate facilities to protect classified data during military Shuttle missions. Although the Sunnyvale Satellite Control Facility handled classified satellite missions, concerns about its capacity and its vulnerability to sabotage or earthquakes made another location desirable. While planners studied the possibility of constructing a new control facility for both satellite and shuttle operations, they decided to establish a temporary secure, or "controlled mode," facility at NASA's Johnson Space Center in Houston, Texas. On 1 June 1979 Space Division activated the Manned Space Flight Support Group to handle Defense Department Shuttle missions and master the complexities of Shuttle operations in preparation for establishment of the Air Force's own Shuttle Operations and Planning Complex. The latter would join a new Satellite Operations Center in what planners referred to as the Consolidated Space Operations Center (CSOC). Meanwhile, the Sunnyvale Satellite Control Facility would report through Space Division's Deputy Commander for Space Operations.[56]

Hans Mark led the Air Force effort to consolidate Shuttle and satellite operations under a new, single management headquarters and, along with NORAD commander General Hill, he emphasized the operational advantages of having the CSOC located in the Colorado Springs area close to Cheyenne Mountain. With strong support from the Colorado congressional delegation, they proved successful in having the Colorado site selected. Construction would begin during fiscal year 1982, with a scheduled completion date of 1985.[57]

General Hill also succeeded in another area of space control. Since early 1978 he had lobbied to replace NORAD's Space Defense Center with a more ambitious, operationally-oriented control center in Cheyenne Mountain. Late in 1979, he received permission to form the Space Defense Operations Center (SPADOC), which would serve as the sole "focal point for national space defense functions." When completed by the mid-1980s, officials expected it to handle not only expanded SPADATS operations, but also to control potential antisatellite countermeasures. With the creation of SPADOC, the Air Force took a another step toward an operational focus for military space.[58]

A final organizational change on the road to a space command occurred when, in September 1981, the Air Staff created the Directorate for Space Operations within the office of Lieutenant General Jerome F. O'Malley, the Deputy Chief of Staff for Operations, Plans and Readiness. Since his arrival on the Air Staff in the spring of 1979, he had been a tireless champion of normalizing space operations in the Air Force. Complementing Hans Mark on the civilian side of the Air Force, General O'Malley provided a crucial, high-level, uniformed voice on operational space issues. His new office, he declared, was to "provide an intensified space focus...and to help reorient USAF philosophy toward an operational approach" by advocating the operational use of space systems at the highest levels of the Air Force. It would, he said, "provide a renewed emphasis that the Air Force plans to stay in the lead in military space operations."[59]

General O'Malley's new office resulted from a recommendation by a study prepared in the summer of 1980 under the auspices of the Scientific Advisory Board. Although another in the long string of studies on space, it proved remarkably influential, even if its immediate impact disappointed the authors. Under the chairmanship of former Air Force Secretary John L. McLucas, fourteen distinguished civilian and military space authorities met in July 1980 to conduct what became known as the *Scientific Advisory Board Summer Study on Space*. The group concluded that, while the Air Force had done good work over the past fifteen years in evolving experimental systems into reliable operational ones, its leaders had only begun to recognize the capability of these systems for military operations. The study, which appeared in August 1980, focused on general deficiencies in the Air Force's ability to perform the space roles outlined in the current draft manual on space doctrine. Technology in space, the *Summer Study* asserted, "does not provide" support to commanders; operational space objectives "are not clearly defined;" space systems are "not integrated" into force structures; and space requirements and employment strategy for operations "are [neither] clearly understood nor fiscally obtainable."[60]

"Inadequate organization for operational exploitation of space" accounted for much of the problem. This, the insightful study asserted, resulted from "a continuing perception that major Air Force commanders do not generally believe that the space program is an Air Force program in which all can take pride, can use to their advantage, can count on, and thus can support." The authors concluded their analysis with a series of recommendations ranging from the importance of operational priorities to the need for inclusion of space systems in an integrated forces architecture of the future. They also urged the Air Force to embrace a long-term "mixed fleet" launch strategy rather than to rely entirely on the Shuttle. This, they argued, would be the best way of ensuring reliable, timely military launches in the likely case that NASA's ongoing Shuttle management and budget problems caused schedule delays and diminished capabilities.[61]

How influential was the *Summer Study*? In the near future the Air Force would develop an integrated space systems architecture that projected developments to the year 2000. A year after the study appeared, the Air Force officially adopted the mixed-fleet approach of using the Shuttle and expendable launch vehicles for the 1980s. General O'Malley found the *Summer Study*'s findings especially important. Not only did he act to establish the Directorate of Space, he worked to have Air Staff space program element monitors transferred from the research and development office to his own Directorate of Operations and Readiness, headed by Major General John T. Chain. General Chain and his staff of space reformers proved to be the driving force for organizational change in support of General O'Malley's efforts. As O'Malley confidently declared, "I believe we can gain MAJCOM support by transferring space systems…into the operational community. If steps are taken at the Air Staff level to normalize space systems, these efforts will eventually permeate the MAJCOMs and the desired pride of ownership will take form."[62]

On the other hand, reflecting on events after the *Summer Study*'s appeared, Secretary McLucas ruefully observed that when he and fellow group member General Bernard Schriever approached the Chief of Staff about presenting the study's findings at the October 1980 Corona conference, General Allen remarked that an already overcrowded agenda precluded discussion of their space analysis.[63] Later, the Chief of Staff lauded the report as part of several efforts, including several recent symposia on space and an important ongoing *Space Policy and Requirements Study*, for helping to focus Air Force efforts on near-term publication of an Air Force doctrine for space. Yet, in the fall of 1980, General Allen, whose strong research and development experience included positions as Deputy Commander for Satellite Programs in the Space and Missile Systems Organization, and Chief of Staff for Air Force Systems Command, again declined to authorize major organizational changes in the area of space operations. He cited the divergent views that emerged from the various studies and symposia,as well as among senior leaders during high-level discussions and conferences.[64]

At the end of the decade, the space community appeared sharply divided over how best to proceed toward more effective space organization. Three positions claimed the broadest support. One favored the status quo and found its strongest supporters among the research and development community centered in Air Force Systems Command and its subordinte organizations. Space Division commander Lieutenant General Richard C. Henry, for example, argued that the organizational changes involving Space Division and the Shuttle were sufficient to insure proper operational space leadership. He stressed the close relationship in the space community between research and development and operations.

> [S]pace is different. Certain functions have to be kept together, specifically the development and building of a spacecraft, the integration of that spacecraft onto its launch vehicle, whether…an Orbiter, or a Titan, or an upper stage of some kind—its launch or orbit, and its on-orbit support. We have…the teamwork within Space Division…that gets the job done. I would be sad to see us forced into, for organizational reasons, the customer-developer relationship that we have today on the airplanes.[65]

Another group, often including the Air Force Chief of Staff, seemed to favor a more centralized operational focus, including establishment of a major Air Force space command, but that group preferred to take a more deliberate, evolutionary approach. A third group, represented by General O'Malley, Hans Mark, and other missionaries for space operations, thought an operational space command long overdue and favored immediate action. At various times General Allen seemed to favor the views of the first or second group but, under considerable pressure, he eventually would come to support General O'Malley's position.[66]

The question by decade's end seemed to center less on whether change should occur than on the proper pace of change. Although the series of organizational changes in 1979, 1980, and 1981 heartened space advocates, momentum for change had yet to achieve a level that promised immediate success. The "organizational prelude" had moved space further along the path toward normalization, but much work remained for General O'Malley and his supporters before they could expect to achieve broad agreement on space policy, doctrine, organization, and operations. Nevertheless, space proponents from the operational side of the house were closer to success than many realized. By the end of the Carter administration, the Air Force and the nation were much farther along the road to achieving consensus on policy and requirements for the nation's space program than surface changes might have suggested. Presidential Directive 37 established declaratory national policy, and Air Force leaders had made substantial progress on policy and doctrinal issues, as well as on determining requirements for future actions with respect to satellite survivability and antisatellite development. The Air Force had moved far toward what Colonel Moorman had termed "socialization for the issues." It remained for the incoming Reagan administration to provide the necessary final momentum. Given the new president's defense agenda, an overly cautious Air Force reaction to change might have found important decisions for space dictated by outsiders.

The Reagan Administration's Plan for Space in 1981

The new president took office in 1981 determined to upgrade the nation's military posture. His Strategic Modernization Plan, which called for major improvements in all offensive and defensive areas, would provide important focus and momentum for change in the Air Force space community. For its part, the Reagan administration initiated the now traditional White House policy review for space, and the Defense Department followed suit.

Meanwhile, General O'Malley spearheaded yet another study to investigate means of broadening Air Force space policy to emphasize operational support. Shortly after the new administration took office, O'Malley assembled thirty people from four major commands and five Air Staff organizations for an intense four-month assessment of the advantages of, and requirements for, using space as a warfighting support medium. Building on the issues raised in the Scientific Advisory Board's 1980 *Summer Study*, General O'Malley's approach stressed the broad operational needs of user commands and the means of providing the necessary capabilities. When published in May 1981, the *Space Policy and Requirements Study* represented the most comprehensive analysis of the Air Force space program to date.[67] The influence of the *Space Policy and Requirements Study* on future organizational and doctrinal developments surprised many who worked on it, particularly since General Allen initially chose not to endorse its recommendations, which asserted Air Force space leadership and advocated a space-based "military" capability. By the end of the year, however, events would compel him to change his mind.[68]

Already by the fall of 1981, General O'Malley had his Directorate of Space functioning under Brigadier General John H. Storrie as the Air Staff focal point for space and space-related plans and operations. By the end of 1981, the directorate was hard at work on an Air Force Space Master Plan and a detailed space surveillance architecture report that would set the stage for the ambitious *Space Systems Architecture 2000* study published in 1983. The influence of the *Space Policy and Requirements Study* also would be seen in the Air Force doctrinal publications issued in 1982 and, even earlier, in the administration's assessment of space policy and programs that commenced in the summer of 1981.[69]

Within the Air Force, the events of late 1981 took on a momentum that threatened to outpace the desire or ability of Air Force leaders to control them. Organizationally, they resulted in agreement to form the Air Force's Space Command as the operational and management focus for Air Force space interests. Clearly, the Defense Department and White House policy reviews, together with the Air Force *Space Policy and Requirements Study*, created widespread support for major policy and organizational changes. Moreover, President Reagan's Strategic Modernization Plan, issued in October 1981, included provisions which called attention to space systems. Throughout the final weeks of 1981, Air Force leaders would seldom refer to President Reagan's October 1981 Strategic Modernization Plan in terms of its importance for or impact on space. Yet, the plan served as a call to action for the Air Force to get its house in order by agreeing on a reorganizational roadmap for future space activities.

From that point momentum increased for an Air Force decision on military space reorganization. Along with the interplay of several important events, the actions of a number of key individuals compelled General Allen to take action leading to creation of an Air Force major command for space operations. For one, Under Secretary of the Air Force Edward C. "Pete" Aldridge, who also served as Director of the National Reconnaissance Office, raised the strong possibility of creating a space command within the Air Force in a speech to the National Space Club in November 1981. Referring to the need for better coordination of space activities, he declared, "I believe the right answer may be some form of a 'space command' for the operation of our satellites and launch services." He proceeded to add that "the Air Force is moving in that direction now." For another, Congress now entered the scene more directly. Not only did Senator John Warner's subcommittee on strategic and theater nuclear forces express more interest in space organizational alternatives, but Representative Ken Kramer from Colorado Springs made Air Force leaders extremely uncomfortable by introducing a resolution calling for the Air Force to rename itself the "Aerospace Force." His House Resolution 5130 proposed that such a force "be trained and equipped for prompt and sustained offensive and defensive operations in air and space, including coordination with ground and naval forces and the preservation of free access to space for U.S. spacecraft." He also called on the Air Force to create a separate space command.[70]

At the same time, the Defense Department policy review committee completed its initial draft space study, which raised the question of an Air Force response on organizational issues affecting space. Could the Air Force continue to remain on the sidelines? Taken together, these events of late 1981 prompted General Allen to act, but General O'Malley took the key step. In December O'Malley called together Generals Jasper Welch, Howard Leaf, Bernard Randolf, and John Storrie, along with four action officers, to discuss a course of action. Acting on their recommendation, General Allen directed the Air Staff to develop a "Space Policy Overview" paper as a "think piece" in preparation for the final Defense Department study results. Air Staff planners responded with a matrix, termed the "Navajo Blanket" because of its color-coded format, which outlined the complete range of space programs, costs, functional responsibilities, the impact of the Shuttle, future implications, and various organizational options. Significantly, the organizational alternatives centered on the five proposed in the 1979 *Space Mission Organization Planning Study*, including an ADCOM initiative to "dual-hat" Space Division's Deputy Commander for Space Operations as Vice CINCAD for Space. The Air Staff had its report ready for General Allen's review in early January 1982.[71]

The Air Force Forms a Space Command

As the President's Strategic Modernization Plan went before Congress in the spring of 1982, Air Force leaders continued to wrestle with the difficult management and policy decisions for the future Air Force role in space. A cautious General Allen steadfastly determined to control the process of centralizing space efforts as much as possible. At the same time, it had always been imperative that key commanders, as well as a broad spectrum of the Air Force community, agree on the course ahead. By early 1982, although the reformers could rely on widespread support for the establishment of an Air Force operational space command, outside pressures made it difficult for Air Force leaders to control events.

In January a General Accounting Office report, produced at the behest of Senator and former astronaut Harrison Schmitt (R-New Mexico), brought additional pressure to bear by castigating the Defense Department for poor management of military space systems. Referring to space as a mission rather than a medium, it called on the Defense Department to create a single manager for space activities and to develop a comprehensive plan for the military exploitation of space. The report specifically suggested that funding be withheld from the CSOC project until the Defense Department presented a logical, effective organizational plan for military space operations. Understandably, Senator Schmitt favored Kirtland Air Force Base in his own state as the permanent site of the CSOC. The Air Staff responded by criticizing the report for considering space a mission, which might require special "space forces," rather than a place where the Air Force could carry out missions and special activities. Although Air Force officials had consistently viewed space as a medium for operations, it had become increasingly difficult to maintain a clear distinction between medium and mission. One might even argue that the establishment of a centralized command for space operations would blur the separation further.[72]

In February 1982, a crucial event occurred that opened the door permanently for operational space advocates. At the Corona South commanders' meeting that month, General Robert T. Marsh, commander of Air Force Systems Command, proposed an evolutionary reorganization concept centered around his Space Division. According to his scheme, which stopped short of a new major command for space, the Space Division commander, General Henry, would also serve under NORAD/ADCOM Commander in Chief General James V. Hartinger as ADCOM's deputy commander for space and would maintain direct links to the Under Secretary of the Air Force by means of a special assistant on the Air Staff. If supporters of a greater operational focus for space disagreed with the proposal for a greater Air Force Systems Command role in space operations, they welcomed the new prominence of the issue in high-level discussions. In the months ahead, the move to establish an operational space command would also benefit from the good personal relations among West Point classmates Generals Hartinger, Marsh, and Henry.[73]

General Allen, fortified with the conclusions of the comprehensive report prepared by General O'Malley's study group, directed Generals Marsh and Hartinger to work together to develop reorganization proposals. Soon thereafter an ad hoc working group composed of representatives from Air Force Systems Command, Aerospace Defense Center, Space Division, and the Office of the Secretary of the Air Force began a two-month effort to reach agreement on potential management initiatives. ADC officers strongly favored a space command that would centralize space management, which at that time was spread among twenty-six major organizations, and would provide a focus for advocacy and wartime use of space systems. With representatives from the research and development community unwilling to sanction a space command, working group discussions proved difficult. Meanwhile, in testimony before Congress, Strategic Air Comand commander General Bennie Davis declared that, "Unless the emphasis regarding operational systems is reoriented, the future prospects for coordinated and meaningful space systems development do not appear favorable. Operational requirements must begin to drive the direction of our technology efforts." In December 1981 General Davis had initiated discussions with ADCOM, the specified command, with the objective of returning defensive space systems to ADCOM's control. Then, in March, Under Secretary of the Air Force Aldridge became directly involved when Air Force Secretary Verne Orr charged him to examine the options for reorganization. In a letter to Vice Chief of Staff General Robert C. Mathis, Aldridge warned of "outside pressures" to establish a space command. In order to control the process, he declared, the Air Force should develop its own alternatives leading to an Air Force space organization.[74]

In mid-April, with the momentum intensifying for an operational management focus, General Allen and senior Air Staff officers received the briefing from the working group. The Air Force Systems Comand representatives presented a more elaborate version of their original plan, which had been outlined at the Corona South meeting. Senior Air Staff officers, notably General O'Malley, objected. During a lull in the discussion, General Hartinger offered the alternative of an Air Force major command for space, at which point General Allen decided that further planning efforts should be directed toward a separate space command. The plans for establishing the new organization would be handled by the Air Staff's Space Operations Steering Committee, then chaired by General Chain, who had succeeded General O'Malley as Deputy Chief of Staff for Plans and Operations when the latter was selected as Air Force Vice Chief of Staff.[75]

By early May 1982, although general understanding existed on centralizing space functions, Air Force leaders had yet to reach formal agreement. In a sense, time seemed to be running out for General Allen if he were to retain control of the decision-making process. As he noted later,

> I concluded...that [a Space Command should be formed] before I left office...since I did have a fairly conservative view of this, that it might be appropriate for me to go ahead and then have as much influence as I could in having the command structure not be overblown but get it underway.[76]

In mid-May Air Force Secretary Orr appeared before Congress to reject Representative Kramer's proposal to rename the Air Force the Aerospace Force. He mentioned that a space command study was underway and nearing completion. Moreover, given the growing congressional and Defense Department consensus for consolidation of space activities, the Joint Chiefs of Staff might become involved and take over the issue. As long as Air Force General Jones remained chairman this would be unlikely, but current planning called for him to retire on 1 June. Then, the more aggressive Navy could very well mount a strong campaign for it to receive the space defense assignment. Furthermore, defense observers expected both Secretary Weinberger and the President to announce the findings of their respective space policy review groups in late June or early July. President Reagan planned to make a major policy address on space, and speculation suggested he might even announce the creation of a separate space service equal to the existing three.[77]

On 21 June 1982, the day before the Defense Department announced the results of its space study, and a few days before his own retirement, General Allen appeared publicly with Under Secretary of the Air Force Aldridge to announce the formation of an Air Force space command that would become effective on 1 September 1982. General Hartinger would become commander of the new Space Command, while retaining his responsibilities as commander of the specified ADCOM and binational NORAD. The Space Division commander would serve as the vice commander of Space Command. As General Hartinger explained, the command would provide

> a focus for centralized planning, consolidated requirements and an operational advocate and honest broker for USAF space systems. We will provide the operational pull to go along with the technology push which has been the dominate factor in the space world since its inception.

At the same time, the Air Force established a Space Technology Center at Kirtland Air Force Base, New Mexico, which consolidated the functions of three key Air Force Systems Command laboratories dealing with space-related research on geophysics, rocket propulsion, and weapons.[78]

The official news release concluded with the intriguing statement, "It is the Air Force's hope and belief that Space Command will develop quickly into a unified command." Why, in view of consistent Air Force opposition to such a move for more than twenty years, did the service suddenly reverse itself? Clearly, in the early 1980s, the argument made by Admiral Arleigh Burke in 1959 and 1960 appeared more valid. At the same time, the price for the Navy's acceptance of an Air Force space command apparently was Air Force agreement to form a unified command. General O'Malley's Space Operations Steering Committee had provided what most observers assumed would be the alternative to ensure Air Force dominance in military space operations while satisfying Army and Navy concerns. The plan for an Air Force space command developed by General Chain's Space Operations Steering Committee called for a joint command, but it purposely left open the question of whether the new organization should be a specified or a unified command. The committee assumed that General O'Malley would understand that a specified command would permit representation from the other services, similar to Aerospace Defense Command in its role as a specified command, while preserving Air Force management and command prerogatives. According to committee members, the proposed public announcement of the new Space Command first went to General O'Malley for approval without any mention of a unified command. They assumed that he understood a specified command would be the logical choice. After a meeting between General O'Malley and Vice Admiral Gordon R. Nagler, Director of Command and Control in the Office of the Chief of Naval Operations, however, the announcement included the final sentence calling for the prompt formation of a unified command. Afterward, committee members lamented that they had not been more specific about the need to select the specified command option. On the other hand, it is equally likely that General O'Malley found that Admiral Nagler's price for cooperation was the unified command alternative.[79]

President Reagan used the occasion of July 4th to announce his new national space policy, which appeared in his administration's National Security Council Decision Directive 42. According to the White House policy statement, the basic defense objectives would embrace strengthening the nation's security, creating a Defense Department-NASA cooperative effort to ensure the Shuttle's use for national security and accord such missions launch priority, and deploying an operational antisatellite weapon. The latter received special mention as a specific program. The president also stressed the importance of satellite survivability and durability, once again highlighting his oft-repeated concern for command, control, communications, and intelligence (C3I) effectiveness. The new policy initiative also created a senior interagency group to implement space policy and "to provide a forum...for orderly and rapid referral" of policy matters to the president. The president's space policy clearly gave more attention to the national space program and the military significance of space.[80]

The new Defense Department space policy complemented the President's national policy by stressing the need for a warfighting capability in space, and it had as its major theme the view of space as a theater of operations rather than a mission. General Charles A. Gabriel, the new Air Force Chief of Staff, incorporated this long-held viewpoint of many space advocates into the Air Force's manual on basic doctrine, AFM 1-1, published in March 1984 and, perhaps more importantly, in the first manual on space doctrine, AFM 1-6, issued in October 1982. Begun as early as 1977, the new doctrinal statement for space seemed to solve, once and for all, the issue of whether space should be considered a medium or a mission. As General Gabriel proclaimed, "space is the ultimate high ground...the outer reaches of the Air Force's operational medium—the aerospace, which is the total expanse beyond the earth's surface. Space, then, is an operational environment that can be used for conducting

Air Force missions." Reflecting the increased emphasis on military space operations, he declared that "the nation's highest defense priority—deterrence—requires a credible warfighting capability across the spectrum of conflict. From the battlefield to the highest orbit, airpower will provide that capability." He then proceeded to outline military interests in space, along with Air Force functions and missions. The latter included various means of performing warfighting missions by way of ground- or space-based weapon systems "consistent with national policy and national security requirements." Emphasis throughout centered on space as a medium that contributed to all Air Force mission areas. Although General Gabriel did not accompany the doctrinal publication with a formal Air Force policy statement to update General Jones' July 1977 letter, Air Staff officers expected one to appear the following year.[81]

At the time the new Air Force space doctrine appeared, Air Staff planners were working with Defense Department officials on several comprehensive plans for space requirements and systems, using as a basis the *Space Policy and Requirements Study* of the previous year. In the spring of 1982, for example, General O'Malley had charged the Air Staff's Space Directorate to chair a study of space surveillance systems, termed *Space Surveillance Architecture Study 2000*, which they expected to complete the following spring. More important, however, proved to be General O'Malley's initiative of December 1982, a comprehensive approach to space policy, the space threat, and systems and technologies titled *Space Systems Architecture 2000*. Chaired by the new Space Command and including participants from the Air Staff, Air Force major commands, the Joint Chiefs of Staff and the Defense Communications Agency, this effort focused on preparation of a plan that would provide the basis for the administration's entire defense modernization initiative.[82]

The Air Force had reached a milestone on the space issue by the end of 1982. On the eve of President Reagan's announcement of the Strategic Defense Initiative in March 1983, which would center on space capabilities, Air Force leaders could rely on doctrine as developed in AFM 1-6, a new Space Command to centralize operational space activities, and broad-based planning efforts then underway to chart the nation's future course in space. At long last, the Air Force had committed itself to an operational space future.

An Ending and a Point of Departure

The creation of an operational space command represented the victory of the "space cadets," those intrepid believers in the central importance of space to the Air Force future. A decade earlier they could not be considered in the mainstream of Air Force policy, plans, and operations. Space, declared its supporters, constituted a medium, a place for enhancing established mission elements, rather than a mission in itself. While this approach proved important for preserving the Air Force claim on space, it failed to attract strong advocacy for space programs or loyalty throughout the Air Force to space as an Force element equal to aeronautics. Indeed, the "aerospace" concept contributed to the fragmentation of a military space community that seemed more comfortable within the world of research and development than in an operational environment.

Over the course of the 1970s the forces for change propelled space into the operational arena. Pressure came from many areas, both in and outside the Air Force. The altered perception of the Soviet military threat compelled even reluctant leaders, like President Jimmy Carter, to focus on space in order to make the nation's defensive posture credible. When political leaders encountered a space community in disarray, they moved to promote strong policy and organizational reforms. The Air Force could not but respond to this kind of outside pressure, but building consensus internally for space required time, patience, and far greater understanding among senior leaders than was forthcoming in the 1970s. By the end of the decade, the studies, the conferences, the committee deliberations, the space symposia, all contributed to the necessary "socialization" process that created an appreciation for space previously absent in the Air Force. While these forces were at work, the space systems themselves demonstrated that they no longer should be considered experimental and part of the research and development side of the Air Force. Above all, the imminent arrival of the Space Shuttle made space operations a leading issue, and the various organizational initiatives of the research and development community could not halt the momentum toward centralized space operations. Pushed from without and buffeted from within, Air Force leaders acted as much to avoid external dictation as they did to direct the elements for change within the service.

If the creation of the Air Force's Space Command served as an end point to the long struggle of the space reformers, it also represented a point of departure, a major step on the road to making space an integral part of the service. The challenges ahead appeared formidable. Possessing few resources, the new command would have to acquire space systems scattered among various Air Force commands, not all of which would gladly relinquish their forces. Effective command relationships would have to be established between Space Command and the research and development organizations, the other services, and most likely in the near future, a new unified space command. Above all, the new Air Force command would have to prove itself worthy of the space enthusiasts who saw in its formation the best means of institutionalizing space within the Air Force.

CHAPTER 6
From Star Wars to the Gulf War:
The Air Force Moves to Create an Operational Capability for Space

Consolidation and crisis marked the decade of the 1980s for the Air Force in the space arena. On the one hand, the newly-created Space Command led the development of an operational focus that involved the shift from consolidating control over space systems to making space systems central to the needs of the warfighter. On the other hand, the space launch crisis at mid-decade led to reexamination of the Space Shuttle's promise and the future military agenda in space. Both developments contributed to the growth and maturity of the operational mindset needed to apply space assets effectively under wartime conditions. By the end of the decade, champions of space could, with justice, point to what they termed the new "operationalization" of space. War in the desert would provide the test.

Buoyed by the new Reagan administration's emphasis on building a strong defense, Air Force leaders anticipated a major effort to develop and apply space systems to meet operational requirements. The Air Force's Space Command would chart the course. Created in late summer 1982, the fledgling command would face a difficult path over the next decade. Although designated the focal point for operational space issues, its experience proved that traditional interests and a fragmented space community could not be overcome immediately. Research and development authorities were especially reluctant to relinquish management responsibility for space systems that they considered best operated by their own more experienced units. Establishing consensus on proper space roles and missions both within and outside the Air Force presented a challenge for space operators—one they had yet to completely achieve by decade's end. The victory of the operators in 1982 provided only an initial achievement in the struggle to move space out of the shadow of research and development and into the realm of the warfighter.

Ironically, the crisis produced by the *Challenger* tragedy in early 1986 created further momentum for an operational space focus. The explosion of the Shuttle led to a nearly three-year hiatus in the nation's space program, during which leaders quickly realized the old truth that one could not have a space program without the means to get to space. The immediate concern centered on space launch, as military officials reexamined the policy of relying on the Shuttle for military space requirements. Their investigations led to reemergence of expendable boosters as the primary launch vehicles for military space systems, and to the end of the Shuttle's promise of routine access to space with manned, reusable space vehicles. A return to the dependable booster, however, did not mean a return to business as usual.

Beyond the issue of space launch, the Shuttle disaster precipitated a widespread crisis of confidence in both the civilian and military space programs. In the atmosphere of self-doubt during the last half of the decade, a variety of studies and reports reassessed the objectives and capabilities of the nation's space program. Of these, the most important for military space proved to be the Air Force Blue Ribbon Panel investigation in late 1988. Distinguished panel members representing all segments of the Air Force gave the panel's recommendations a degree of credibility absent in earlier studies. Their assessment of space policy, the role of the Air Force in space, and of space in the Air Force established a firm basis for the broad process of "normalizing" space, or for gradually establishing the view that space activities were operational rather than developmental in nature. As operational activities, space operations contributed to achieving Air Force missions just as much as more traditional service activities.[1]

Strongly supported by the Blue Ribbon Panel, the movement to normalize and operationalize space in the late 1980s centered on Air Force Space Command.* By the end of the decade, this newest Air Force major command had acquired a considerable number of space-based and ground-based space systems, as well as control of the infrastructure to support them. It appeared well on its way to establishing an effective relationship with the unified space command as well as with other civilian and military agencies in the space arena. Above all, Air Force Space Command achieved a landmark victory in its struggle to assume operational responsibilities performed previously by the research and development community when, in 1990, it won operational control of the space launch mission. Almost equally important proved to be the incremental transfer of satellite control activities to the new

---

* Space Command was redesignated Air Force Space Command on 15 November 1985.

command, which began in 1987. The transfer of space launch and satellite control to the operational command represented a crucial victory in the process of institutionalizing space within the Air Force.

At the end of a decade of consolidation and crisis, the Air Force space program had reached a major milestone in the evolution of military space systems from the developer to the operator. Air Force leaders directed their attention to the needs of the warfighter as they sought to make space launch more responsive and space systems more applicable for tactical commanders. Their achievement would soon be put to the test in regional conflict.

Space Command Sets an Operational Agenda

The formation of Space Command on 1 September 1982, the first major command created by the Air Force in thirty years, represented both an end and a beginning. At long last space advocates had convinced the Air Force community that space deserved representation among the operational commands. In an increasingly complex arena, the ad hoc management methods that had resulted in a fragmented space community could no longer be justified. On the other hand, establishing a space command proved only a point of departure. In late 1982 the new command faced the daunting challenge of acquiring ground- and space-based systems, providing an operational focus for the use of space, and serving as the organization best suited to "sell" space to the Air Force. More specifically, the command's initial mission statement, as described in Air Force Regulation 23-51, dated 25 July 1983, included responsibility to manage and operate space assets, consolidate planning, define requirements, provide operational advocacy, and "ensure the close interface between research and operational users." Generally, the command sought to achieve its agenda by expediting the transition of space systems from research and development to operations, and by increasing the evolution of space system applications from national or strategic requirements to those most appropriate to support theater or tactical warfighters.[2]

Space Command began auspiciously with the transfer from the Strategic Air Command (SAC) in 1983 of fifty space and missile warning systems, bases, units, and upgrade projects.* The initial list included Peterson Air Force Base, Colorado, location of the command's headquarters, as well as Thule and Sondrestrom Air Bases in Greenland and Clear Air Force Station in Alaska. Space Command also would own Falcon Air Force Station, located near Peterson and designated the future home of the Consolidated Space Operations Center (CSOC). By early 1984 SAC also had relinquished four major space systems, two operational—the Defense Meteorological Satellite Program (DMSP) and Defense Support Program (DSP)—

---

* See Appendix 6-1.

and two in the development and acquisition phase—the Military Strategic and Tactical Relay System (Milstar) and Navstar Global Positioning System (GPS).[3]

DMSP. The transfer of the Defense Meteorological Satellite Program from Strategic Air Comand to Space Command in 1983 in itself represented an evolutionary shift from strategic to tactical operational applications. In December 1982, shortly after creation of Space Command, the trouble-plagued program achieved a new level of performance with the successful launch from Vandenberg Air Force Base, California, of the first block 5D-2 satellite on an Atlas booster. An Atlas and apogee kick motor launched a second 5D-2 satellite into proper orbit on 17 November 1983, where its Operational Linescan System telescope performed flawlessly in scanning a swath 1600 nautical miles wide thereby covering the globe in nearly 12 hours. Imagery of cloud cover picked up by the optical and infrared detectors, as well as moisture content, temperature, and ionospheric monitoring data, could be stored for later transmission or immediately downlinked to Air Force Global Weather Center at Offutt Air Force Base, Nebraska, or readout stations, one at Loring Air Force Base, Maine, and the other under construction at Fairchild Air Force Base, Washington, as well as numerous tactical terminals deployed worldwide on land and aboard ships. Real-time data received by the terminals reached field commanders to support tactical military operations. Down-linked transmissions passed directly to the Navy Fleet Numerical Oceanography Center at Monterey, California, prior to their merging with Commerce Department satellite data and, then, went to Defense Department users through a global network of weather stations. Ongoing improvements in subsequent 5D-2 satellites included plans for more reliable inertial measurement units and celestial sensor assembly units, computers with larger memories, and more efficient solar array panels.[4]

At the same time, planners looked ahead to a new type of satellite, referred to as Block 5D-3, which would be designed for launch either on the Space Shuttle or on an expendable booster. Hoping to begin development in 1986, officials worried that designing and building a Shuttle-compatible satellite would delay delivery of the first Block 5D-3 spacecraft by a year. Moreover, funding constraints threatened to delay the next two in the series, which could leave an additional gap in orbital coverage. Planners thus began considering use of refurbished Titan II missiles as launch vehicles. As with the other satellite programs, future progress would depend on the Space Shuttle's development and the solution of ongoing technical and budget challenges.

DSP. Space Command also gained operational control of the Defense Support Program, the central element in the nation's space-based early warning system that monitored missile launches and nuclear detonations. The three operational satellites, each measuring 21 feet high by 10 feet in diameter, contained a telescopic infrared sensor for detecting missile launches, an additional (RADEC I) sensor for nuclear detection, and star sensors for attitude determination. Signal processing electronics within the infrared sensor helped to discriminate between signals representing missile launches and other radiation sources. Computers housed in the system's two ground stations completed the process of signal discrimination. An improved satellite, designated #12, had received a modified star sensor, new power supplies for command decryption units, and an upgraded nuclear detection package. Following deployment of a DSP satellite in early 1984 aboard a Titan 34D/ Transtage combination, future satellites of this kind would be configured for launch by the Space Shuttle.[5]

Milstar. In 1983, Space Command received management responsibility from SAC for the extremely high frequency (EHF) joint-service Military Strategic and Tactical Relay System (Milstar) program, then in the early stages of satellite concept defini- tion and communications terminal development. Defense Department officials planned for Milstar to provide worldwide jam-resistant voice communications for the National Command Authorities and, ultimately, to serve as the main element in the Military Satellite Communications System (MILSATCOM), replacing the Navy's Fleet Satellite Communications System (FLTSATCOM), the Air Force Satellite Communications System (AFSATCOM), and multiuser Defense Satellite Communi- cations System (DSCS) networks. The Air Force contracted through Lockheed Missiles and Space Company for development of the satellite and control system, while MIT's Lincoln Laboratory prepared a Milstar-compatible device for use on Fleet Satellite Communications spacecraft in order to support operational testing of terminals. The Navy supervised terminal development by each of the services to ensure commonality and sufficient logistical support. Air Force planners looked to the Space Shuttle as the future launch vehicle for Milstar in the late 1980s or early 1990s.[6]

Meanwhile, the Defense Department's main long-haul moderate-to-high-data- rate communications satellite system, the super high frequency (SHF) Defense Satellite Communications System (DSCS), also progressed by means of the launch in late 1982 of the first DSCS III satellite, which joined the three DSCS II satellites in geostationary orbit and achieved full operational status in May 1983. The new satel- lite benefited from improved physical and electronic survivability measures, while 21 new AN/GSC-39 medium terminals replaced the obsolete ground terminals, and work continued to convert the entire system from analog to digital transmission by the end of the decade. Although the Air Force retained responsibility for the space or satellite segment, overall management responsibility remained with the Defense Communications Agency rather than being transferred to Space Command. The Army continued its responsibility for the ground segment, which planners expected to improve with the addition of five fixed and six mobile operations centers.

Eventually, DSCS would join the other defense support satellite systems which depended on the Space Shuttle for launch.

Navstar GPS. When turned over to Space Command in early 1984, the Navstar Global Positioning System project was nearing the end of its successful validation phase, during which a limited constellation of five to seven prototype Block I satellites, orbiting at an altitude of 10,900 nautical miles, provided navigation signals transmitted from atomic clocks through a 12-element antenna array to various types of user equipment. The GPS control segment consisted of several monitor stations, a master control station, and ground antennas. Improved Block II satellites for the operational system would have nuclear-protective hardness, longer and more accurate navigation signals, and measures to prohibit unauthorized use. Although Rockwell International had experienced problems building and testing the new satellites, the company still planned to meet the schedule, which called for the initial launch aboard the Shuttle in October 1986 with a Payload Assist Module (PAM-DII) upper-stage vehicle.[7]

By the end of 1987, planners expected GPS to provide worldwide, two-dimensional coverage 24 hours daily and, when fully deployed as a 21-satellite constellation (18 operational spacecraft and 3 spares) in December 1988, full worldwide three-dimensional coverage that would enable users to determine their position to within 15 meters fifty percent of the time and 27 meters ninety percent of the time. By then, the master control station would be functioning in the Consolidated Space Operations Center, which the Air Force began to construct in 1983 at Falcon Air Force Station, Colorado, while a monitor station would be installed at nearby Peterson Air Force Base. The deployed system would rely on three types of user sets already undergoing testing in aircraft, on naval surface vessels, in wheeled and tracked vehicles, and by foot soldiers. The Defense Department hoped future funding would permit the purchase between 1984 and 1997 a total of 23,000 improved user sets that relied on more sophisticated software programming. Although the Air Force served as resource (program) manager, GPS continued as a joint-service program. There were deputy program managers from the Army, which handled the ground segment, and the Navy and Marine Corps, as well as the Defense Mapping Agency, Department of Transportation, and the North Atlantic Treaty Organization (NATO).[8]

Together with the space infrastructure transfers, the four satellite programs provided Space Command a strong initial space system foundation to build upon over the course of the decade. As demonstrated by the command's early experience with the Strategic Air Command (SAC), the effort proved difficult. Despite its willingness to divest itself of missile warning and space surveillance systems, SAC sought to retain a strong operational voice in the control of space systems in the period prior to formation of the unified space command in the fall of 1985. SAC's attempt to preserve an operational hand in the Navstar GPS program, for example, had delayed its transfer to Space Command until the first month of 1984. In fact, during the two years after creation of Space Command, SAC commander General Bennie L. Davis and his staff proposed that resource management for future space-based systems be divided between operational resource management and support resource management. While General James V. Hartinger's command would retain responsibility for support management, the operational issue would be determined by a particular system's mission. Arguing that systems are independent of the "basing" mode and that unity of command should not be violated, Davis and his staff believed offensive-oriented space systems should be subject to SAC's direction while Space Command should retain resource management responsibility for defensive strategic systems. SAC also turned to the traditional Air Force view of the nature of space to argue its case. If space represented a place and not a mission—hence a medium where space assets could satisfy a variety of missions for a number of commanders—Space Command should not attempt to own all space assets in order to perform a space "mission." According to General Davis, SAC, as an operational user, should be accorded basic responsibilities to "advocate, deploy and employ strategic offensive systems in the space environment" through operational resource management.[9]

General Hartinger countered by arguing that the SAC proposal would further fragment the space operations structure, confuse the wider Air Force community, and heighten the "current level of ambiguity." Although the close personal ties between Generals Davis and Hartinger, along with formation of United States Space Command in September 1985, served to alleviate the immediate problem between the two commands, the controversy suggested the difficulties Space Command would continue to face as it moved to consolidate its position as the operational command for the space "mission."[10]

Air Force Systems Command proved to be a more challenging obstacle to Space Command's pretensions to operational space leadership. In this case, the historical role of the research and development command in space operations made it a reluctant participant in the movement to transfer operational control of space assets to the fledgling command. Space Command's mission statement included its responsibility to "ensure close interface between research and operational users," and the appointment of Air Force System Command's Space Division commander as vice commander of Space Command until 1 October 1985 contributed to this end. Yet the larger issue of when the point arrived at which a space system moved from "experimental" to "operational" remained open to debate. Given the complex, unique nature of the space environment and the systems functioning in the medium, Air Force Systems Command questioned the competence of the "inexperienced" operational command and favored lengthy on-orbit checkout procedures and repeated use to achieve "commonality" and consistency of operations before turning over systems to Space Command.[11]

As a result, Air Force Systems Command proved reluctant to hand over satellite control and space launch responsibilities. Not until late 1987 would Air Force Space Command acquire the Air Force Satellite Control Network. The Consolidated Space Operations Center (CSOC) represented the network's primary operational element. Although construction of the CSOC began in May 1983, it seemed an inordinately lengthy process to Air Force Space Command before it became operational in March 1989, two years after the projected date for initial operational capabilities for GPS and DSP. Air Force Systems Command argued that funding, management, and technical problems, together with evolving requirements, accounted for the "delayed" turnovers. In 1986 and 1987, studies of the CSOC's capabilities determined that current and programmmed CSOC facilities and equipment could not support intensive launch recovery operations forecast for the early 1990s. As a result, planners decided to build a new mission control center in space made available when construction of the CSOC's Shuttle Operations and Planning Complex was canceled after the *Challenger* disaster. This requirement further delayed completion of the CSOC. While Air Force Space Command became the resource manager of the Air Force Satellite Control Network in 1987, Air Force Systems Command retained several important responsibilities, including operation of the Satellite Test Center at Onizuka Air Force Station, California. Not until 1993 would Air Force Space Command receive final turnover of the CSOC, thus completing the transfer of all Air Force Satellite Control Network elements and responsibilities.[12]

Space launch would remain the responsibility of the research and development command until the fall of 1990 when Air Force Space Command gained authority to begin a phased takeover. Even then, only strong pressure from Air Force headquarters and Defense Department officials compelled Air Force Systems Command and its Space Division to comply. The space launch issue represented the most intriguing and important element in the development of Air Force Space Command as the operational focal point of Air Force and Defense Department space operations. From the vantage point of 1990, official studies and histories note that Air Force Space Command had focused on acquiring the space launch mission since its activation in 1982. Before the *Challenger* catastrophe, the launch issue created little controversy between Air Force Systems Command and Air Force Space Command. In November 1982 the new operational command received responsibility for Space Shuttle contingency operations. When completed, the CSOC would provide Air Force Space Command not only control of satellite operations through its management of the facility's Satellite Operations Complex but also an active role in Defense Department Shuttle operations through its participation with Air Force Systems Comand in the operation of the collocated Shuttle Operations and Planning Complex (SOPC). Concerned about Air Force System Command's deliberate approach to turning over space systems, Space Command sought and obtained an agreement in 1984 whereby the two commands recognized that Space Command would assume more responsibility for space systems. Air Force Systems Command argued, however, that expendable launch vehicles should not be considered "operational" because each launch involved unique payload and mission demands. As such, space launch did not represent an operational task and should be omitted from the agreement. Space Command concurred. With the Shuttle designated as the primary space launch vehicle for all future Defense Department missions, Space Command expected to gain responsibility for the bulk of the space launch mission with activation of the CSOC. Later, when expendable launch vehicles gained a new lease on life after the *Challenger* tragedy, Air Force Space Command would reopen the issue of space launch responsibility.[13]

A United States Space Command Joins the Space Community

On 23 September 1985, Space Command's position in the military space arena received an additional challenge with the creation of United States Space Command, a unified command for space operations directly responsible to the Joint Chiefs of Staff. When the Air Force announced its intention to establish its own Space Command in May 1982, the official statement expressed the view that "it is the Air Force's hope and belief that Space Command will develop quickly into a unified command."[14] By early 1983, all signs pointed to the imminent creation of a unified operational command for the military space activities of all the services as the "next logical step" to centralize and maximize space operational effectiveness. Yet twenty-five years earlier, Air Force leaders had strongly opposed the Navy's repeated attempts to diminish the growing Air Force space mission by proposing a unified command. Now the two sides had reversed positions. What had happened? Clearly the world of military space had undergone remarkable changes in the previous quarter century. While Air Force responsibility for space by the mid-1980s embraced 70 percent of all Defense Department space systems and 80 percent of the budget, program management had to be shared with the other services, as well as Defense Department and civilian agencies. Moreover, the increasing reliability and effectiveness of second- and third-generation space systems created greater support from a growing user communityfor a single Defense Department organizational focus for space operations. In the final analysis, establishment of a unified United States Space Command proved to be a prerequisite for Navy and Army approval of an Air Force Space Command.[15]

The Defense Department's space policy of June 1982 and President Reagan's national space policy of July 4th of that year focused on ready access to space and the importance of military space by stressing the need to integrate into operational commands survivable space assets that supported tactical applications. The initial impetus of the administration's new policy led to the creation of the Air Force's Space Command on 1 September 1982. At the same time, the Joint Chiefs of Staff polled the warfighting commanders-in-chief (CINCs) for their views on space requirements, while joint exercises in 1982 and early 1983 involved elaborate space scenarios for the first time. Meanwhile, shortly after activation of Space Command, General Hartinger and his staff developed procedures and a rationale for a unified space command that would involve his Air Force major command as the "core" component of the unified command. As such, the Air Force would take the lead in coordinating all American military space operations, and he would serve as commander of both the unified and major commands, as well as the North American Aerospace Defense Command (NORAD).[16]

Above all, the planning and support for a unified space command received a crucial boost from President Reagan's Strategic Defense Initiative (SDI). On 23 March 1983 President Reagan concluded a dramatic speech on national defense by proposing a major national—and later international—program to develop technologies capable of defending against ballistic missiles. In ringing tones he declared, "I call upon the scientific community in our country, those who gave us nuclear weapons, to turn their great talents now to the cause of mankind and world peace, to give us the means of rendering these nuclear weapons impotent and obsolete." If achieved, gone would be the 1960s doctrine of Mutual Assured Destruction which relied on massive nuclear retaliation as the ultimate deterrent. In its place, Reagan proposed Mutual Assured Survival, a "positive" alternative strategy based on strategic defensive systems capable of destroying ballistic missiles in flight, leading to the objective of eliminating the threat of ballistic missiles entirely. The proposed change in the nation's space policy represented an enormous break with past developments because, if accepted and funded by Congress, it would permit weapons in space.[17]

To some, the Strategic Defense Initiative, as the administration eventually termed the President's proposal, appeared visionary. Others found it naive and more suitable to the "Star Wars" label it quickly received, suggesting a saga out of science fiction, as in the 1977 motion picture of the same name. In any case, SDI clearly turned the spotlight of attention on strategic aerospace defense in unprecedented fashion. President Reagan's speech had an electric effect on the space community. Because SDI would clearly be dependent on space-based systems, it compelled officials to review the entire role of space in military operations. In effect, SDI provided additional incentive and broader support to proceed with a unified space command, which seemed the sensible organization to become the operational focus for SDI planning and systems operations.

When the Joint Chiefs of Staff in April 1983 requested suggestions for the best organizational means of supporting SDI, General Hartinger responded with his proposal for a unified command. He immediately realized the potential of SDI to enhance the importance of his command, and he hoped Space Command would become responsible for Air Force participation in the test program. In June, Air Force Chief of Staff General Charles Gabriel concurred on the need for a unified command. The Navy's decision to activate its own space command on 1 October

1983 served to increase support, although the Navy itself remained generally unenthusiastic about a unified structure that would be dominated by the Air Force.[18]

Early in 1984 General Gabriel and Air Force Secretary Verne Orr, in a joint statement, reaffirmed Air Force support for a unified command by asserting that "no single military organization exercises operational authority over military space systems in peace, war, and the transition period from peace to war." Late that year President Reagan approved the recommendation from the Secretary of Defense and the Joint Chiefs of Staff. Following extensive studies on roles and missions, the United States Space Command was activated on 23 September 1985. Appropriately, on hand for the ceremony was retired Admiral Arleigh Burke, who had unsuccessfully championed the cause of a unified command in 1959 and 1960.[19]

The Growing Conflict Over Space Roles and Missions

As proposed by General Hartinger, the arrangement also called for the unified commander-in-chief to serve as commander of Air Force Space Command and commander-in-chief of NORAD. From the start the command structure created tension and raised issues similar to those that earlier beset the Air/Aerospace Defense Command.[20] As NORAD commander-in-chief, General Hartinger needed to deal with a Canadian partner that had never been comfortable with SAC's control of "defensive" space assets from 1979 to 1982 and, now, had grave reservations about its own role in the Strategic Defense Initiative. Moreover, the unified command received operational control of the missile warning and space surveillance missions, which meant that its personnel exercised peacetime as well as wartime control over Air Force space assets in the Cheyenne Mountain Complex's Space Surveillance and Missile Warning Centers. The issue of peacetime control remained relatively unimportant as long as the same individual headed the unified and major Air Force commands; but in October 1986, the Air Force elected to separate leadership of the commands, leaving Air Force Space Command with a two- rather than four-star general and without responsibility for day-to-day operation of crucial space resources. As a result, the space roles and missions debate would resurface with a vengeance in the last half of the decade as the Air Force sought to redefine its institutional commitment to space.[21]

The saga of the Air Force Space Plan, as well as various other doctrinal and mission statements, also reflected tension between Air Force Space Command and Air Force Systems Command, specifically, and within the Air Force, generally, as the space community attempted to develop a uniform approach to space operations. Just over a year after its activation, on 18 November 1983, Space Command accepted custodianship of the Space Plan, the first approved by the Air Staff since the early 1960s. This seemed entirely appropriate given the command's mission responsibility to "consolidate planning…, define requirements…, and provide advocacy…for Air Force space issues." The Air Staff viewed the document as a comprehensive, integrated long-range planning effort involving space activities, missions, and operations. It would serve to justify a future space investment strategy that would ensure continued procurement and funding support. Air Force leaders also considered the plan an educational tool that embodied corporate thinking on space and, thus, could help institutionalize space within the service.[22]

Planners hoped to update the Space Plan periodically to reflect the evolving space community. When Space Command received the plan in 1983, it became responsible for 21 of the required 37 actions to implement the document. By the end of 1984 the command had completed 10 actions.* Although most requirements could be completed without difficulty, Space Command repeatedly failed to reach agreement with the Air Staff and U.S. Space Command on interpretation of appropriate mission area functions. What appeared to be minor differences over space operational terminology in fact represented profound disagreement on proper roles and missions, as well as widespread uncertainty on the role of space in the Air Force. The document, which was expected to help unify the Air Force on space, actually became more of a hindrance.[23]

The Air Force Space Plan described the general uses of military space and identified four specific terms for space operations. "Space control" involved maintaining freedom of action in space and denying the same to the enemy. "Space support" referred to the deployment, maintenance, and sustenance of equipment and personnel in space, primarily by means of space launch and on-orbit repair or recovery. "Force enhancement" referred to traditional defense support functions such as communications, navigation, and weather designed to "enhance" terrestrial and space-based forces. "Force application" referred to the performance of combat functions from space.[24]

Air Force Space Command asserted that the use of this terminology in Defense Department space policy and in the Air Force as a whole differed in key respects from policy followed by U.S. Space Command, which relied for guidance on JCS Publication Number 1 and the Unified Command Plan. The unified space command focused on two mission areas, space control and space support, and subsumed under these areas force enhancement and force application. According to U.S. Space Command, space control involved all aspects of the space defense mission, including force application, while space support involved force enhancement functions. Air Force Space Command's staff especially opposed the unified command's interpretation of space support. The major command's planning chief, Brigadier General G. Wesley Clark, for example, explained that for U.S. Space Command, the space support function included support to terrestrial forces, an employment function that rightfully fell within its area of responsibility. It

---

* See Appendix 6-2.

also involved, however, the preparation, maintenance, and sustenance of space forces, which was a space service support function that properly belonged to the Air Force and should be assigned to Air Force Space Command. To clarify the situation and to avoid promoting the wrong perception of the nature of space operations, General Clark proposed that space support be subdivided into "space combat support" and "space service support," with U.S. Space Command responsible for the former and Air Force Space Command the latter, which would also involve coordination with NASA.

In effect, Air Force Space Command proposed modifying the traditional mission functions with special terminology to account for the unique nature of operating in space. Neither U.S. Space Command nor the Air Staff, however, proved amenable to the changes. By 1985 the Air Force Space Command staff successfully incorporated into its draft revision the results of various studies, such as *Space Systems Architecture 2000*, operational intelligence and antisatellite plans, satellite control architecture, and a military man-in-space plan that examined military roles for the Shuttle's Spacelab program. Nevertheless, the Air Force Space Command's Space Plan repeatedly failed to gain Air Staff approval. Likewise, disagreement over space terminology plagued every effort by Air Force Space Command to achieve consensus on space operational doctrine and a revised command mission statement. The different interpretations of space terminology reflected the larger issue of appropriate command responsibilities that continued to divide the parties. Indeed, throughout the 1980s all attempts to update the Space Plan, revise the command's mission statement, and publish operational space doctrine floundered. The failure to produce a revised Space Plan suggests the difficulty of reaching consensus within the Air Force space community, which sought to make space an accepted "mission" throughout the service.[25]

Nevertheless, by mid-decade, space operators could point to major achievements in the establishment and growth of both Air Force Space Command and the U.S. Space Command. To be sure, command relationships needed sorting out and a reluctant Air Force Systems Command would require considerable prodding before relinquishing its traditional hold on space systems. Even so, Air Force space leaders had good reason for optimism in the era of the Space Shuttle. After 1986, however, Air Force space issues would be played out against the background of the *Challenger* tragedy, which forever altered the landscape of future national space operations.

The *Challenger* Disaster Creates an Uncertain Launch Future

NASA had expected a triumphant but routine mission of the orbiter *Challenger* on 28 January 1986 in celebration of the Space Shuttle's twenty-fifth flight. Initiating use of the nation's second Shuttle pad at the Kennedy Space Center, Mission 51-L was to launch the "first teacher in space," Christa McAuliffe, perform unprecedented observations of Halley's Comet, and deploy one of the space agency's

Tracking and Data Relay Satellites. After cold weather delayed the flight for several days, the *Challenger* rose from its launch site that January morning at 11:39 a.m. Eastern Standard Time. Just 73 seconds after liftoff, a massive explosion destroyed the spacecraft, killing all seven crew members and plunging the nation's space program into the greatest crisis in its young history.[26]

While the nation justifiably focused on the *Challenger* tragedy, military space officials had additional worries. In early 1986 the Air Force had only begun to recover from the failure in August 1985 of its Titan 34D rocket, which had to be destroyed when one of its engines shut down after liftoff and the rocket veered off course. Then, in April 1986, another Titan 34D exploded over its launch pad at Vandenberg, and in May NASA lost a Delta rocket. After those launch vehicle failures, space leaders effectively grounded the space program by prohibiting further flights of the Shuttle and expendable launch vehicles (ELVs) until the problems could be solved. The nation confronted an ailing space industry and a space program in disarray. President Reagan appointed a commission chaired by former Secretary of State William P. Rogers to investigate the *Challenger* accident. Among other findings, the commission's exhaustive report, issued on 6 June 1986, concluded that defective seals between two solid-rocket-motor sections sparked the chain of events that produced the explosion. NASA had much work to do before confidence in manned spaceflight could be restored.[27]

Without an assured heavy-lift launch capability, the military space program also found itself in crisis. The Shuttle had been designated the primary launch vehicle for all future Defense Department payloads, and the Titan 34Ds had been scheduled only until the Shuttle achieved its full flight schedule in the late 1980s. The Air Force expected to run out of expendable boosters sometime in 1988. Programs most immediately affected by the grounding of the Shuttle would be the Navstar Global Positioning System (GPS) and the early warning Defense Support Program, although others would suffer from launch delays and the related "ripple" effect. Payloads previously manifested for the Shuttle would remain in storage rather than replenish aging satellite constellations. There, while expensive investigations continued, they would generate a high cost while officials worried about potential atrophy and projected booster replacements.[28]

The *Challenger* accident proved to be a watershed in the nation's space program. The moratorium on Shuttle flights, which extended for 31 months, forced civilian and military leaders to investigate not only the future of space launch but the nation's entire space program. During the hiatus Air Force officials led the way in reassessing the military space program. By the time the Shuttle resumed operations on 29 September 1988, the Defense Department's relationship with NASA had been transformed and the Air Force had immersed itself in a searching self-examination of its commitment to space.

The *Challenger* tragedy had not caught the Air Force totally unprepared. Several years earlier, doubts about relying exclusively on four very complex space launch vehicles had prompted Air Force officials to pursue a "mixed fleet" concept of complementary expendable boosters. Indeed, the Air Force had never been comfortable with the decision to rely entirely on the Shuttle for space launch. Back in the mid-1970s, writers on Air Force issues noted that earlier "resigned acceptance" of the Shuttle as the space transportation system for both civilian and military users had evolved into "cautious enthusiasm." After all, Shuttle proponents predicted routine, high-capacity, fast-turnaround access to space with a schedule of 60 flights per year (40 at the Kennedy Space Center and 20 at Vandenberg Air Force Base) at half the cost of expendable boosters. The Shuttle also promised to preserve a manned, military presence in space and achieve the long-sought goal of normalizing space operations through standardized, reusable launch vehicles. To maintain funding and political support for the Shuttle, NASA officials insisted the Defense Department commit to a "Shuttle-only" policy and phase out its fleet of expendable launch vehicles. The Defense Department agreed.[29]

A 14 January 1977 Memorandum of Understanding (MOU) between NASA and the Air Force, as the Defense Department's executive agent for the Shuttle, formally confirmed Shuttle program responsibilities. NASA would be responsible for Shuttle development, flight planning, operations, and control, regardless of the user, as well as landing-site arrangements at the Kennedy Space Center and overall financial management. The Air Force, for its part, would develop a controlled node at the Johnson Space Center for classified missions, and supervise integration of military flights, construct a second launch facility at Vandenberg Air Force Base, and build an inertial upper stage (IUS) vehicle, a two-stage solid-propellant upper stage carried into orbit in the Shuttle cargo bay, to lift payloads from the Shuttle to higher altitudes and inclinations. NASA expected to use the IUS for its ambitious planetary missions. For all intents and purposes, military space launches would be accomplished exclusively by the Space Shuttle. The reusable Shuttle would make the expendable launcher truly expendable once and for all time.[30]

NASA initially expected to begin test flights in 1980. By the spring of 1979, however, agency officials had slipped the initial operating date to early 1981 in light of technical problems and related cost increases. The technical challenges associated with the Shuttle's complex design and payload configuration proved more difficult to master than expected. The Defense Department became alarmed that further delays would result in an unresponsive space launch program and a diminished operational flight schedule in the next decade. Critics increasingly faulted NASA's research and development mentality and called for more military involvement in Shuttle management. Military concerns prompted Carter administration officials in 1978 and 1979 to conduct high-level policy reviews, which led in March 1980 to a modification of the 1977 NASA-Defense Department agreement. The revised accord sought to accommodate the military by assigning priority to the Defense Department in Shuttle mission preparations and flight operations, and by integrating Defense Department personnel more directly into NASA's line functions.[31]

Despite a tilt in the Defense Department's favor, the Air Force remained uneasy about its commitment to a Shuttle-only policy. In 1980 both the Air Force Scientific Advisory Board and the Defense Science Board addressed the space launch issue. Citing Shuttle delays, the likely lack of an "on-call" launch capability, and the general austerity of space launch assets, the two boards proposed a "mixed fleet" policy of using both the Shuttle and expendable boosters for military payloads. At this time officials remained uncertain whether the mixed fleet concept should become a permanent policy or only be pursued until the Shuttle proved capable of fulfilling its early promise of routine spaceflight.[32]

Meanwhile, the Air Force had decided to use the Titan 34D as its heavy-lift booster during transition to the Shuttle, while the IUS would be configured for both Titan and Shuttle vehicles. By 1982, however, NASA had backed out of the IUS joint purchase arrangement with the Air Force, which meant higher costs for the Air Force vehicle. Worried that the IUS two-stage vehicle would be underpowered for planetary missions, NASA expressed renewed interest in the liquid-propellant Centaur G, the most powerful upper-stage vehicle in the space arsenal. NASA's flip-flop on its commitment to the IUS provided ammunition for critics of the civilian agency's competence and management practices.[33]

By the early 1980s, NASA had further lowered its Shuttle flight predictions from a planned 14 launches in 1984 and 24 per year by 1986 to 5 in 1984 and 13 in 1986. A General Accounting Office (GAO) investigation in 1982 noted that the earlier 1977 projected schedule of 487 flights during the first twelve years of operation had been reduced by more than 50 percent to 234. Although the successful maiden flight of the Shuttle in April 1981 eased some of the tension between NASA and the Defense Department, Air Force leaders still were concerned about phasing out expendable launch vehicles once the Shuttle became operational.[34]

In October 1981 Air Force Chief of Staff General Lew Allen formally identified as a problem the total reliance on the Shuttle and called for study of a "mixed fleet" strategy. The following month Under Secretary of the Air Force and NRO director Edward C. "Pete" Aldridge, who would become a central figure in the space launch arena throughout the decade, appeared before the National Space Club in Washington, D.C., to give a "my views only" assessment of military space issues. Calling for a "new management structure for our space operations," he asserted that the Air Force "cannot continue to look to NASA as our country's Launch Service Organization in the Shuttle era." Although he cited as positive the appointment of Major General James A. Abrahamson as NASA's Associate Administrator for Space Transportation Systems, he argued that the space agency should focus on "developing civilian space assets and transportation systems" and consider leaving operational responsibilities to others. The under secretary also appeared to favor retention of expendable launch vehicles even after the Shuttle became fully operational. He observed, "It…seems illogical that our only 'truck' to deliver our goods to space be in the form of 3, or 4, or 5 highly complex launch vehicles. Fleet grounding, launch failures, or both could severely limit our access to space." Aldridge noted that new presidential science advisor Jay Keyworth had undertaken a study of the need for a mixed-fleet concept.[35]

Although President Reagan's national space policy statement of 4 July 1982 reaffirmed the Shuttle as the primary launch vehicle, the Air Force sought in 1983 to ensure a sufficient supply of expendable boosters. It officially proposed a mixed-fleet program based on commercial production of the Titan III, along with the purchase of additional Titan 34Ds and refurbishment of Titan II ICBMs. The latter would be used for launching DMSP payloads. The Titan 34D, nearing the end of its scheduled availability, however, could provide only an interim solution, because it could not match the Shuttle in launch weight and volume capacity. Moreover, NASA elected to modify only two of the four Shuttles to handle heavy Defense Department payloads. By the end of 1983, Under Secretary Aldridge, proclaiming the need for "assured access to space," outlined growing Air Force support for the additional step of developing an upgraded Atlas, termed the Atlas II, and a more powerful Martin-Marietta Titan. The latter vehicle would consist of a 200-inch payload fairing to handle a Shuttle-configured Centaur upper stage and a Shuttle-configured payload; it would possess the capability of launching 10,000 pounds into geostationary orbit. Initially referred to as the Titan 34D7 because of its 7 rather than 5° segmented, solid-rocket motors, it soon became known as the Complementary Expendable Launch Vehicle (CELV), then later the Titan IV.[36]

By early 1984 the Defense Department had accepted the Air Force position. A "Defense Space Launch Strategy" statement, issued on 23 January, declared that:

> while affirming its commitment to the STS [Space Transportation System], DoD will ensure the availability of an adequate launch capability to provide flexible and operationally responsive access to space, as needed for all levels of conflict, to meet the requirements of national security missions.[37]

In support of an "assured access to space" policy, the defense secretary approved the Air Force plan to procure 10 Titan 34D7s, or Complementary Expendable Launch Vehicles. The Air Force hoped to see the CELVs enter the inventory by 1988 to support a schedule of two launches per year.

NASA officials found themselves on the defensive, pleased with neither the prospect of a competitive booster nor the growing criticism of its relationship with the military. Critics inside and outside Congress had been castigating the "militarization" of the Shuttle program for several years. The civilian agency, they asserted, had signed a "pact with the devil" by according the military priority on the Shuttle manifest, by placing active military officers in key NASA posts, and by supplying the bulk of development funding. In response, however, NASA defended its relationship with the Defense Department. Glynn Lunney, manager of the Space Shuttle program at Johnson Space Center, even favored strengthening the already close ties. In late 1983 a General Accounting Office report examined NASA-Defense Department funding disparities and recommended Congress withhold support for the Shuttle Operations and Planning Complex (SOPC) at Falcon Air Force Station near Colorado Springs until the Defense Department and NASA developed effective, long-term operational objectives. In response, Lunney defended the SOPC as strengthening the "separateness" of military and civilian space activities. He also saw nothing amiss in NASA's funding of the "national" space transportation system. From his standpoint, the military earlier had gained experience with unmanned space systems but had neglected manned spaceflight. "It is now time," he asserted, "for the DoD to fully embrace and exploit the manned spaceflight capabilities which NASA has developed for our nation." Doing so would put the military squarely behind the Shuttle. In early 1984 NASA officials fervently lobbied against the Complimentary Expendable Launch Vehicle because, they said, it would result in lower Shuttle flight rates and higher costs.[38]

NASA had another reason for concern when Under Secretary Aldridge called for commercial production of expendable launch vehicles as a means of providing the Defense Department more affordable backup boosters. Commercial ELV production would infringe on NASA's Shuttle marketing operation. In the early 1980s, when the European Space Agency's successful marketing of the Ariane rocket threatened to corner the commercial satellite market, NASA received permission to promote the Shuttle commercially at artificially low prices. The American ELV industry, meanwhile, had been blocked from commercial competition and, subsequently, had suspended production in light of the military's Shuttle-only policy. NASA expected to recoup its costs later in the decade through cost-effective commercial operations, but it had based its planning on erroneous estimates of yearly flights, without accounting for such vagaries as mechanical difficulties, weather delays, and slow turnaround procedures. After four orbiters and six years of operation, *Challenger*'s January 1986 mission had represented only the twenty-fifth orbiter flight. At the same time, the producers of satellites had proceeded on the assumption that future flights would be cheap and frequent. By 1984 the Reagan administration had become sufficiently concerned about the likely shortfall in NASA's commercial operations to pass the 1984 Commercial Space Launch Act, which sought to ease the cumbersome, bureaucratic launch process by centralizing all commercial launches under the Secretary of Transportation. At the same time, the act also tended to move NASA out of the private launch business.[39]

Despite NASA's objections, the Air Force went ahead with a contract in February 1985 for development of the Titan IV. As Under Secretary Aldridge declared, "we cannot have our access to space as 'fragile' as it will be without ELVs complementing the Shuttle." In August 1985, the administration confirmed the decision through a National Security directive titled "National Space Strategy," which authorized a limited number of ELVs as part of the mixed-fleet approach to support "assured access to space." By that time NASA had dropped its objection to the Air Force's procurement of ten Titan IVs in return for a Defense Department commitment to book one-third of all forthcoming Shuttle flights.[40]

Before a Joint Subcommittee on Space Science and Applications in July 1985, a number of prominent military space figures addressed the subject of "Assured Access to Space During the 1990s." Congressional officials wanted to know whether the space leaders favored production of a fifth Shuttle orbiter. General Abrahamson, now head of the SDI program; Lieutenant General Donald J. Kutyna, Air Force Director of Space Systems and Command, Control and Communications; and General Robert T. Herres, commander-in-chief of NORAD and commander of Space Command, argued for a limited ELV program and against an additional Shuttle orbiter. Noting that the early decision to rely on the Shuttle had left little funding over the years for launch-related technology, they supported an advanced launch system technology program to replace the Shuttle by the turn of the century. Air Force Under Secretary Aldridge agreed when he testified before the subcommittee. He expressed concern about the Shuttle's ability to support all scheduled Defense Department flights in addition to NASA's domestic and foreign commitments. Aldridge declared that, assuming no major delays, four orbiters could likely meet the Defense Department's expectations—but only with programmed Titan IV and Titan II payloads as part of the launch plan.[41]

Moreover, the precise heavy-lift requirements for the Strategic Defense Initiative and NASA's proposed space station were yet to be determined. The technological initiative for development of a new expendable launch system drew increasing support following an Air Force Space Command study, the *Space Transportation Architecture Study*, which concluded that payload requirements involving SDI and the space station would likely exceed booster capabilities in the late 1990s. The new launcher proposal, referred to as the Advanced Launch System (ALS), incorporated the Strategic Defense Initiative Organization's requirement for a heavy-lift vehicle capable of launching 150,000 pounds into low-earth orbit. The Air Force also expressed interest in such a vehicle, which would have three times the lifting capacity of the Space Shuttle. By the end of the decade, the ALS program would be restructured to promote new booster technology for a variety of requirements.[42]

On the eve of the *Challenger* disaster, the Shuttle remained the centerpiece of America's space launch program. Although the Air Force's commitment to the Shuttle as its primary launch vehicle had been tempered by diminishing expectations, it hoped that the addition of a limited number of "mixed fleet" expendable boosters would aid in realizing the Shuttle's lofty promise. The foresight and concern of Air Force leaders helped cushion the shock of the *Challenger* and Titan losses at mid-decade.

The Response to the *Challenger* Shock Waves

In the aftermath of the *Challenger* disaster and the expendable booster failures, the nation's launch activities came to a near standstill while officials awaited the Rogers Commission report amidst widespread soul-searching and public criticism. A variety of "experts," with the benefit of hindsight, claimed to have foreseen the disaster and the policy failure that led the nation to rely solely on the Shuttle for America's space launch future.[43]

During the moratorium on Shuttle flights, NASA conducted political damage control and turned to the military for assistance. As part of its recovery plan, NASA appointed Admiral Richard H. Truly as Associate Administrator for Space Flight, Space Division Commander Lieutenant General Forrest S. McCartney as Director of the Kennedy Space Center, and also turned for advice to its former deputy director of the Apollo program, General Samuel C. Phillips. Not only did NASA specifically request help in a variety of areas, it agreed that military missions should take precedence on future Space Shuttle flights. It also agreed to a temporary mixed-fleet space launch policy. At the same time, the administration ordered NASA out of the commercial launch business, which opened the door to a resurgence of the expendable launch vehicle business.[44]

Moreover, when the Rogers Commission report appeared in June 1986, it advocated a Space Shuttle with lower weight and payload capabilities and a conservative launch schedule. The Air Force interpreted this as more reason to focus on dependable unmanned boosters, and worked to find launch vehicles for its delayed inventory of satellites. As the Shuttle launch schedule showed increasingly lengthy delays, the Air Force estimated that as many as 25 payloads would be affected and that the launch backload could not be overcome before 1992. As the situation unfolded, satellites currently in orbit would help by functioning well beyond their original design lifetimes. Nevertheless, the launch delay created a major challenge that would leave nearly a three-year gap without alternative launchers and would raise important questions about the future of the nation's space industrial base.[45]

Most seriously affected were the operational Global Positioning System (GPS) satellite constellation, the early warning Defense Support Program (DSP), and the satellites controlled by the National Reconnaissance Office. Defense Department planners had programmed these payloads exclusively for Shuttle launches. The Air Force moved immediately to reinforce its expendable launch arsenal. By July 1986 the Air Force had recommended producing an additional 13 Titan IVs, as well as 12 new medium-launch Delta II vehicles to help perform GPS flights beginning in 1989, two years behind schedule. The Delta II proved to be the only booster that resulted directly from the Shuttle crisis. The Air Force expected to launch DSP satellites on the Titan IV, Defense Meteorological Satellite Program payloads on Titan IIs, Defense Satellite Communications System satellites on Atlas IIs, and the future Milstar on Titan IVs. At the same time, the service strongly supported Advanced Launch System studies designed to determine a successor launch vehicle to the Shuttle and Titan IV.[46]

The Air Force's decision to focus on expendable launch vehicles seemed more credible when NASA announced in May 1987 that Shuttle flights would resume in June rather than February of 1988 and would be limited to 14 instead of 24 per year. Moreover, only lighter payloads would be flown. Under Secretary Aldridge responded by calling for an additional 25 Titan IVs, Titan launch pads, and 5 to 10 more Delta II medium launch vehicles. The under secretary also defended his new space launch budget that would be doubled by the early 1990s. Although military missions would receive priority once the Shuttle resumed flying, eighteen of thirty-six previously manifested payloads for the Shuttle would be reprogrammed for expendable launchers. After 1992, however, the Defense Department would use the Shuttle only for SDI or research and development missions. In effect, the Air Force would abandon the standardized Shuttle, the "airliner to space," for the diversification represented by expendable boosters. At the same time, no one wanted to resort to business as usual and to the practice of linking specific satellites to particular launch vehicles, which required months of prelaunch preparation. Emphasis now would be on developing an "assured launch strategy" highlighted by lower costs and greater launch responsiveness.[47]

While space launchers remained grounded and public questioning of the future direction of the space program continued into 1987, the White House initiated a new review of national space policy. The Air Force also undertook a comprehensive reassessment of its role in space. In the spring of 1987 the Secretary of the Air Force produced an important "White Paper" on Air Force space policy and space leadership. The paper took as its point of departure the 1983 policy letter from then Chief of Staff General Gabriel that claimed Air Force responsibility for most of military space. This claim, according to the White Paper, had not been fulfilled, and the defense community perceived that the Air Force only grudgingly supported space activities. As a result, the nation faced a void in space leadership at a time of growing Soviet space presence, and the Air Force had failed to "exhibit a sense of institutional purpose or responsibility toward space." In short, space had been relegated to fourth priority in the service behind the strategic, tactical, and airlift missions.[48]

Because outsiders perceived a lack of support for space within the Air Force, they raised challenges to the Air Force's role as executive agent for military space. The Office of the Secretary of Defense, for example, retained a dominant voice in the acquisition area through the DDR&E, while U.S. Space Command and the Strategic Defense Initiative Organization advocated space survivability and surveillance requirements, and the Army and Navy worked on space master plans of their own. The White Paper's authors posed a central question: did the Air Force wish to act as the lead service for space? They declared that the answer should be "yes" because of the service's space expertise and especially the potential of Air Force Space Command for operational leadership. At the same time, however, the Air Force had neither a mission statement for space nor a current space operations doctrine, and its operational space command could not play a strong advocacy role throughout the corporate Air Force and Defense Department because its leader was only a two-star commander.[49]

The White Paper suggested specific actions the Air Force should take to lead the military space community. It should develop a new policy statement that reasserted the Air Force claim as "lead" service for space and should work to revise Defense Department Directive 5160.2 on service space responsibilities. Leadership did not mean an "exclusive" Air Force space role, the paper said. Rather, the service should establish a formal structure to ensure that it met the needs of the other services. Within the Air Force, a corporate commitment could be developed by means of expanding space infrastructure and supporting the SDI and "military-man-in-space" missions. Finally, the Air Force should upgrade the commander of Air Force Space Command to three-star rank and work to increase the interaction among the operational command, Air Force Systems Command, and the Air Staff. The Air Force secretary's White Paper reached a wide audience and provided important impetus to the establishment the following year of the important Blue Ribbon Panel on Space Roles and Missions. Meanwhile, a few months before the White Paper appeared, space operations advocates received a new champion in the person of General John L. Piotrowski, appointed to head U.S. Space Command on 6 February 1987.[50]

General Piotrowski Champions Operational Space

The arrival of General Piotrowski signaled the advent of three years of strong leadership in a variety of operational space areas. His initiatives and actions had significant impact on the thinking and development of Air Force space activities. As commander-in-chief of the unified command, Piotrowski sought to bring an operational focus to the space mission, much of which was accomplished by involving Air Force Space Command, the unified command's largest component. He represented as well a symbolic shift in leadership of the unified command. While his predecessor, General Herres, focused primarily on developing an effective organizational framework, General Piotrowski made it his mission to stress the needs of the warfighter and the importance of normalizing military space operations. As he explained, it was absolutely essential that the unified and specified commanders-in-chief, the Joint Chiefs of Staff, and Defense Department leaders develop an "operational mindset for the use of space." This would reflect the "natural process of maturing space operations from a research and development orientation to an operational mode for the employment of US space-based resources."[51]

General Piotrowski used as a springboard the new Defense Department Space Policy that Secretary of Defense Caspar Weinberger signed on 4 February 1987. The new policy affirmed that the Shuttle would no longer be designated the primary launch vehicle for military missions. The nation must develop an assured space mission capability through balanced launch assets and more survivable systems. Moreover, the military should develop an operational antisatellite weapon system, take advantage of civil and commercial space assets, and promote advanced launch technology. Above all, the Defense Department must "provide operational capabilities to ensure the US can meet national security objectives" by focusing on the mission areas of space control, space support, force enhancement, and force application. The Joint Chiefs of Staff called on the new commander of U.S. Space Command to assess current programs and required actions. Although Piotrowski used his position to advocate a variety of improvements in space infrastructure, his attention centered on space launch and future operational payload requirements that would support theater and tactical commanders.[52]

General Piotrowski believed that the Air Force needed to make radical changes in two areas of space launch—payload manifest procedures and launch responsiveness—in order to make operational priorities the driving force. U.S. Space Command and Air Force Space Command, for example, played only a minor role in launch manifest arrangements. From the 1970s the Defense Department Space Shuttle User Committee had essentially "rubber stamped" payload manifest schedules determined by Air Force Systems Command's Space Division. In September 1985, the redesignated Defense Department Space Launch User Committee began addressing expendable-booster manifest requirements, but the *Challenger* accident interrupted its work. When NASA and Space Division reviewed the Shuttle recovery schedule in the fall of 1986, they did not contact the services or the unified and specified commands for their inputs. Piotrowski considered this situation a prime example of the "technology push" rather than the "requirements pull," whereby space assets and needs traditionally reflected the concerns of the technologists rather than the warfighters. As he explained,

> I believe it is vitally important for the operational requirement to be present in the decision-making process....[O]ur role should be to act as an operational consultant to ensure the risk-vs-requirement discussion is not based solely on technical and programmatic concerns. I recommend for future launches of DoD systems, by either NASA or Systems Command, that US Space Command perform that consultant role.[53]

Specifically, he proposed that U.S. Space Command be accorded formal voting membership on the user committee, now termed the Space Launch Advisory Group. His proposal, however, became part of the thorny issue of "normalizing" the relationship between his command and his component Air Force Space Command. By 1987, he agreed with Under Secretary Aldridge and Major General Maurice C. "Tim" Padden, commander of Air Force Space Command, that the Air Force component should represent U.S. Space Command interests at user meetings. In any event, now operators would be more directly involved.[54]

General Piotrowski also spearheaded the effort to achieve a more responsive space launch capability. The problem with manifesting space payloads led him to reassess the issue of responsiveness in the context of deterrence and warfighting. Current policy, he argued, only guaranteed a return to a peacetime capability and a gradual recovery from the launch standdown. This would mean a relatively rigid "launch on schedule" policy that often required as much as six months of preparation by contractor personnel before each launch. Such practices did not provide the responsive space infrastructure needed for warfighting. Moreover, "deliberate" on-orbit checkout procedures by Air Force Systems Command's Space Division meant that space systems remained under control of the research and development community too long before transfer to operational users. Piotrowski believed that the best way to ensure a launch system responsive to the warfighter would be a complete transfer of the launch mission from Space Division to Air Force Space Command. He formally proposed the transfer in a letter to Chief of Staff General Larry D. Welch on 28 September 1987. Launch transfer, he argued, would represent a natural evolution as Air Force Space Command matured in its operational role and would enable the commander-in-chief of U.S. Space Command to use his component directly for launch-related activity in wartime. He also advocated an Air Force "blue suit" launch operation managed by the operational commands. He proposed that Air Force Space Command immediately assume operational responsibility for either the test ranges or upcoming Delta II/GPS launches.[55]

General Piotrowski and his fellow space operators believed that developments in the wake of the *Challenger* tragedy supported their argument. For one, a special Defense Department commission on defense management practices led by former Deputy Defense Secretary David Packard called for acquisition commands to concentrate on research, development and acquisition by divesting themselves of "operational" responsibilities. This led to the transfer in 1987 of the Air Force Satellite Control Network, including the remote tracking stations, from Air Force Systems Command to Air Force Space Command. Piotrowski hoped that this transfer would provide sufficient incentive for reconsideration of the launch issue. At the same time, recent Defense Department policy relegating the Shuttle to second priority behind expendable boosters effectively sealed the fate of Air Force Space Command's expectation to control military space launch through its Shuttle responsibilities. By February 1987 the Defense Department had decided to cancel funding and development of the Shuttle Operations and Planning Complex (SOPC) at Falcon Air Force Station and to mothball the Shuttle launch complex at

Vandenberg Air Force Base. As a General Accounting Office report suggested, cancellation of the SOPC also represented an end to dedicated military manned spaceflight efforts for the foreseeable future.[56]

Most affected was the Military-Man-in-Space project supported by Under Secretary Aldridge. When it began in 1985 this program embraced a study of potential tests aboard the Shuttle and military uses of on-orbit satellites. After President Reagan announced support for NASA's Space Station *Freedom* in January 1984, the Air Force also examined the possibility of participating in certain space station experiments. The problem, however, remained the Air Force's traditional inability to specify requirements that could be achieved only by military personnel aboard a space station. As a result, the Defense Department continued to question Air Force involvement in manned spaceflight, while political support in the late 1980s threatened to eliminate the space station altogether. Nevertheless, the Air Force persisted with the low-priority Military-Man-in-Space project, directed largely from Air Force Space Command, which had established an office in 1987 to provide "centralized focus for all Air Force military manned activities in space." Air Force officials hoped that reorienting the program's objective from earth observation to more technically demanding uses of military astronauts involving analysis and processing of data would prove more worthy of funding support. With the decline of military interest in the Shuttle and the space station's future in doubt, however, planners developing experiments for Shuttle flights in the early 1990s had no certainty they would be flown.[57]

With the return to expendable launchers and no provisions for turning over to Air Force Space Command the new Titan IV and Delta II boosters, the Defense Department's shift to expendable launch systems revitalized Air Force Systems Command's central role in launch operations and reinforced the status quo. Piotrowski's initial effort with the space launch issue proved unsuccessful. In denying his request in December of 1987, Air Force headquarters argued that the disruption involved in such a transition would adversely affect the launch recovery process. At the same time, even an Air Force Space Command study had raised questions about the lack of expertise within the command to handle a rapid rather than evolutionary transition. Further progress would have to await the renewed momentum in late 1988 created by the Blue Ribbon Panel on Space.[58]

General Piotrowski's initiatives on space manifesting and space launch should be considered as part of the U.S. Space Command and Defense Advanced Research Projects Agency (DARPA)-led "space in transition" movement that involved all elements of the space community in the late 1980s. From Piotrowski's perspective, the Air Force had to transition its force posture from one of remoteness to the concerns of the commanders-in-chief to one that ensured integration with warfighters' requirements. It should do this by emphasizing the interrelationship among survivable space systems and quick-reaction launch capabilities. These issues surfaced in early 1988, when Piotrowski surveyed the commanders-in-chief and theater commanders on their dependency on space systems. In response, the commanders declared that they had found weather, intelligence, and communications satellite information increasingly necessary for their operations, but they bemoaned their inability to control these assets. The unified space command chief's survey also revealed that without having access to weather and communications from satellites in a crisis situation, the commanders-in-chief did not conduct training to use this information. Piotrowski focused on the satellites themselves, particularly the trend toward multimission, multiuser satellites. They had proven cost-effective and capable of satisfying a broad spectrum of requirements, but had they met user needs? Piotrowski and his counterparts thought not.[59]

Piotrowski's "responsive" proposal called for developing many small, low-cost, single-mission satellites that could be launched on short notice and receive early on-orbit checkout. As such, they would be readily available for theater commanders. DARPA, which did not favor the practice of hardening satellites and producing more complex spacecraft, had long advocated cheaper, lighter satellites (LIGHTSATs) and a survivable launch capability through its Advanced Satellite Technology Program. In the early 1980s, however, an assessment by the Office of the Secretary of Defense recommended retaining high-altitude deployment of multi-mission satellites. Over the course of the decade theater commanders, the Strategic Air Command, and the Strategic Defense Initiative Organization increasingly looked to so-called cheap satellites (CHEAPSATs) as the best means of satisfying theater weather, communications, reconnaissance, and intelligence requirements during a crisis. The Air Force became most interested in the possibility of lightweight communications satellites to complement existing networks in a "communications by the yard" approach to fulfill theater needs not met by current systems. Piotrowski and others saw small satellites as a key means to transition from the existing peacetime situation to a more responsive warfighting posture and, thus, to realize the objective of assured access to space. Moreover, a quick-reaction "on-call" launch response would meet operational needs and help institutionalize space inside and outside the Air Force. Such a capability would involve simpler, smaller, short-life payloads launched aboard a standardized bus by quick-reaction launchers from multiple launch sites across the country. Short-term tactical satellites from a mixed-fleet arsenal could meet important surge requirements of wartime commanders.[60]

The Blue Ribbon Panel Provides a Space Agenda

Like General Piotrowski's other space launch concerns, discussion of responsive light satellites became overshadowed in 1988 by deliberations of the Blue Ribbon Panel on Space Roles and Missions, which proved to have the most far-reaching influence of the many space panels and studies over the years. In the spring of 1988 Air Force Chief of Staff General Welch formed a Blue Ribbon Panel consisting of senior representatives from all major Air Force commands to assess Air Force space issues. The Vice Chief of Staff of the Air Force chaired an Executive Steering Group that included Lieutenant General Donald J. Kutyna, commander of Air Force Space Command, and vice commanders from the other Air Force major commands. The main work would be done by the Panel Study Group, headed by Major General Robert Todd, vice commander of Air University. Echoing the 1987 White Paper on space, the chief of staff justified another study on space in terms of major changes in the space landscape that resulted from new policy statements by the Defense Department and the White House, technical advances, and the potential of SDI, as well as friction and funding problems with the other services. He worried above all about the ambivalence toward space in the Air Force. While the service had played a leading space role for thirty years and continued to garner 50 percent of the national space budget and 75 percent of the Defense Department's space funding, it remained uncertain about its future space role. The commitment of Air Force leaders to the institutionalization of space, he asserted, was not shared throughout the service. This resulted from misunderstanding about the potential of space systems, a multiuser approach to systems that placed space at a disadvantage in the budget process, and the historically closed nature of the space community.[61]

General Welch charged the panel to examine the role of space for the warfighter, responsiveness of space systems, and organizational relationships. After deliberating over the summer, the panel issued a report in August 1988 that dealt with three broad areas. First, Air Force space policy should be revised to reflect realistic capabilities and pretensions. This meant an Air Force role as principal, but not exclusive, agent for military space activities and a major effort to achieve the capability of performing warfighting missions in and from space. Secondly, the panel assessed the Air Force role in space in terms of the four mission functions described in the 1983 Air Force Space Plan. For these, the panel recommended a reasoned approach involving acquisition, operation, and support of military space systems. Finally, the panel investigated the organizational, institutional, and personnel issues associated with the role of space in the Air Force. The panel asserted that Air Force Space Command must continue its central role as advocate, operator, and single manager for space support, while U.S. Space Command should normalize its relationship with its Air Force component by returning to it operational control of peacetime space assets. The institutional challenge had occurred because many viewed space systems as vulnerable during conflict, without an assured mission capability of providing ready space system replacements. Generally, there continued to be a lack of broad institutional involvement in the space program, an absence of space expertise in the various commands, and overall minimal appreciation of the value of space throughout the Air Force. The panel concluded its evaluation by specifically recommending that doctrinal manuals be revised to include space in combat operations, and that space expertise be spread throughout the service.[62]

After receiving the Blue Ribbon Panel's report in late 1988, Air Force headquarters in February 1989 issued an implementation plan designed to realize the panel's twenty-nine specific action recommendations. The implementation plan declared in ringing words that "the Air Force is and will be responsible for the global employment of military power above the earth's surface." The plan expected to lay the groundwork for establishing a decisive space role in combat operations. The Air Force must foster among itself and the other services a "broader institutional view of how military power is applied above the surface of the earth." It charged Air Force Space Command with developing a "Space Roadmap" for updating the Air Force Space Plan and integrating all existing Air Force space operations. The Space Roadmap, projecting space into the 21st century, would link space systems to warfighting requirements, global strategy, and the four mission areas. The implementation plan asserted that "spacepower" would assume an importance equal to airpower in future combat and that the Air Force must ready itself for the "evolution of spacepower from combat support to the full spectrum of military capabilities." Above all, the road-map had to lead to a "coherent Air Force role in space."[63]

The Blue Ribbon Panel report and the Air Staff's implementation plan provided necessary momentum on a number of important space issues. They helped the Air Force space community, primarily Air Force Space Command, pull the rest of the Air Force along the path to an improved and clearer understanding of, and vision for, the space mission for the Air Force. Although Air Force Space Command's revision of the Space Plan continued to face opposition at Air Force headquarters, prospects for approval had brightened in light of the various ongoing studies. These included the Space Roadmap, an Air Force Investment Strategy for Space directed by the Assistant Air Force Secretary for Space, and an Assured Mission Support Space Architecture Study led by U.S. Space Command. In addition, doctrinal statements that had long been controversial faced good prospects for approval given the Panel's recommendation that the Air Force promote the "direct integration of space operations with the Air Force's more traditional roles." Moreover, the Panel's call for "normalization" of space led to a change within the Air Force board structure, whereby Air Force Space Command received a "home board" for space in order to effectively advocate space systems for several users.[64]

The Blue Ribbon Panel's findings also led to important changes in the relationship between U.S. Space Command and Air Force Space Command. The Panel called on the unified command to establish a more effective relationship with its component commands, especially Air Force Space Command, by relinquishing peacetime operational control of the space surveillance and missile warning functions. U.S. Space Command personnel had been exercising operational control over these Air Force assets since the separation of the two commands on 1 October 1986. Air Force Space Command leaders argued that as a component command it should serve as the focal point for the management and operation of Air Force strategic defense space assets and command, control, and communications systems in support of NORAD and U.S. Space Command. Under pressure from Air Force headquarters, General Piotrowski's command, by November 1988, had agreed to transfer to Air Force Space Command the Space Surveillance Center functions through the formation of a new organization, the Air Force Space Surveillance Element. The unified command, however, proved less forthcoming in transferring to its Air Force component command the three other Cheyenne Mountain Complex operations centers: the Missile Warning Center, the Space Defense Operations Center, and the Intelligence Operations Center. Air Force Space Command expected to gain responsibility for these remaining centers in the early 1990s, when the commander of Air Force Space Command once again would be dual-hatted as commander-in-chief of U.S. Space Command.[65]

Panel recommendations also supported the major effort to develop effective new launch technology through the Advanced Launch System (ALS) program. By the end of 1989 the ALS had evolved from a technological initiative to produce a heavy launch vehicle for the Strategic Defense Initiative and future space station to a multivehicle technology-oriented project. Not all participants approved of the restructured program's objectives, which eliminated production of the vehicle itself.

Space architecture studies during 1984 and 1985 based on Strategic Defense Initiative requirements had identified the need for a launch vehicle capable of placing at least 200,000 pounds into low-earth orbit. By 1987 the Strategic Defense Initiative Organization called for a capability of nearly 400,000 pounds per year into low-earth, polar orbit by 1993, with an expected increase to 5 million pounds per year by the end of the 1990s. Air Force officials, including Under Secretary Aldridge expected ALS also to meet future Air Force requirements for large multi-user satellites that could not be handled by the Shuttle or Titan IV, although General Piotrowski and Air Force Space Command planners feared that ALS furthered peacetime rather than wartime objectives by undermining their initiatives to produce tactical satellites of less size, weight, and complexity. Meanwhile, NASA had joined the competition by proposing an unmanned derivative of the Shuttle, termed Shuttle-C (Cargo). By the late 1980s, however, the climate of fiscal austerity and strong opposition to the prospect of space-based missile defense systems raised doubts about proceeding with an ALS program aimed only at producing a large new booster to support the Strategic Defense Initiative.[66]

Air Force Space Command led the effort to restructure the program to support development of a new "family of vehicles" that by the late 1990s could provide responsive, reliable, low-cost access to space for a variety of payloads. But with funding in short supply, might the launch dilemma be better addressed with a technology-only program directed toward improving the existing fleet of expendable boosters? This recommendation emerged from a 1989 Defense Science Board study of space launch. Board members argued for limiting ALS to a study and technology program without a full-scale development phase, because upgraded expendable launch vehicles would meet operational requirements for the foreseeable future. Air Force Space Command wanted ALS, now termed the Advanced Launch Development Program (ALDP), to address requirements for an operational launch system rather than merely focus on upgrading existing launch vehicles. The larger issue had become the classic development dilemma of whether to continue investing in improvements to systems based largely on 30-year-old technology or, instead, to support promising but unproven technology that might result in a family of launch vehicles that Air Force Space Command argued could provide "responsive, reliable, flexible, low cost access to space for the broad range of expected payload sizes, orbits and launch rates…essential to satisfy…requirements in the late 1990s and beyond."[67] By the end of the 1980s, the uncertainty of space launch for the future compelled the vice president's National Space Council to schedule a major assessment of the issue in 1990 or 1991.[68]

Air Force Space Command Gains the Space Launch Mission

If the Blue Ribbon Panel's findings did not lead to clarification of the Advanced Launch System program, they nevertheless helped produce major changes in the Air Force space launch mission. By the time the Blue Ribbon Panel's *Implementation Plan* appeared on 3 February 1989, the country had just completed its "year of recovery" for "assured access to space." The Titan 34D had returned to service with launches from both coasts; the first of the refurbished Titan II's for Defense Meteorological Satellite Program flights began operations in September; and the new Titan IV would enter the inventory with projections of three to five flights per year. Additionally, the new Delta II medium launch vehicle would make its first flight with Global Positioning System satellites in early 1989, and the Air Force had issued a contract for a second medium launch vehicle, a stretched version of the Atlas-Centaur for Defense Satellite Communications System launches.[69] The Blue Ribbon Panel had applauded the recovery of the expendable launch vehicle industry and mission. It also created momentum for transfer of the space launch mission from Air Force Systems Command to Air Force Space Command, and led to a revised Air Force Space Policy in December 1988 that declared that the Air Force would "consolidate space system requirements, advocacy, and operations, exclusive of developmental and, for the near term, launch systems, in Air Force Space Command." Although the policy stopped short of reassigning the launch function, it clearly reflected a central objective of the Blue Ribbon Panel, namely to institutionalize the role of Air Force Space Command as the focal point for operational space activity. Increasing awareness of Air Force Space Command's responsibilities and the importance of space in the Air Force set the stage for action on the launch transfer issue.[70]

After General Piotrowski failed in late 1987 to convince Air Force leaders to transfer the launch mission, he relinquished the burden of advocacy to Lieutenant General Donald J. Kutyna, the new commander of Air Force Space Command. In February 1988, General Kutyna provided Air Force Chief of Staff General Welch a lengthy rationale for transferring launch responsibility that became the command's basic position in the months ahead. Space boosters, he argued, while complex and costly vehicles, represented operational rather than developmental systems, yet Air Force Systems Command's research and development personnel performed operational tasks involving range and launch pad operation, supervision of contractor personnel, and execution of launch countdown checklists. These could, and should, be handled by "operators" who could boast of considerable experience with current boosters over the years. General Kutyna favored a "clean stroke" transfer similar to that involving the Satellite Control Network rather than a piecemeal change. At the same time, Kutyna and his staff had always understood that such a transfer would require resolution of difficult budget, manpower, and contractor issues, as well as interface challenges with NASA and the classified programs, along with responsibilities for upper-stage vehicles.[71]

General Welch, however, reaffirmed his earlier opposition, and the launch transfer issue joined a number of other concerns that would have to await Blue Ribbon Panel deliberations. In the new climate for change following publication of the implementation plan, General Welch directed the Air Staff in late May 1989 to review responsibilities of Air Force Space Command and Air Force Systems Command in order to recommend "a more normal relationship between developers and operators." Subsequently, Air Force headquarters directed both commands to prepare and discuss with each other their positions on space launch. By the end of the year, the two sides continued to differ fundamentally on the nature and control of space systems. Air Force Systems Command proposed a lengthy, phased turnover of individual launch vehicles, but only after sufficient improvements had been made to make them "operational." Space Command, by contrast, favored immediate transfer of space launch, represented at this time by the Space and Missile Test Organization, as well as all residual satellite control operations. In his presentation to General Welch in March 1990, General Kutyna declared that the transfer would enhance operational effectiveness in four ways. Making a single command responsible for the entire space support function would ensure unity of command, render systems more responsive to the warfighter, improve methods for the formulation of operational requirements, and assist the acquisition community by freeing it from performing operational functions. The Air Force Space Command chief also countered the objections of Air Force Systems Command representatives which centered on potential disruption to classified reconnaissance programs and contractor arrangements, and especially on what they considered the specialized, nonoperational nature of space systems.[72]

Although General Welch agreed with General Kutyna's basic position, he preferred to forego an immediate transfer and, instead, appointed a Launch Operations Transfer Steering Committee to examine various options for an effective transfer with minimal disruption. The goal would be to produce a plan "to bring launch operations into line with the normal division of roles and missions between operational commands and the acquisition command." Included among the committee members were Lieutenant General Ronald W. Yates and Major General Thomas S. Moorman, Jr., who would soon assume command of Air Force Systems Command and Air Force Space Command, respectively. In the spring of 1990 the committee examined sixteen options that in one way or another compared Air Force Space Command's position, which supported a direct transfer leaving launch systems to become more "operational" in the future, and Air Force Systems Command's argument, which favored an incremental transfer after first improving the launch systems to make them "operational." In mid-May General Welch agreed to the committee's compromise recommendation, which clearly favored the operational command. On 1 October 1990, Air Force Systems Command would transfer to Air Force Space Command its launch-related centers, ranges, bases, and the Delta II and Atlas E missions. The remaining Atlas II, Titan II, and Titan IV missions would be turned over later on a phased schedule. Approving the transfer on 12 June, Secretary of the Air Force Donald B. Rice declared that the "change in assignment of roles and missions further normalizes space operations and pursues our corporate commitment to integrate space power throughout the full spectrum of Air Force operational capabilities."[73]

It was left to General Moorman, in ceremonies on 1 October at Patrick Air Force Base, Florida, marking the transfer, to best describe the "landmark event."

> I believe this transfer is part of the natural evolution of the Air Force space program. It is a testimony to how our thinking about space operations has matured....[O]ver the past several years our leadership has been examining the role of the Air Force in space as well as that of space in the Air Force. The result of this review is an Air Force policy which has two basic tenets—that the future of the Air Force is inextricably tied to space, and that spacepower will be as decisive in future conflicts as airpower is today. The policy also states that we will make a solid corporate commitment to integrate and normalize space throughout the Air Force....[T]his transfer of launch responsibility is the tangible result of the Air Force's desire to fulfill these policy objectives. The decision to transfer the launch mission was based on the beliefs that placing satellites into orbit has matured to a point where it should be considered an operational task, and that Air Force Space Command had sufficiently matured where it could assume the responsibility....The transfer...is intended to be virtually transparent to both the users and operators. That transparency will help guarantee continued smooth

operation of launch activities and will establish a foundation for moving
forward toward normalizing our military access to space.[74]

The transfer of space launch represented not only the "most significant operational
milestone" in the command's brief history, but a major step on the road to an
operational, warfighting perspective for space.[75]

The Decade in Retrospect

By the end of the 1980s, the Air Force was well on its way toward achieving the institutionalization of space that enthusiasts had long envisioned. Space activity no
longer seemed primarily developmental in nature but, rather, an operational
element whose systems could fulfill Air Force missions in a manner comparable to
the service's traditional activities. Over the course of the decade the space launch
issue remained central to every aspect of the space program. Without assured access
to space there could be no space program. In the atmosphere of self-examination
following the *Challenger* tragedy and the Titan booster failures, the Air Force at the
highest levels moved to reassess not only its investment in the Shuttle but its entire
commitment to space.

The *Challenger's* shock waves generated a variety of space studies that attempted
to understand the present and chart the future. Of these, the Blue Ribbon Panel far
and away provided a realistic sense of the potential of space through its policy
analysis, and its examination of the Air Force role in space and the role of space in
the Air Force. It called on the Air Force to undertake sober leadership, and it set the
stage for the Space Roadmap. The Blue Ribbon Panel's recommendations served as
the linchpin for the broad process of "normalizing" space within the Air Force that
gained momentum in the late 1980s.

To be sure, much remained incomplete at decade's end. While the return to
expendable boosters enabled the service to continue launching communications,
weather, navigation, and early warning satellites, it would be 1992 before the three-
year Shuttle delay would be overcome. At the same time, roles-and-missions issues
continued to demand accommodation between the United States Space Command
and its component Air Force Space Command, as well as among the latter and other
Air Force and Defense Department organizations with space responsibilities. Likewise, the future of space launch also persisted unresolved. A return to the diversity
of reliable space boosters did not alleviate troublesome questions about the feasibility and necessity of developing a standardized launch vehicle for the new century.

Nevertheless, the end of the decade offered more hope than pessimism. Through
all the turmoil surrounding space launch in the movement away from the Shuttle,
the focus remained centered on operational requirements and the needs of the warfighter. In this regard, Air Force Space Command provided the focus as it moved to
consolidate operational responsibilities. Its victory in garnering the space launch
mission represented a final shift in the long struggle to move Air Force space from

the research, development, and acquisition community to the operational arena.
The Air Force had proclaimed itself the lead service for military space. Only a policy,
infrastructure, and institutional commitment wholly oriented toward space operations could provide the conditions to achieve the claim in reality. At the close of the
decade, General Kutyna, commander of Air Force Space Command, best described
the promise and potential of the Air Force space challenge. He said the Blue Ribbon
Panel had determined that:

> spacepower will assume an increasingly decisive role in future combat
> operations, and the future of the Air Force is inextricably tied to space.
> We are at the forefront in the evolution of spacepower from combat
> support to an actual warfighting capability. Spacepower is important
> today, but it will be absolutely critical in the future for effective military
> operations.[76]

General Kutyna's prediction would soon be put to the test in a major regional
conflict in Southwest Asia, Operation Desert Storm.

CHAPTER 7

Coming of Age:

Operation Desert Storm and Normalizing Military Space Operations

On 2 August 1990, Iraqi leader Saddam Hussein shocked the world by invading and rapidly overrunning the small, oil-rich country of Kuwait, sending the Kuwaiti government into exile. The Iraqi action threatened vital Western oil reserves in Kuwait and neighboring Saudi Arabia, which for many years had served as the basis for American policy in the Arab world. American President George Bush reacted promptly by convincing the United Nations to condemn the invasion, implement economic sanctions, and demand an unconditional Iraqi pullout. At the same time, the world body authorized President Bush to forge a multinational military alliance and force compliance if the Iraqis did not withdraw by 15 January 1991. On 7 August, under the operational name Desert Shield, allied forces began a five-month-long buildup in the Persian Gulf region. America now faced its first post-Cold War crisis.[1]

The Iraqi challenge found the United States already well on its way toward adjusting to new political and economic realities. The so-called New World Order that emerged after the 1989 political revolutions in eastern Europe and the impending demise of the Soviet Union, which finally occured in late 1991, precipitated a major reassessment of a military force structure designed to meet the threat of global nuclear war. In the new "multi-polar" world, regional crises and conflicts seemed more likely to test the United States. To prepare itself for new responsibilities as the only remaining superpower, America's national security strategy now focused more directly on revitalizing its domestic economy as a basis for strong international leadership, supporting emerging democracies, and maintaining traditional alliances. The Gulf War would show that joint and coalition warfare, rather than unilateral action, represented the wave of the future.[2]

The Gulf conflict also would demonstrate that national strategy could no longer support large prepositioned forces in a "forward defense" role. Rather, leaders envisioned restructured, smaller armed forces in the future, characterized by high readiness, technological superiority, and extensive mobility. Such forces would be capable of rapidly projecting power anywhere in the world without benefit of a supporting infrastructure already in place. Moreover, the armed forces of the 1990s should be able to emerge victorious from two simultaneous regional conflicts. Air Force Chief of Staff General Merrill McPeak termed the Air Force role in the new force structure as one of providing "global reach and global power." A central element in this Air Force vision would be the application of military space capabilities. Space forces would lead the way by providing global coverage, a nonthreatening forward presence, and inherent flexibility that guaranteed real-time and near-real-time support across the spectrum of military conflict.[3]

The Gulf War, fought under the operational name Desert Storm, represented the first major trial by fire for space forces, whereby military space systems could fulfill their promise as crucial "force multipliers."[4] By all accounts, space forces provided the vital edge in ensuring the victory of the U.N. Coalition. Their contribution proved more impressive because of the difficulties that had to be overcome. Space systems, up to this point, had focused primarily on strategic rather than tactical requirements. Some embryonic planning and testing of tactical uses of space capabilities had emerged by the late 1980s; however, ensuring nuclear warning and monitoring arms control agreements had been more important than supporting tactical operations. As a result, Coalition planners had to make important adjustments in both the satellite and ground segments of their space forces in order to meet tactical contingencies. Although remarkably successful, a number of persistent deficiencies could only be minimized, never overcome. In their many postwar assessments of space system performance, military authorities attempted to use the lessons learned from the desert conflict to ensure that space systems would better support the tactical warfighter in the future. The Air Force saw in its Gulf War experience a springboard for charting the future of the nation's military space program and assuring its own leadership role in space for the century ahead.[5]

To be sure, military space systems had provided important operational wartime support long before the Gulf War.[6] As early as the Vietnam conflict, weather and communications satellites furnished useful data and imagery to commanders in Southeast Asia and linked them with Washington, D.C. More recently, satellite communications had proven important in the British Falkland Islands campaign and in Urgent Fury, the Grenada invasion of 1983. In 1986, during Operation Eldorado Canyon, space systems provided a vital communications link and supplied important mission planning data to aircrews that bombed targets in Libya. In 1988, Operation Earnest Will witnessed the first use of GPS test satellites to support ships and helicopters during mine sweeping operations in the Persian Gulf. During Operation Just Cause in Panama in 1989, DSCS satellites provided long-haul communications links and DMSP supplied important weather data.[7]

These operations, however, involved only portions of the military space community for a relatively brief period of time, and the contribution of space systems was not widely understood or appreciated. Desert Storm, by contrast, involved the full arsenal of military space systems. Nearly sixty military and civilian satellites influenced the course of the war and helped save lives. Communications satellites established inter- and intra-theater links to support command and control requirements for an army of nearly 500,000 troops. Weather satellites enabled mission planners to keep abreast of constantly changing atmospheric conditions, while early warning spacecraft supplied crucial data on enemy missile launches. Navigation satellites furnished precise positional information to all elements of the armed forces. Then, too, commercial satellites not only assisted in filling coverage and system gaps, but broadcast the war over television to a worldwide audience. Desert Storm was, indeed, the first large-scale integration of space systems in support of warfighting.[8]

Operation Desert Shield—Preparation

At the outset of Desert Shield in early August 1990, communications satellites served only an American administrative unit in Bahrain and two training groups in Saudi Arabia, while the priorities lay elsewhere for weather, navigation, early warning, and remote sensing satellites. Much time and effort would be required to reconfigure satellite systems and overcome shortfalls before the Coalition's enormous space potential could be decisively marshaled. As for Iraq, it possessed no space assets of its own and had to rely on the international Intelsat and Inmarsat networks as well as the two Arabsat regional telecommunications satellites. During Desert Shield, the Iraqi leadership made little or no effort to integrate space into their military planning. Coalition forces, on the other hand, took advantage of a five-month "grace period" before the onset of Desert Storm to orient their space systems for maximum support to the warfighter. They faced a formidable challenge in all areas of space support.

Communications—Defense Satellite Communications System (DSCS). Analysts of the Gulf conflict found it difficult not to overplay the role of communications satellites (COMSATs) because of their vital importance to the success of every aspect of Desert Shield and Desert Storm operations. Although Saudi Arabia possessed a modern communications system, it did not service key areas of the potential battlefield or possess the circuit capacity needed to support the requirements of a half-million personnel. During Desert Shield preparations the Coalition put into place ten different COMSAT systems, which carried over 90 percent of U.S. communications to and from the Gulf area. Of this, commercial satellites accounted for 24 percent of the traffic. Communications satellites also furnished tactical links within the theater and served as relays for terrestrial radio systems suffering from line-of-sight limitations. They provided total communications to air, sea, and ground forces, and brought the war to television screens around the world.[9]

Establishing effective communications during Desert Shield presented a major challenge. Satellites needed to be repositioned or activated from standby status. To meet the high demand for communication circuits, military leaders reallocated circuits from other users to U.S. Central Command, leased civilian COMSAT circuits, and deployed thousands of terminals to the operational area. Coalition forces received communications support from the Defense Satellite Communications System, Fleet Satellite Communications (FLTSATCOM), NATO III, and Skynet systems, as well as commercial satellites. The variety of systems required considerable coordination with individual agencies and extensive integration before the communications system could function smoothly and efficiently.[10]

Of the various communications satellite systems involved in the Gulf operations, Air Force attention centered on the super high frequency (SHF) Defense Satellite Communication System, which had long met the bulk of global, long-haul communication requirements for all branches of the armed forces. At the outset of Desert Shield, the DSCS constellation appeared in good shape. In 1989 the launch of DSCS II and DSCS III satellites ended the marginal status of the constellation in the wake of the *Challenger* disaster. In August 1990 the DSCS network consisted of two DSCS II and three DSCS III operational satellites, together with one DSCS III reserve and two DSCS II limited-use spacecraft.[11]

Nevertheless, from the outset of Desert Shield Lieutenant General Thomas S. Moorman, Jr., commander of Air Force Space Command, expressed concern that military satellite communication links between the United States and the Middle East, and especially DSCS circuit capacity, would prove insufficient. Although his command was responsible for the space segment, overall DSCS management remained in the hands of the Defense Communications Agency (DCA). On 15 August General Moorman requested from the DCA the status of network allocations in support of U.S. Central Command. In a reply two days later, the agency did not seem worried, because the dedicated circuits had received little use. Circuits on two DSCS satellites had been earmarked to support Gulf operations. On an older DSCS II satellite positioned over the Indian Ocean (*DSCS IO-II*), 85 percent of channel 3 had been reserved for U.S. Central Command, but only 30 percent was in use. As for the DSCS III Eastern Atlantic spacecraft (*DSCS EA-III*), DCA, as system manager, had allocated 100 percent of channel 2, 85 percent of channel 4, and 100 percent of channel 3. Yet, usage figures for the three channels amounted to 10 percent, 20 percent, and 0 percent, respectively.[12]

Air Force Space Command remained troubled about potential saturation of the communication links. By late August *DSCS IO-II* usage had risen to 85 percent of allotted capacity under pressure from the growing buildup of forces in the region. Moreover, Air Force Space Command's staff became concerned with the vulnerability of DSCS satellites and other Coalition spacecraft to Iraqi jamming of satellite transponders, either by using the fire control radar for SA-6 surface-to-air missile batteries or by overrunning the U.S. Diplomatic Telecommunications Terminal at the American embassy in Kuwait City. In the latter event, Coalition forces might have to rely on naval ultra-high-frequency communications provided by the Navy's FLTSATCOM and additional commercial satellite support. Unfortunately, FLTSATCOM also was susceptible to jamming, and its large terminals made its bandwith unsuitable for tactical requirements.[13]

By mid-September, the problem of circuit loading could no longer be overlooked. Although use of the Eastern Atlantic spacecraft had reached only 60 percent of capacity, the figure for the Indian Ocean satellite's channel 3 had skyrocketed to 98.4 percent. The staffs of Air Force Space Command, DCA, the Joint Chiefs of Staff, U.S. Space Command, and U.S. Central Command considered several possible solutions, including moving users to different channels on the Indian Ocean satellite or entirely to the Atlantic Ocean satellite. Doing so, however, would temporarily sever connections with deployed forces in the theater, and U.S. Central Command declined to take this risk. Instead, DCA authorized the unprecedented step of re-aligning *DSCS EA-III*'s high-gain multibeam antennas to enhance its Middle East communications traffic capacity. Officials also continued to emphasize "bandwidth discipline" to users to prevent misuse of the strict channel allocations. But what if the current DSCS satellites experienced partial or complete system failure? The DSCS II satellite, after all, had been operating well beyond its projected "lifespan."[14]

Another potential solution to bolster Gulf support involved repositioning one of the satellites in orbit or launching a new DSCS satellite. This became a subject of serious debate in the fall, especially after U.S. Central Command realized that available funding for launch vehicles could not support a "launch on demand" requirement for additional DSCS satellites. Even with additional funding, however, it immediately became clear that a new DSCS launch could not take place before the spring of 1991. On 26 September Air Force Space Command initially recommended the activation of an orbiting NATO III Flight D spare satellite to relieve the two DSCS satellites of non-Desert Shield communications, but by the end of October growing political opposition to the use of a NATO asset for non-European operations ended consideration of this alternative.[15]

Meanwhile, Air Force Systems Command's Space Systems Division proposed a "spare-DSCS alternative," which involved using an old spare DSCS II, Flight D-14, and newly developed portable terminals to furnish SHF communications for Coalition war orders. The technical specialists also believed that Flight D-14 could be moved closer to the Gulf to provide the narrow coverage required for the use of small terminals in the theater. The first contingent of U.S. troops brought with them 48 tactical terminals, and by late September, with 200,000 additional troops on their way, the number of sets had increased to 58. The spare-DSCS option gained momentum, and by mid-November DCA requested that U.S. Space Command obtain permission from the Joint Chiefs of Staff to reposition Flight D-14, a "West Pacific Narrow Coverage Reserve" satellite, from its parking slot at 174 degrees east to 65 degrees east, over the Indian Ocean. On 17 November Air Force Space Command's 3rd Satellite Control Squadron initiated the move, and, at a drift rate of 4 degrees per day, Flight D-14 arrived at its new location on 19 December. After three days of intense testing, the second "Indian Ocean" DSCS satellite, began providing direct support to Desert Shield users on 22 December.[16]

By this time, Coalition authorities had arranged to use the British Skynet 4B satellite positioned at 53 degrees for additional SHF links, along with several categories of UHF satellites. These included two small experimental MACSATs (Multiple Access Communications Satellites) which had been launched by Scout boosters just prior to the Iraqi invasion. The minisatellites used "store-and-retrieve" procedures to relay logistics information between the United States and Saudi Arabia. U.S. forces also relied on geostationary FLTSATCOM spacecraft, largely for nontactical naval communication requirements, and commercial satellites. At the time of Desert Shield, the Hughes Corporation operated an orbiting network of five leased satellites (Leasats), although one had failed. For Gulf contingency support, Hughes moved one Leasat to provide better coverage of Iraq. At the same time, DCA contracted for the launch on 9 January 1991 of another commercial satellite, "Syncom" IV-5 1990-002B. Even if the commercial circuits were to prove unnecessary, Air Force Space Command felt more secure realizing that the commercial satellites could provide "redundancy" should the military systems fail.[17] The satellite communications network established during Desert Shield reflected considerable system flexibility and cooperation among the military, civil, and commercial space sectors.

Navigation—Navstar Global Positioning System (GPS). The Global Positioning System became the best known space system during the war. It proved capable of answering the age-old questions of "where am I" and "where am I going" in the featureless desert. Ironically, the military had been slow to accept GPS, partly because of reluctance to forego existing navigation systems. As for the Air Force, it seemed at times unable to dedicate itself to a multi-service system. As a result, a lower funding priority for GPS translated into delays, made worse by the *Challenger* Space Shuttle disaster. Not until 1989 did the Air Force launch the first five Block II operational satellites aboard the new Delta II booster to join the test satellites that had been supporting user equipment evaluations at Yuma, Arizona, since the late 1970s. Planners did not expect the full configuration of 21 operational and three spare satellites to be in orbit providing 24-hour, worldwide three-dimensional coverage and positioning information before late 1992 or early 1993. Meanwhile, to meet an interim deadline of April 1991 for 24-hour, global two-dimensional coverage, authorities decided in early 1990 both to begin repositioning the satellites from their test locations to their operational positions and to launch additional GPS satellites.[18]

Desert Shield sparked urgent efforts to provide navigation coverage to forces in the field. In early August, GPS rephasing already was well underway. By the 22nd of the month, when U.S. Space Command notified U.S. Central Command that GPS II-8 had been activated just twelve days after launch, the fourteen-satellite constellation consisted of six prototype and eight Block II operational satellites. With the buildup continuing, the previously scheduled launches of GPS II-9 and II-10 on 2 October and 26 November, respectively, increased the configuration to sixteen satellites on the eve of Desert Storm. Planners also decided to alter the orbit of GPS II-9 to optimize its coverage over Baghdad, especially at night. With the re-phasing process completed by year's end, program managers expected the constellation to provide 24-hour two-dimensional and 19-hour three-dimensional coverage of Gulf operations.[19]

The optimistic plan threatened to go awry in December, however, when prototype satellite #6 failed. First launched in 1980 with a projected lifespan of only five years, the test satellite had finally succumbed to old age. Although the launch of GPS II-11 scheduled for 30 January 1991 would compensate for the loss of the prototype satellite, its launch was delayed after engineers discovered a flaw in the solar array drive-control electronics unit. By late December, work to revive the test satellite continued, while General Donald J. Kutyna, the commander-in-chief of U.S. Space Command, and General Moorman, commander of Air Force Space Command, convinced Space Systems Division to give high priority to solving GPS II-11's design flaw. Meanwhile, Air Force Space Command delcared GPS II-10 ready for operations on 15 January, well below the normal check out time of 30 to 60 days, and one day before the air campaign began.[20]

Problems with the GPS ground segment proved more alarming than the challenge of achieving the optimum satellite configuration. In August planners found themselves woefully short of receiver terminals. The system relied on two main types of receivers at the outset of Desert Shield. Rockwell International produced some 550 manpack/vehicular sets weighing approximately 18 lbs. each at a cost of $45,000 per set. Their design called for "Selective Availability," the capability of receiving encrypted precision-coded signals from the satellites that resulted in positional accuracies to within thirty feet. Because of their weight, most of these sets had been vehicle-mounted, while the high cost had limited their production to only 550 sets.[21]

Meanwhile, other companies had begun to produce commercial receivers to meet the growing demand in the private sector. One of these, Trimble Navigation of

Sunnyvale, California, had introduced a Small, Lightweight, GPS Receiver (SLGR) in 1989. The SLGR provided positional accuracy of between 50 an 100 feet. Less accurate than the Rockwell military receiver, the SLGR weighed only 4 lbs., cost about $3400, and proved small enough to fit into a soldier's pocket. The Army had purchased 500 demonstration SLGRs for testing before Desert Shield, and it distributed the sets to forces that deployed in August. Although every available military and commercial set quickly found its way to the Middle East, the dramatically increased demand could not be met. In an area of difficult terrain, few landmarks, and poor maps, GPS quickly assumed vital importance. After the Army Space Command expressed an "urgent need" for additional SLGRs, the GPS joint program office initiated two emergency requisitions for the commercially produced SLGR. In September, Trimble began shipping to the theater 1,000 SLGRs and 300 vehicle installation kits. Later, in December, Trimble contracted for an additional 7,178 SLGRs, with most earmarked for Army use.[22]

Along with a shortage of receivers, authorities also had to face the issue of protecting GPS signals broadcast by means of Selective Availability. In contrast to the military P-code sets, the commercially produced SLGRs received a coarse acquisition signal, which could be further degraded to deny precise navigational data. On 10 August Selective Availability was turned off to permit use of the Army's initial batch of commercially produced SLGRs. It remained off when it became apparent most sets would be commercially procured and, also, at the request of U. S. Special Operations Command. At the same time, GPS authorities made an additional compromise by permitting GPS course acquisition transmissions to occur without introducing additional error. The "open" signal resulted in the relatively high accuracy of approximately 100 feet for the commercial receivers, in contrast to half that figure with Selective Availability turned on.[23]

Beyond the difficulties of receiver shortages and signal usage, GPS users faced considerable challenges in establishing sets and terminal networks, ensuring effective linkage between field units and commands and, especially, in familiarizing themselves with the space equipment and its capabilities. Fortuitously, on the eve of Desert Storm, GPS officials had nearly finished the major improvements to the satellite constellation and had "solved" the receiver shortfall problem.

Environmental Monitoring. Defense Meteorological Satellite Program (DMSP), Landsat and SPOT. Coalition leaders understood the importance of weather satellite data and imagery for mission planning from the beginning of Desert Shield. The United States averaged six meteorological satellites in orbit simultaneously during the Gulf conflict. These included the military's three polar sun-synchronous DMSP satellites, the two National Oceanic and Atmospheric Administration (NOAA) TIROS polar-orbiting spacecraft, and the civilian GOES geostationary satellite. In addition, coalition forces received weather information from Japan's GMS system, positioned at 140 degrees east, two European METEOSATs located at 50 degrees west and 0 degrees west, and twelve Russian polar-orbiting METEOR satellites. Both civilian and military satellites played important roles, because they frequently complemented each other's coverage.[24]

DMSP proved to be the most useful system in providing cloud-cover imagery and temperature and moisture data in its twice daily sweeps of the Gulf. At the beginning of Desert Shield, the DMSP configuration consisted of a standard, two-satellite constellation. Flight 8 had been launched on 19 June 1987 and Flight 9 on 2 February 1988. Both had been performing without major incident. Air Force officials, in fact, had stressed improvements to the control segment of the program by creating a new operations center at Fairchild Air Force Base, Washington, to replace the Offutt facility, as well as upgrade the Thule, Greenland, tracking station to replace the site at Loring Air Force Base, Maine. By August 1990 most of these changes had been completed.[25]

Traditionally, the weather satellite program had followed a "launch on need" strategy, which required at least a 90-day wait before the request from the Air Weather Service resulted in a launch. In mid-1990, prior to Desert Shield, planners decided to replace the two orbiting DMSP satellites with Flights 10 and 11, scheduled for launch aboard Atlas E boosters in October 1990 and July 1991, respectively. After the Iraqi invasion of Kuwait, Air Weather Service officers met with their Air Force Space Command counterparts to determine the need for launching Flight 10, now rescheduled for 21 November. Problems with the Operational Linescan System (OLS) on Flight 9 convinced them to proceed with the launch. As it turned out, the launch took place on 1 December, but it proved far from routine. Apparently, the apogee kick motor exploded leaving the satellite in an incorrect, lower orbit. Although it would be replaced in less than a year, it nevertheless supplied useful information for the Gulf operations. By year's end, three healthy DMSP satellites stood ready for weather support.[26]

The main difficulty with the DMSP system proved to be its lack of tactical mobility. A system originally designed to fulfill strategic requirements only it recently had stressed tactical applications. Successful operations depended on establishing high-quality communication links between field units and the central DMSP Mark IV van. This 26,000-lb vehicle required a C-130 for transport. On 20 August the first of six Mark IV "tactical" terminals deployed to the theater to receive downlink imagery from the satellites. Five supported marine aviation and amphibious operations; one Air Force van was positioned at Riyadh together with the theater Tactical Forecast Unit. The latter transmitted weather data to more than thirty sites within the region by means of secure weather fax. The Navy also had DMSP terminals on its carriers and command and flag ships.[27]

Given the focus on tactical operational requirements, however, officials preferred a more mobile DMSP terminal. During Desert Shield, the Air Force moved rapidly to acquire three so-called Rapid Deployable Imagery Terminals (RDIT), which could be transported by two people with a weapons-carrier vehicle. But the search for more tactically-oriented terminals continued. On 7 November, Military Airlift Command proposed to Air Force Space Command the development of a light-weight, rapidly-deployable "Small Tactical Terminal" for each unit to receive near-real-time data. General Moorman approved the concept, and developers hoped the tactical terminals could replace the single DMSP van early in 1991.[28]

From the beginning of Desert Shield, Coalition forces supplemented DMSP forecasting with weather data form the civilian satellites. NOAA's TIROS spacecraft, for example, provided useful transmissions at 2 P.M. local time on the late-afternoon jet stream that affected evening weather patterns in the desert. For the Army, the geostationary civilian satellites proved to be more useful than the military polar-orbiting satellites. The civilian satellites re-imaged the same portion of the earth every thirty minutes in contrast to the twelve-hour cycle of the polar satellites. Moreover, although DMSP provided a smaller-scale weather picture, which proved useful for identifying fog and sandstorms, the Army suffered from limited access to DMSP data because of the small number of receivers in the theater. Since the Mark IV terminal did not meet Army mobility requirements, Army units below theater level relied on a commercial weather receiver, the German-made WRAASE terminal. Available prior to Desert Shield, the civilian weather receiver could obtain imagery from civilian satellites of four nations, but not from U.S. military DMSP satellites. Most DMSP readouts did not reach Army units, which prompted the Air Force to develop a small, high-resolution tactical terminal for direct satellite-to-user trans-mission of imagery and data. In mid-December a small experimental DMSP receiver had proven successful in at the Army's Central Command headquarters. At year's end, it remained uncertain whether the tactical terminals could arrive in sufficient quantity in time to be useful in Gulf hostilities.[29]

Weather satellites represented one avenue to gaining a better understanding of the battlefield environment. Another involved the use of wide-area and multi-spectral imagery (MSI) from space systems to prepare accurate maps and to support terrain analysis requirements. MSI images depict features beyond human visual capability, including spectral change resulting from ground disturbances. In the Persian Gulf region, where accurate mission planning maps did not exist, multi-spectral imagery could benefit all the services.[30]

For multispectral imagery and wide-area coverage of the area of operations, Coalition forces relied on two civilian sources, the U.S. Commerce Department's *Landsat 4* and 5 satellites, and the French-owned earth resources satellite system, SPOT (Satellite Probatoire d'Observation de la Terre/Exploratory Satellite for Earth Observation). Both of these earth resources imaging satellites provided essential wide-area surveillance of the battlefield area not available from the high-resolution sensors aboard the U.S. national reconnaissance satellites. Landsat satellites imaged each part of the earth every sixteen days in seven different bands of the spectrum with a spatial resolution of thirty meters. Their ability to image a 185-kilometer wide area on each pass produced a wide field of view that enabled mapmakers to create products with a scale of approximately 1:80,000. SPOT satellites, although not MSI-capable, performed in like fashion. The primary SPOT sensor used three different bands at a 20-meter resolution, and one 10-meter panchromatic band to achieve an image map scale of 1:25,000. In contrast to Landsat, SPOT satellites viewed each part of the earth every twenty-six days, and the width of its pass measured approximately sixty kilometers. Owners of both systems required users to purchase the requested imagery and refrain from sharing it with third parties. Coalition leaders, worried that Iraq might acquire multispectral imagery, convinced Landsat and SPOT officials not to make it available to Saddam Hussein's regime between August 1990 and March 1991.[31]

The U.S. Army's topographical battalion arrived in the Gulf in August with maps based on 1987 Landsat imagery. By November, however, it had established three operational MSI workstations and had begun receiving updated imagery. Yet officials did not have an effective courier system to deliver the imagery to users until late January 1991, and they worried that units might not have sufficient time to fully exploit it.[32]

The U.S. Air Force relied on Landsat imagery for a variety of purposes, including construction of large airfields in Saudi Arabia. Specialists converted Landsat imagery of existing airstrips into engineering drawings for use in planning and building some of the world's largest air bases. For mission planning and rehearsing, however, both the Air Force and the Marine Corps preferred using SPOT imagery because of its 10-meter resolution. In September the Air Force purchased a sizable amount of SPOT images and offered to share the data with the Army. But the Army had no funding available to pay the royalty fees and, as a result, Army units did not have access to SPOT data during the Gulf War.[33]

During Desert Shield, Coalition forces became convinced that multispectral imagery could provide direct warfighting support. Beyond its use in preparing accurate maps, terrain analyses, and strike planning, it enabled U.S. Central Command leaders to keep abreast of Iraqi activity. By comparing Landsat imagery taken in August and December along the Kuwaiti-Saudi Arabian border, planners could determine the changes that had occurred. Bright spots appeared when the data was displayed on a single image, which indicated ground cover had been altered. At the same time, Coalition authorities realized that multispectral imagery could also give the Iraqis insight into Allied planning. Shortly before the ground war, the Defense Intelligence Agency intervened to prevent U.S. news media from obtaining Landsat data of the Kuwait-Iraq-Saudi border, which might have revealed the Coalition buildup in preparation for the "left hook" offensive maneuver at the war's start. Although users of multispectral imagery during Desert Shield would have preferred a system more timely, accurate and responsive to tactical requirements, the imagery clearly provided Coalition forces an important advantage in preparing for the conflict ahead.[34]

Early warning—Defense Support Program (DSP). Early warning Defense Support Program satellites would play a crucial role in detecting tactical ballistic missiles. Well aware that Iraq possessed Scud tactical missiles, United States military planners worked from the start of Desert Shield to optimize the strategic early warning satellite system for coverage of the tactical threat. In August 1990, the DSP network consisted of three operational satellites and two spares in geostationary orbit. Originally designed for strategic requirements, DSP's primary mission continued to be warning of ballistic missile launches. Secondary missions included detection of space launches and nuclear detonations for test ban monitoring. One satellite positioned over the Indian Ocean at 70 degrees east monitored intercontinental ballistic missile (ICBM) launches from the Asian landmass, while the other two focused on the sea-launched ballistic missile (SLBM) threat from their positions in the South Atlantic at 70 degrees west and over the eastern Pacific at 135 degrees west. Although in 1989 and 1990 the Air Force had launched the first two of its new-generation DSP satellites, DSP-1/Block 14, it elected to use its two oldest orbiting spacecraft to monitor Scud launches, because they were better positioned to support Gulf operations.[35]

During the Desert Shield buildup officials worried about timeliness of detection and warning. Technically, the spacecraft's infrared telescope scanned an area larger than it could observe at a particular time because it was mounted off-angle to its rotational axis. Rotating six times per second, the satellite required ten seconds to re-image an area where a missile had been detected. As a result, the longer time needed to determine launch site, trajectory and impact point became especially troublesome for Coalition forces challenged to destroy short-range tactical missiles. Air Force planners sought to increase early detection of Scud launches by repositioning the Indian Ocean satellite farther westward in order to maximize its coverage of the Gulf region.[36]

DSP program managers also attempted to reduce the time it took to process and relay warning data to Patriot antiaircraft missile crews in the Gulf. DSP procedures called for the Indian Ocean satellite to transmit data first to the Air Force Space Command ground station at Woomera, Australia, then up to a Defense Satellite Communications System (DSCS) satellite for relay over the Pacific to the ground terminal at Buckley Air National Guard Base, Colorado, and onward from there to the U.S. Space Command's Missile Warning Center, Air Force Space Operations Center, and Space Command Center at Cheyenne Mountain. There, analysts determined the likely impact zone of the missile and transmitted this information to U.S. Central Command and air defense commanders by way of a DSCS satellite over the Atlantic Ocean. The entire process from launch detection to warning took up to five minutes. With a Scud flight time of only seven minutes, Patriot crews and civilian target areas would receive precious little warning.[37]

To enhance the Coalition's ability to defeat the Scud threat, personnel at U.S. Central Command, Air Force Space Command, and U.S. Space Command emphasized training and straightforward procedures—and they worked hard to improve warning and response efficiency during Desert Shield. At the same time, Air Force Space Command sought, unsuccessfully, to reduce the warning time by deploying DSP terminals to the theater. By doing so, the delay could have been reduced from five minutes to as little as 90 seconds, thereby increasing up to five minutes the warning time given to Patriot batteries. Although satellite early warning had been in place since early August, it required the full five and half months of Desert Shield to establish the secure communication paths, develop alert procedures and train all involved to achieve the fine-tuning Air Force Space Command believed necessary for success.[38]

In mid-January 1991, on the eve of combat, the military space community could look back on over five months of intense effort to adjust space forces for tactical operations in the Middle East. Often resorting to innovative solutions to support the tactical warfighter, they relied on the inherent flexibility of space systems and their own ingenuity to overcome the limitations of a Cold War space posture. Although confident of success, the planners realized that the system remained fragile in crucial aspects and vulnerable to potential Iraqi counter actions. It remained to be seen whether space would prove decisive in the unfolding conflict.

Operation Desert Storm—Combat

When the United Nations ultimatum on Iraqi withdrawal from Kuwait expired on 15 January 1991, the Coalition decided on immediate military action. Desert Storm began on the night of 16–17 January with a massive air campaign led by F-117 Stealth fighters firing laser-guided weapons at targets in Baghdad. Following the radar-evading fighters came a series of coordinated air strikes in Iraq and Kuwait in conjunction with the launching of Navy Tomahawk cruise missiles. Coalition leaders thought the air assault would last only a few days, but it continued for another six weeks. Observers found the course of the ground war equally surprising. Expected to last several weeks, the land campaign ended after only four days. No one would deny, however, that the "blitzkrieg" ground offensive resulted from the length and effectiveness of the air campaign. Because the combat phase of Desert Storm has been discussed in detail elsewhere—as well as shown to millions of television viewers worldwide—this study will focus on the performance of space systems and Air Force operations in the Gulf War.

Communications—DSCS. As Chairman of the Joint Chiefs of Staff General Colin L. Powell observed near the end of Desert Shield:

> When we started our deployment, we had only the most rudimentary communications infrastructure in Southwest Asia and the challenge of distance was daunting. Thanks to good planning and to our understanding of the importance of satellites, we quickly and smoothly transitioned to a mature tactical theater network.[39]

The military satellite communications network quickly proved its worth. When Coalition air forces began on 16–17 January what would become a 39-day air assault, they had the unprecedented advantage of access to a single data base, or Air Tasking Order (ATO). Communications satellites also made possible immediate updates of target assignments and provided "positive" control of combat operations from pre-mission planning to post-mission aircrew debriefing. Executing over 700,000 transactions daily, satellites made Desert Storm air operations the most efficient and accurate to date. Once Desert Storm started, satellite communications more than doubled. By this time, the Coalition communications satellite network transmitted to more than 1,500 satellite communications terminals in the theater. More than three quarters of these were single-channel "manpack" military and commercial receivers.[40]

DSCS proved to be the most important intra-theater long-haul, multichannel communications system. DSCS satellites carried over 50 percent of communications traffic during the war and ensured effective command and control for both strategic and tactical operations throughout the conflict. DSCS provided the daily tasking order to every air base in the theater, and continued to link air and ground units to their bases in the States. It handled a 75 percent increase in intelligence relay to the United States for analysis, then back to the U.S. Central Command for use by deployed warfighters.[41]

By the end of Desert Shield, more than 120 DSCS tactical terminals had been delivered to the Gulf. When the ground forces initiated their "left hook" attack, many units moved with tactical DSCS terminals, some on flatbed trucks to avoid reassembling the satellite antenna during relocation. At the conclusion of hostilities, 33 DSCS terminals supported the warfighters in Kuwait and Iraq. In short, DSCS helped guarantee the command and control vital to the success of the war effort.

Navigation—GPS. By mid-January 1991, Air Force Space Command had declared the problem-plagued satellite #6 fit for operations, but engineers could not alleviate GPS II-11's design flaw in time to affect the conflict. As a result, the GPS constellation remained at sixteen satellites during Desert Storm. U.S. Space Command's postwar evaluation would characterize GPS as "perhaps the most visible example of space systems support to U.S. troops in Operations Desert Shield and Desert Storm."[42] Most attention has focused on the navigation system's vital contribution to ground forces in the land campaign—and rightly so. After all, its precise positional data and time readouts provided an average accuracy of 7.5 meters. GPS supported every type of ground operation, from large-scale maneuvers to individual soldiers moving through the featureless desert. The Rockwell "man-pack," which troops mounted on trucks or helicopters, and the SLGR, affectionately known as "Slugger" and carried by individual soldiers, enabled units to plot and achieve objectives and relocate tactical operations centers, Special Forces personnel to operate effectively in enemy territory, artillery observers to target enemy positions and direct friendly fire, and troops to clear land mines. In short, GPS gave the Army eyes to operate in the desert and made possible the successful envelopment maneuver that brought the ground war to a rapid conclusion.

GPS also served the other services. Not only did it furnish Marine artillery units with precise positioning data, it helped naval forces clear mines and provide precise coordinates for cruise missile strikes against Baghdad targets. The Air Force bene-fited from GPS in a variety of ways. For one, the system gave B-52 bombers an all-weather flight capability for their missions. For another, in the opening hours of the war, Special Operations Forces Pave Low helicopters with GPS teamed with Army Apache attack helicopters to destroy two Iraqi radar sites and, thereby, create a major gap in the Iraqi air defense network. F-16 aircraft also used GPS for passive navigation to the initial point of the bomb run. Of nearly 200 F-16s in theater, 80 to 90 possessed GPS receivers. Especially valuable in bad weather, the precision navigation system freed the pilot to take care of other business and optimize his use of weapons. Fuel permitting, GPS enabled aircraft to remain in the target area as forward air controllers, who could furnish to later strike aircraft precise coordinates and current battlefield data. GPS also promoted superb close-air-support coordina-tion with GPS-equipped ground units. Possessing a GPS receiver and a laser com-pass, ground troops could triangulate on a target and pass the coordinates to the GPS-equipped F-16 which, in turn, could relay them to strike aircraft. In light of the traditional Army-Air Force divisiveness over the close-air-support mission, space-based GPS foreshadowed the advent of a new era in air-ground cooperation.[43]

Environmental Monitoring—DMSP, Landsat and SPOT. When coalition forces launched the air war, they confronted the worst weather experienced in the Gulf in fourteen years. As Air Force Chief of Staff General Merrill McPeak later recalled, weather conditions were "at least twice as bad as the worst-case estimates."[44] Moreover, Coalition forces learned during Desert Shield that weather in the region proved notoriously susceptible to sudden changes. Heavy coastal fogs, blinding sandstorms, and heavy rains could seriously hinder operations. On 24 January 1991, for example, DMSP imagery depicted a clear Baghdad and overcast Basra, yet a second DMSP readout less than two hours later showed the reverse. Given these conditions, DMSP and supporting civil weather satellites made possible the planning and execution of the most sophisticated air campaign in history. Over the course of the 39-day air campaign, Coalition forces averaged over 2,500 sorties per day. Imagery and data transmitted from the three-satellite DMSP constellation, in particular, helped planners develop real-time schedules, make immediate, accurate retargeting decisions for reconnaissance and tactical missions, and aid in bomb damage assessments. The tactical operator could effectively choose the best weapon for the target based on known weather conditions in the target area. Current weather data proved especially useful for enhancement of night vision and infrared targeting. In this regard, DMSP weather reports proved vital for the success of precision-guided laser and optical ordnance, which depended on clear weather for accurate target designation.[45]

In addition to DMSP's importance for tactical air operations, it aided in the movement of troops during the ground war. Moreover, it helped predict and track rainstorms and sandstorms, oil fires and oil spills, and cloudcover, as well as analyze the potential spread of chemical agents and correlate storms with flood threats. General Norman Schwarzkopf, commander-in-chief of U.S. Central Command, thought so highly of DMSP that he always kept the most current DMSP data within arm's reach for quick reference. On balance, DMSP proved to be a crucial "force multiplier" during the conflict.[46]

Although the Landsat and SPOT remote sensing satellites played a larger role in Desert Storm planning, their multispectral imagery also supported tactical operations during the battle. The special capabilities of SPOT sensors proved very useful for engineers, who could adjust the optical images for off-nadir observing. This allowed any site to be viewed up to almost 1,000 kilometers on either side of the satellite's path. As a result, the normal period before revisiting the particular location could be reduced from twenty-six days to two. Aircrews in the Gulf used a data base of stereoscopic images to prepare flight routes and target attack procedures. One mission that used this system to great advantage involved an attack on Kuwait's Mina al Ahmadi oil complex, which Iraq had used to create a massive oil spill in the Persian Gulf. The U.S. F-111 pilots who bombed the well heads stated afterward that, in effect, they had flown the mission long before taking off.[47]

Although Landsat and SPOT wide-area surveillance contributions received well-deserved accolades, U.S. national reconnaissance satellites also played a key role in the space surveillance and intelligence war. Likewise, these space systems confronted the challenge of adapting their strategic capabilities to meet tactical requirements. It has been widely reported that the sensitive intelligence program directed by the National Reconnaissance Office used both optical imaging KH-11 "Keyhole" satellites and radar-imaging "Lacrosse" satellites for intelligence collection in the Gulf War. With their multispectral optical sensors, the Keyhole spacecraft are reported to be capable of achieving a resolution of nearly ten centimeters during daylight hours in clear weather. The Lacrosse radar imaging satellite gave Gulf forces the benefit of day and night coverage in all types of weather with resolutions reported to be between one and 1.5 meters.[48]

Battlefield commanders usually preferred the wide-angle imagery of the civil satellites to the incredibly large-scale detail depicted by the intelligence collectors. Yet, in many cases these satellites furnished analysts superb battle damage imagery. The ability to determine detailed damage caused by precision, "smart" weapons, for example, made it unnecessary to dispatch of reconnaissance aircraft to overfly the heavily-defended target area. Intelligence satellites also contributed to one of the most challenging missions of the war, locating Iraq's mobile Scud launchers for destruction before the DSP early warning satellites became involved.[49]

Early warning—DSP. Saddam Hussein saw in the surface-to-surface Scud missile a terrorist weapon that could split the allied Coalition and bring Israel into the war. DSP's role was to detect and provide sufficient warning for strikes against the launchers and for the Army's Patriot batteries to intercept incoming missiles. Like the intelligence satellites, DSP had been designed for national strategic objectives rather than battlefield support. Nevertheless, the measures taken during Desert Shield to make the system more tactically responsive proved successful. DSP satellites detected Scuds in time to alert civilians and military defense personnel to don their chemical protection suits and take cover, and for Patriot batteries to engage the missile.[50]

Like DSP, Patriot had not been designed for tactical ballistic missile warning in desert conditions. Limited to a 50-mile range, the system's target radar could not spot a Scud before the missile's terminal phase of flight. At the same time, the fire control electronics system often overheated in the Gulf's desert climate, which led battery operators to keep the radar systems in an inactive, standby mode until DSP satellites detected a Scud. Initially, missile crews had as little as 90 seconds after DSP warning to acquire, track, lock-on, and launch to destroy the Scud. Often the Patriot intercepted the missile at ranges of five miles or less, in full view of ground observers and a worldwide television audience. In cases where television broadcast the attack, the allies worried that Iraqi viewers could adjust to more accurately re-target the site. As the war progressed, the arrival of recently-tested Constant Source terminals in the theater gave Patriot batteries as much as five minutes warning time. In fact, DSP acquired all of the Scud launches. On balance, the missile defense system gave a good accounting of itself. Of the reported forty missiles launched against Israel and forty-six against Saudi Arabia, the vast majority either fell victim to Patriot crews or dropped harmlessly well away from their intended targets.[51]

Lessons Learned and Normalizing Military Space

Military analysts concluded that, in Desert Storm, space systems contributed to victory in the political battle, ensured effective command and control, and helped make the war a short conflict, which saved lives. Speaking for many after the conflict, General Moorman, commander of Air Force Space Command, observed:

> Desert Storm was a watershed event for space systems. Satellites, and the ground systems and people trained to control them, played a crucial role in the outcome of the conflict. Space owned the battlefield. We had a robust on-orbit constellation and the inherent spacecraft flexibility to alter our operations to support specific needs of the terrestrial warfighter.[52]

In many ways the most impressive element of the Gulf War proved to be the ability of space personnel to adapt their systems for the tactical warfighter. At the outset of Desert Shield, few of the space systems were in position to provide the support upon which Coalition forces would come to depend. Most had been designed during the Cold War to satisfy strategic requirements. Planners needed the full five months of Desert Shield to optimize space and ground segments and to create the necessary inter- and intra-theater infrastructure. Fortunately, in Saddam Hussein, the Coalition faced an enemy without significant space assets of his own, and one unwilling to prevent the buildup or seriously menace the vulnerable space network in place by January 1991. At the same time, Coalition members benefited from not having to face another conflict while dealing with the Gulf crisis. They remained well aware that their good fortune could not be guaranteed in future situations.

While basking in the glow of a justly-praised, decisive victory, the U.S. space community sought to learn and improve. The Air Force was at the core of this effort. Postwar analyses correctly emphasized deficiencies and the challenges ahead. Above all, analysts realized that in spite of mounting the largest contingent of space-based forces to date, their systems proved insufficiently designed for tactical use, and ground personnel often lacked the necessary equipment and training to fully exploit space capabilities. Even systems traditionally more oriented toward tactical operations encountered problems. In order to meet the challenge of supporting the warfighter, Air Force leaders realized that they must lead the effort to modernize space infrastructure, continue technical improvements to space systems, and extend space awareness throughout the Air Force and the armed forces as a whole. They expected Desert Storm to provide the momentum in the early 1990s for improvement in every area of space operations. The attention of the military space community focused on systems and capabilities represented by the four mission areas first established in the mid-1980s: space control, force application, force enhancement, and space support.

Space control referred to operations to maintain friendly use of space and deny the same to potential enemies. Over the years authorities in this area, had been most successful in developing protective measures for satellite systems by means of hardening, increasing the number of satellites, applying better tactics, and deploying mobile ground segments. Likewise, space surveillance capabilities had improved with more effective radar, electro-optical, and passive radio frequency sensors. Although attempts to deploy an operational antisatellite system had proven unsuccessful, Desert Storm renewed interest in developing such a capability.

Ever since the beginning of the space race, the Air Force had stated a need for an antisatellite system. By the mid-1980s, the service began testing an F-15-launched heat-seeking antisatellite weapon, termed the Miniature Homing Vehicle. Concerned about expense and the system's potential impact on the arms race, however, Congress in 1988 banned further testing. The Air Force subsequently canceled the program. Although the congressional decision addressed orbital testing specifically, the Army proceeded with tests of a ground-based kinetic energy interceptor as part of the Strategic Defense Initiative. Program managers remained unenthusiastic about the Army's antisatellite requirement, even as they continued to develop the weapon in the early 1990s.[53]

The Gulf War convinced commanders of the importance of satellite reconnaissance and the need to deny it to potential enemies. General Charles A. Horner, commander-in-chief of U.S. Space Command and commander of Air Force Space Command after Desert Storm repeatedly argued for the capability to destroy foreign satellites, even those belonging to allies if they were aiding an enemy. Other Air Force leaders agreed on the need to control space. As Air Force Secretary Sheila Widnall asserted in the fall of 1994, "Part of the Air Force mission is control of space, our ability to deny the use of space if necessary." Despite the pleas from Desert Storm leaders, the antisatellite program was confined at mid-decade to a research effort by all three services.[54]

The force application area confronted similar roadblocks to the use of military weapons in space. This element comprised fire support operations from space against enemy forces by means of ballistic missile defense and "power projection" operations against terrestrial targets. The latter represented only a theoretical application, and no plans existed to include power-projection space weapons in the force structure. Ballistic missile defense, which had developed under the auspices of the Strategic Defense Initiative, called for layered defenses comprised of both terrestrial and space assets. Space-based elements included Brilliant Eyes and Brilliant Pebbles. Both programs had experienced considerable change over time as a result of the SDI's uncertain fortunes. In the early 1990s, Brilliant Pebbles envisioned satellite interceptors designed to demolish ballistic missiles in their midcourse and terminal phases of flight. But, like the antisatellite program, the Strategic Defense Initiative Organization's space-based interceptor was limited to a technology base program, with the objective only of developing technologies as security against potential future threats.[55]

Brilliant Eyes, however, drew more interest from Air Force space officials who sought to improve theater space surveillance capabilities after Desert Storm. The Brilliant Eyes concept called for a "distributed" satellite network consisting of several hundred spacecraft with infrared and laser sensors orbiting at 700 kilometers, capable of tracking missiles in their midcourse phase, discriminating among reentry vehicles and decoys, and predicting impact points. Like national reconnaissance assets, these satellites would also perform an arms control monitoring function. Air Force Space Command had been interested in Brilliant Eyes because of its relationship with plans to upgrade the DSP early warning satellites. On the other hand, it seemed that the project did not address the command's space surveillance requirements. Following the Gulf War, the issue became more complicated when the Strategic Defense Initiative Organization decided to refocus the program to emphasize theater missile defense. The new concept called for development in stages to provide both national and global protection against ballistic missiles. The reoriented Brilliant Eyes distributed sensor program, now managed by Air Force Space Command, led Air Force space authorities to reexamine the operational implications for their own missile warning and space surveillance requirements.[56]

In the mid-1990s both the ballistic missile defense and antisatellite programs continued, but only as research efforts that reflected considerable debate about their necessity, feasibility, and cost-effectiveness. By contrast, postwar interest in improving the military space posture centered on programs embraced by the mission areas of force enhancement and space support.

The elements of the force enhancement area involved space combat support operations, which had proven the most visible and important during the Gulf conflict. Dating back to the Eisenhower era, they represented the traditional defense support functions of communications, tactical warning and attack assessment, navigation, environmental monitoring, reconnaissance, and surveillance.

Space-based reconnaissance and surveillance systems. These systems operated under the auspices of the National Reconnaissance Office, which remained a sensitive agency in spite of the first public discussion of its activities in September 1992. Following Desert Storm, in which "national" space sensors proved capable of providing outstanding resolution but little wide-area surveillance, overcoming this coverage limitation became part of a broad effort to increase space-based wide-area surveillance. In the post-Cold War arena, intelligence and surveillance imaging systems required more flexibility to respond to rapidly-developing tactical intelligence needs. Some critics recommended a distributed system of satellites that could be "everywhere, all the time." But did this mean replacing the large, heavy strategic-oriented reconnaissance and radar-imaging satellites with smaller spacecraft carrying lower-resolution sensors? The latter would also have the advantage of not requiring launch by the expensive, heavy-lift Titan IV booster. By mid-decade, planners at the National Reconnaissance Office were reported to be weighing a new system of numerous smaller, lighter satellites against retaining the existing configuration that could be improved to image eight times the area of current satellites.[57]

Nevertheless, the national intelligence space community continued to operate largely independent of the broader military space sector. This had been a major complaint of Air Force space agencies which worried that the intelligence community too often failed to coordinate its space requirements to best support the warfighter. A report by the vice president's National Space Council late in the Bush administration called for reduction and, where possible, elimination of security constraints that continued to work against effective integration of the military, civil, and commercial space sectors.[58]

Environmental monitoring, including weather and earth-sensing satellites. Evaluators of satellite performance in the Gulf War concluded that while DMSP "exceeded expectations," data must be made available in greater volume and more frequently. Criticism centered on the lack of sufficiently mobile receivers in the field. Because Army units below theater level had no access to the non-mobile Mark IV receiver, they had no direct means of using DMSP data and relied, instead, on a commercial receiver for satellite data. Only late in the war did the Air Force introduce two prototype terminals small enough for use in the rear of the Army's High Utility Multipurpose Vehicle, or "Humvee." Air Force analysts agreed that they needed to develop and field sufficient numbers of mobile high-resolution DMSP receivers capable of obtaining imagery and data from both military and civil satellites.[59]

Planners expressed satisfaction with the evolutionary development of the weather satellites and, together with deploying more mobile terminals, expected to replace the current DMSP Block 5D-2 spacecraft with the more capable 5D-3 series early in the new century. On the other hand, the widespread use of military and civil polar-orbiting satellites for weather data precipitated reconsideration of the old issue of combining programs to save money and avoid redundancy. With post-Cold War budget austerity looming, proponents of "convergence," or merging the systems, optimistically forecast success in this eighth major examination of the question. Air Force officials appeared more favorable as long as service requirements could be satisfied. Yet the postwar plans for convergence of the civilian and military systems foundered on the same hurdles that had prevented the success of earlier efforts. NOAA's international commitments meant including the Japanese and Europeans, who balked at participating in a military system and at agreeing to U.S. control of weather data during hostilities.[60]

Postwar reports also examined the need for an advanced multispectral imaging (MSI) capability. Landsat and SPOT, they noted, did not always provide timely or accurate data for mission planning, bomb damage assessment, or use of precision guided weapons. Neither system combined sufficient spectral and spatial capabilities, while efforts to merge MSI data to obtain the required resolution proved slow. National systems, on the other hand, provided superb spatial resolution, but little overall view of the battle area. They lacked sufficient MSI capability and proved awkward to handle because of their security classification. A future system, analysts concluded, should possess a wide-area MSI system, high-resolution and spectral coverage, direct links to users, and timeliness. Meanwhile, in the aftermath of Desert Storm, the Defense Department worked with NASA to achieve improved multispectral imaging capabilities through acquisition of the *Landsat 7* spacecraft.[61]

Navigation—GPS. Evaluations of the GPS navigation system expressed little criticism of the system's performance, especially after three-dimensional coverage could be supplied. Like DMSP, the assessment of GPS centered on the shortage of mobile receivers. Planners simply had not foreseen the need for sizable quantities of mobile receivers for large maneuvers or operations like Desert Storm. With so few military Selective Availability receivers on hand at the outset of Desert Shield, Coalition forces requisitioned less accurate, but more functional commercial sets. It took up to six months for their arrival, which limited training effectiveness. Above all, use of commercial receivers caused military authorities to forego encryption, which would have provided data with even greater accuracy. Throughout Desert Shield and Desert Storm, managers of the GPS program worried that Iraqi forces might take advantage of the "open" signals. Understandably, their primary postwar recommendation involved securing sufficient crypto-capable GPS receivers and providing thorough training in their use. Even so, Selective Availability remained a source of debate because space officials knew that a process termed Differential-GPS allowed users to work around Selective Availability to achieve highly accurate position information. Air Force priority following the Gulf War involved achieving initial operational status with a full constellation of 24 satellites, which occurred in 1993, and completing installation of receivers in aircraft by the end of the decade.[62]

Tactical warning and attack assessment. The integration of DSP warning and Patriot anti-tactical ballistic missile capability proved to be one of the great achievements of the Gulf conflict. Although the anti-Scud warning system exceeded every expectation, it took months to develop the necessary coordination with a space system that originally had been designed and configured for strategic warning, not tactical defense. Warning time for Patriot crews remained uncomfortably brief, while the satellite sensors could provide only general launch site and impact point identification. Desert Storm had clearly demonstrated the need for improved ballistic missile tactical warning and assessment and midcourse tracking capability. For years the Air Force had sought to improve the warning satellite network. In the early 1980s, the service studied a replacement program called the "Advanced Warning System," which the Strategic Defense Initiative Organization absorbed in 1984 and renamed "Boost Surveillance and Tracking System." By the end of the decade, pressure from a Congress worried about violating the anti-ballistic missile treaty led instead to the Organization's adoption of Brilliant Eyes and the return of the Boost Surveillance and Tracking System to the Air Force as a potential DSP replacement. By 1991 a modified version of the program reemerged as the Follow-on Early Warning System (FEWS).[63]

In the wake of Desert Storm, leaders like General Horner made FEWS their highest priority. The new system would retain its strategic capability but be far more effective against short-range tactical missiles. It would possess greater on-board processing capability and more flexible communications. Mobile DSP receivers would also become part of the improved system. U.S. Space Command's postwar assessment made its number one recommendation the normalization of tactical warning support through provision of the necessary equipment, including requirements in operational plans, and thorough training of personnel. Evaluators went on to declare that "what was said about warning is true for all space systems. All space support should be normalized…[and] institutionalized." In short, space must be integrated into all preexisting and new plans and become part of the mainstream for all services. Air Force space leaders had been promoting this theme for years, and would soon return to it in their renewed campaign for military space leadership.[64]

High costs and technological challenges, however, led FEWS down a rocky road in the early 1990s. A number of studies compared FEWS with various alternatives, including a combination of Brilliant Eyes and DSP satellites and a modified DSP system. Air Force officials also eliminated a number of FEWS sensors to reduce the expense and weight, which made it a candidate for the less costly Atlas II booster instead of the Titan IV. A replacement for DSP still had not been determined by the time the Clinton administration assumed office in early 1993. The new defense team made the issue a major focus of its so-called Bottom Up Review of defense acquisition programs. On the basis of the review, in early 1994 Defense Secretary Les Aspin canceled FEWS, declaring it too expensive. Meanwhile, the search continued for a more effective system than DSP. Replacement options for DSP included an Alert Locate and Report Missiles system (ALARM), that was proposed in the summer of 1994. After assessing several ALARM proposals consisting of both geosynchronous and low-earth orbiting satellites, congressional opposition led planners to settle on a new Space-Based Infrared System, termed SBIRS, that would involve a configuration of four geosynchronous-orbit satellites in its first phase.[65]

Communications. Along with tactical warning and attack assessment, military satellite communications (MILSATCOM) received special attention from the planners and postwar analysts. Not only did the Gulf War highlight both its importance and shortcomings, but for many years projected replacement systems had been among the most controversial space programs under development. Changing requirements, technological challenges, and high costs had led to delays and restructuring—and growing doubts about their operational future. To be sure, communications satellites provided Desert Storm forces the support they needed in spite of the variety of systems used and the fact that they had not been designed to provide intra-theater communications between commanders in the field. Moreover, they were stretched to the limit, highly exposed to jamming, and far less mobile than ground forces desired. DSCS, for example, provided rapidly moving ground forces a multichannel terminal with an 8-foot satellite dish, but available power and bandwidth limited the system's capacity, and prompted use of the higher-gain 20-foot antenna. The trade-off became less mobility and greater exposure to enemy forces.[66]

Military communications authorities looked forward to upgrading existing networks like DSCS, developing more-mobile receiver terminals, and introducing subsequent systems like the Navy's UHF "follow-on" system and Milstar to ensure reliable, global communications support—especially for the tactical operator. Most MILSATCOM concerns centered on the troubled Milstar (Military Strategic and Tactical Relay) program. In 1989 Congress had directed the Pentagon to restructure the program from strategic communications support to tactical requirements at considerably less cost. The Gulf War occurred in the middle of a major effort to reorient Milstar and save the program from outright termination. Desert Storm might well have rescued Milstar from certain cancellation.[67]

Milstar had emerged in the late 1970s from an Air Force proposal for a strategic satellite system (Stratsat). Stratsat was to consist of a four-satellite constellation designed entirely to support nuclear forces. It would avoid potential antisatellite threats by orbiting at supersynchronous orbits of about 110,000 miles, and operate in the extremely high frequency (EHF) range to provide more bandwidth for spread-spectrum anti-jam techniques. Considered too ambitious for so limited a mission, Stratsat gave way in 1981 to Milstar. Planners viewed Milstar as capable of both strategic and tactical operations, and they proceeded to add to the design numerous additional missions and requirements. In 1983 President Reagan accorded Milstar "highest national priority" status, which allowed the program to proceed with little regard for funding restrictions.[68]

From the perspective of systems development, Milstar represented a throwback to "concurrency" procedures which had characterized the Atlas ballistic missile development program of the 1950s. Program leaders tried to develop Milstar's "cutting edge" technology and system procurement concurrently, which led to delays, redesigns, and cost overruns. Thirty years earlier these kinds of problems seemed relatively unimportant to an Eisenhower administration desperately determined to produce an operational missile as quickly as possible. Over the course of the 1980s, however, Milstar's difficulties increasingly drew the ire of a budget-conscious Congress.[69]

Initially designed to provide EHF low-data-rate (LDR) communications, the eight-satellite constellation would benefit from crosslink capabilities and extensive hardening against radiation. The EHF range had the advantage of allowing use of antennas as small as six inches in diameter, which suited highly mobile special operations forces. Launched by the Titan IV, four of the satellites would operate in various polar orbits and the other four in geosynchronous orbits. Because Milstar's chief goal was survivability and not high performance, planners did not design it for high data rates or for each satellite to serve more than fifteen users simultaneously. As a result, it would supplement rather than replace existing satellites like DSCS and FLTSATCOM.[70]

Increasingly plagued by cost overruns and schedule delays, Milstar's original strategic orientation seemed anachronistic in the post-Cold War world. Desert Storm, however, reinforced interest in promoting Milstar's tactical capabilities, and the program underwent several alterations in the early 1990s after Congress demanded its restructure. Milstar also became a subject for the Pentagon's Bottom Up Review in early 1993. By early 1994, the program envisioned six rather than eight satellites, without the vast array of survivability features, and with fewer ground control stations. The first block of two satellites, referred to as *Milstar I*, would retain the limited-use LDR capability, but the next-generation *Milstar II* satellites would be equipped with a medium-data-rate (MDR) package to support tactical forces. On 7 February 1994, seven years after its projected launch, the first *Milstar I* satellite achieved a successful launch and on-orbit checkout. A second Milstar satellite was launched in November 1995, and the following month system operators achieved crosslinking. Plans called for launch of the first *Milstar II* satellite in 1999, and, by the year 2006, Air Force program managers expected to see a transition to a less expensive, lighter *Milstar III* Advanced EHF satellite. In the future, Milstar would supplement the Navy's FLTSATCOM and Ultra High Frequency (UHF) Follow-on (UFO) satellites but coexist with the successor to DSCS III. Nevertheless, in the mid-1990s the jury remained out on Milstar's ability to provide survivable, jam-resistant, global communications to meet the needs of the National Command Authorities, battlefield commanders, and operational forces at all levels of conflict.[71]

Although the Milstar debate drew most of the attention in the early 1990s, Desert Storm also raised the issue of small satellites for tactical communications that had been championed by Air Force operational leaders in the late 1980s. Although the two Scout-launched Marine MACSATs received considerable publicity, most analysts determined that "the war showed there was no limitation of capacity for tactical commanders, so the word is there is no need for small satellites." The Pentagon trend in the 1990s still seemed against light "tacsats" in favor of higher frequencies and other high-power transmission requirements, survivability, and tri-service satellite systems. Even so, DARPA and the services persevered with their modest light-satellite programs, and the debate continued.[72]

Space support. Space launch completed the triangle of special interest space programs in the Pentagon's early 1993 Bottom Up Review of space acquisition systems. It fell within the mission area of space support, which involved deploying and sustaining military systems in space. Desert Shield and Desert Storm had exposed the Achilles heel of the space program. When personnel from U.S. Space Command and Air Force Space Command reviewed U.S. Central Command's request in the fall of 1990 to launch an additional DSCS III satellite, they quickly determined that the launch needed to await completion of the Atlas II's new Centaur upper stage, scheduled for July 1991. To be sure, from August 1990 to the end of Desert Shield, six military satellites joined the existing network, and all contributed to Desert Storm operations. Yet these spacecraft had been scheduled well in advance of Desert Shield. In effect, the U.S. space launch system continued to reflect a policy of launching on schedule, not on demand. It simply could not respond to short-notice requests.[73]

The familiar conditions that had made space launch the space program's weakest link dated to the Eisenhower presidency. Beginning in that era launch systems and infrastructure had supported research and development rather than operational requirements, then fell into decay following the decision to use the Shuttle in place of expendable boosters. Despite the demand for expendable rockets after the *Challenger* tragedy, the space industry could not retool fast enough to meet rising demand. Moreover, military, civil, and commercial launch needs meant supporting three separate launch teams and related equipment. Aging boosters and range system components, as well as inefficient production lines and launch procedures, resulted in an expensive, operationally limited system. Space leaders had been trying for years to solve the launch dilemma.

On the eve of Desert Storm, the Advisory Committee on the Future of the United States Space Program, known as the Augustine Committee after its chairman, Norman R. Augustine, recommended deemphasizing Shuttle operations and developing "an evolutionary, unmanned but man-ratable heavy lift launch vehicle." On 24 July 1991, President Bush announced a new national space launch strategy that incorporated many suggestions offered by the Augustine Committee. Calling for continued use of improved expendable boosters, the President's National Launch System called for a new, heavy lift vehicle that would reduce launch costs and improve performance. Yet, late the following year, in November 1992, the Vice President's Space Policy Advisory Board's Task Group on The Future of the U.S. Space Launch Capability suggested canceling the National Launch System. Chaired by former Air Force Secretary Edward C. Aldridge, the Aldridge Report proposed instead a new program called "Spacelifter." The Spacelifter would be developed under Air Force leadership, using a single "core" vehicle to meet lift requirements of all three sectors of the space community—military, civilian, and commercial. This seemed to reflect Air Force Space Command's interest in developing a family of vehicles leading to operational systems.[74]

The Clinton administration's Bottom Up Review seemed to favor a return to improved expendable launchers. Its analysis addressed the launch issue in terms of several options. One would be to extend the existing fleet to the year 2030; a second to develop a new family of expendable launch vehicles to replace the current fleet beginning in 2004; a third to promote a technological effort to develop a reusable vehicle; and finally, "austere" variations of the first two alternatives. The Defense Department decided on an austere approach, funding only required improvements to existing launch and range infrastructure. In short, officials decided to proceed with modest improvements to the status quo, which many considered an acceptable solution. In the mid-1990s, the space launch issue remained far from resolved.[75]

In the aftermath of Operation Desert Storm, the Air Force played the central role in evaluating the capabilities of space systems to meet the needs of the warfighter. Air Force leaders realized that they must provide the necessary leadership if military space were to benefit from infrastructure modernization and new technological initiatives and, ultimately, achieve "normalization" of space within the Air Force and throughout the military community. But the momentum for change represented by the performance of space assets in the Gulf War diminished considerably when confronted by the challenges of developing a new generation of space systems and an effective launch capability. Moreover, continued fragmentation of the nation's space community in an era of budget austerity severely hampered efforts to make the changes Air Force leaders deemed essential. The situation called for strong, central direction, and the Air Force responded with another initiative, one designed to chart the course of military space into the post-Cold War future of the 21st century.

CHAPTER 8

An Air Force Vision for the Military Space Mission:

A Roadmap to the 21st Century

In the aftermath of Desert Storm Air Force leaders took significant steps to establish the Air Force as the "lead" service for space. Their motivation stemmed from multiple sources. It reflected the pride of having been the principal steward of military space capabilities for over thirty years. As important, the Air Force space community recognized the crucial turning point represented by space accomplishments in the Gulf War and anxiously sought to apply the operational "lessons learned." Finally, Air Force leaders saw that in a post-Cold War world, Air Force leadership in the space arena remained not only critical for the future of space within the service, but essential to support the new demands of "global power and global presence."

A Generation of Leadership in Military Space Activities

In promoting its leadership role in military space for the 21st century, Air Force leaders relied on the institutional memory and experience acquired over more than a generation, dating back to the second Eisenhower administration. The Air Force role in space proceeded along two broad levels: one involved a number of "campaigns" to convince national leaders that the imperative of national security required assigning the Air Force sole responsibility for military space activities that included development and deployment of weapons in outer space; the other centered on continuation of the effort to institutionalize space within the Air Force and throughout the armed forces by transferring responsibility for Air Force space activities from the realm of research and development to the operational side of the service. Space advocates believed that "normalizing" and "operationalizing" space within the Air Force would also buttress the service's claim to be designated executive agent for space and to lead the national military space program. Air Force space pioneers never achieved the lofty goals they established for the service. Like their aviation counterparts after World War I, ambitious Air Force space agendas did not always receive sufficient support from within the service, or from national leaders who remained opposed to an expanded military space program. Nevertheless, the Air Force achieved a remarkable record as the service preeminently involved in initiating, developing, and applying the technology of space-based systems in support of the nation's security.

The foundation for Air Force space leadership was established before the Eisenhower era, at the close of the Second World War. At that time, the Army Air Forces took two important steps to set the stage for an Air Force future in space. With the publication of *Toward New Horizons* in late 1945, Commanding General of the Army Air Forces Henry H. "Hap" Arnold and his close friend and Chairman of the Scientific Advisory Board Theodore von Kármán provided the service a sound research and development focus and an agenda for the future. Shortly thereafter, in early 1946, the service-sponsored Rand Corporation issued its prescient report on satellite feasibility, *Preliminary Design of an Experimental World Circling Spaceship*, which predicted that an artificial Earth-observation satellite could be launched within five years. Neither *Toward New Horizons* nor the Rand report produced a rush to develop space capabilities in light of Cold War tensions during the late 1940s, tight budgets, and focus on the strategic bomber as the first line of national defense. Even so, if the newly designated United States Air Force proved unwilling to seriously pursue satellite development itself, it was determined to prevent the other services from capturing what it termed the "space mission."

The Air Force renewed its interest in satellites in the early 1950s, when technological progress affirmed the promise of long-range ballistic missiles carrying thermonuclear devices, and the new Eisenhower administration took measures to defend the nation from a surprise attack. Development of an American ICBM represented one means of strengthening national defense; at the same time, development of a reconnaissance satellite, launched into orbit by rocket boosters, offered the prospect of obtaining vital strategic intelligence data on the Soviet Union. In 1954, Rand's landmark Project Feed Back report affirmed the technical feasibility of artificial satellites and recommended the Air Force develop an electro-optical reconnaissance satellite to meet President Eisenhower's requirements.

The administration initially supported the missile and satellite efforts of all three services. The Air Force redesigned and intensified its development of an earlier Convair ICBM proposal, which it renamed Atlas and placed under the direction of the hard-charging Brigadier General Bernard A. Schriever. By 1957 the crash program led by Schriever and his cohorts at the newly-established Western Development Division encompassed not only the Atlas ICBM, but the Thor IRBM and, in conjunction with Lockheed's Missile Systems Division, the military reconnaissance satellite project. The latter would lead to the Agena upper-stage booster, the Defense Support Program's early warning infrared satellite, and the recon naissance satellites managed as a national program by the National Reconnaissance Office.

At the same time, the Eisenhower administration established a "freedom of space" policy that promoted unrestricted overflight to allow the free passage of military reconnaissance satellites. This meant emphasizing civil spaceflight and prohibiting the deployment of space-based weapons. Air Force leaders believed otherwise. They preferred to guard against potential threats, and viewed the self-limiting "space for peaceful purposes" policy dangerous and self-defeating. Space-based weapons would remain restricted to studies only. Consequently, Air Force space efforts centered on what came to be called defense support functions—reconnaissance and surveillance, early warning, navigation, communications, and meteorology. These activities, and the Eisenhower space policy that framed them, would endure largely unaltered for the next 30 years.

The launch of the Sputnik satellites in late 1957 intensified an already heated contest for leadership of the national space program among the Army, Navy, and Air Force. Air Force leaders coined the term "aerospace" to justify their claims, first to lead the national space effort and, when that failed, to be designated the executive agent for all of military space. The Air Force confronted a host of competitors in its bid for space primacy. The creation of NASA in 1958 proved a mixed blessing for the Air Force. On the one hand, NASA acquired its most important space assets from the Army and Navy which, by 1960, left the Air Force the dominant military space service and NASA dependent on the Air Force for support. On the other hand, NASA now would chart the nation's civil spaceflight future and compete for space funding. Moreover, in the military sphere, the Air Force found itself subordinated to the Pentagon's Advanced Research Projects Agency and, later, to the Director of Defense Research and Engineering. A new competitor appeared in 1960 when the administration created the National Reconnaissance Office (NRO) to manage Project CORONA, the sensitive reconnaissance satellite program. Although directed by the Under Secretary of the Air Force, this Central Intelligence Agency-Air Force program would remain outside the control of Air Force headquarters. Finally, for defense support missions, the Air Force often had to share responsibilities with other services and agencies.

By the end of the Eisenhower presidency, the Air Force clearly had not achieved the "independent" leadership position claimed by its most ardent spokesmen. Even so, it could point to an impressive list of achievements that included providing the bulk of space booster and infrastructure support and managing the early warning satellite and the ground-based space surveillance network. Air Force leaders also thwarted two attempts by their Army and Navy rivals to create a unified command for military space activities. Responsible for nearly 80 percent of the military space budget, it clearly found itself the leading service for military space.

In the spring of 1961, Secretary of Defense Robert McNamara designated the Air Force the military service for space research and development. As part of the arrangement, the Air Force reorganized to create a more centralized focus for space by establishing Air Force Systems Command with General Schriever as its first commander. The previous year the Air Force had created the nonprofit Aerospace Corporation to provide needed technical expertise. The beginning of the Kennedy administration was a period of high expectations for Air Force space leaders, who believed they had a "green light" to promote an expanded military space program and gain recognition as "executive agent for military space." The Air Force agenda included making permanent NASA's early dependence on the service. As executive agent for NASA support, the Air Force sought an equal partnership with NASA in the decade ahead. The service also attempted to convince Defense Department officials that the military had a legitimate requirement for a manned space mission apart from NASA's program. Manned spaceflight was also seen as the best means of generating support for space within the Air Force. Finally, despite established national space policy, the Air Force strongly lobbied for permission to develop space-based antisatellite weapons.

Well before the end of the decade, the Air Force campaign had failed all across the board. NASA basked in the glow of the unprecedented Project Apollo moon landing. Meanwhile, Air Force efforts to make military manned spaceflight the focal point of a space-oriented service ended when President Nixon in 1969 canceled its remaining human spaceflight project, the Manned Orbiting Laboratory. The Air Force could never convince the Defense Department that the military had a legitimate requirement for a man-in-space "mission." Likewise, Air Force attempts to move space-based weapons projects beyond the drawing board proved fruitless. Air Force pretensions to lead an expanded space program received further setbacks under the Defense Department's policy of tri-service management and military-civil cooperative efforts designed to reduce costs and service bickering. With the larger Air Force space agenda unrealized, the service's research and development organizations, by default, assumed operational responsibility for space programs and systems. This set the stage for the future contest between R&D and operational elements for control of Air Force space. Meanwhile, by the end of the 1960s, Air Force leaders downplayed space issues and spoke instead of taking care of traditional Air Force aviation needs.

At this juncture, two developments reinvigorated the Air Force space program. One proved to be the rapid growth of unmanned, instrumented spacecraft and their potential importance for military operations. Communications (DSCS) and weather

(DMSP) satellites provided crucial data to commanders in Vietnam, while by the early 1970s, the Air Force had launched its first early warning satellites (DSP) and readied for development the nation's first three-dimensional satellite navigation system (GPS). The National Reconnaissance Office also prepared to launch the successors to Project CORONA reconnaissance satellites. Artificial earth satellites were coming of age.

The other important development was the advent of the Space Shuttle, the NASA-Defense Department project for a reusable launch vehicle that NASA predicted would provide more inexpensive and more frequent access to space. Under pressure to use the Shuttle in place of expendable boosters, the Air Force agreed to assist with development costs, produce an upper-stage vehicle, and construct a West Coast launch facility. In return, NASA accepted an enlarged cargo bay to accommodate military satellite requirements and resolved to give Defense Department missions operational priority. Air Force space enthusiasts could also argue that involvement with the Shuttle preserved a military manned spaceflight mission.

The coming of the Shuttle and artificial satellites compelled Air Force leaders in the 1970s to seriously address organizational issues. Because satellites increasingly provided operational support to a variety of users, the practice of assigning operational responsibility to one particular Air Force command seemed inappropriate. Likewise, by the mid-1970s four Air Force commands promoted themselves as best qualified to manage Shuttle operations. The potential operational impact of space systems prompted Air Force leaders to assess the importance of space for operational commanders and the service's institutional commitment to a space future. The growing debate focused on whether the research and development community should continue to launch and control space systems or relinquish those responsibilities to the operational side of the Air Force. If the latter, should the Air Force create a new, major command for space operations? The decade of the 1970s witnessed a plethora of studies, conferences, and symposia that helped to build consensus for an operational space focus within the Air Force. At the same time, the contributions of the space systems themselves showed that they had moved beyond the experimental stage and could no longer be confined to the research and development realm. By the early 1980s, the Reagan administration's interest in an expanded defense space program provided important momentum for organizational changes already underway within the Air Force. By late summer 1982 the Air Force had an operational Space Command—for the price of a unified space command to follow three years later.

During the 1980s, Air Force Space Command needed to acquire systems, gain the necessary experience, and convince the wider Air Force of the operational importance of space for traditional missions. Becoming operational proved to be a long and difficult process. Not until 1993, for example, did the research and development community relinquish complete responsibility for satellite control and space launch.

Along the way, the new Air Force Space Command had to establish effective relationships with the unified command and deal with the launch crisis following the *Challenger* tragedy. The latter precipitated not only a return to expendable boosters but, also, a reexamination of the Air Force commitment to space. By the end of the 1980s, Air Force leaders referred to the responsibility of the Air Force as the "lead service for military space" to "normalize" and "operationalize" space within and outside the Air Force—in short, to institutionalize space to the point where space systems furnished support essential to the warfighter.

Desert Storm provided the needed catalyst in the "operationalization" of military space systems. In the Persian Gulf conflict, space systems that had traditionally performed a strategic function proved sufficiently flexible in a tactical environment to provide critical support to the warfighters. Space systems helped achieve victory, which served as the springboard for Air Force leaders to assert their vision for the nation's space program and the Air Force's leadership role in achieving it.

An Air Force Vision for Another Generation of Space Leadership

In order to chart the course for the Air Force space program into the next century, Chief of Staff General Merrill McPeak in the fall of 1992 established another Blue Ribbon Panel on space. Led by Lieutenant General Thomas S. Moorman, Jr., vice commander of Air Force Space Command, it included nearly 30 officers and civilians from Air Force headquarters and the major commands. Like the Blue Ribbon Panel of four years earlier, the Moorman Panel addressed space roles and missions issues that affected the Air Force internally. But the new panel, in the aftermath of space contributions to Desert Shield and Desert Storm, expanded its analysis to emphasize the role of the Air Force in the wider military and national arena. Meeting at Maxwell Air Force Base, Alabama, from early September to early November, the panel reviewed existing Air Force space policy, organization and infrastructure, charted the service's future role in space, developed a strategy to achieve that objective, and outlined an action plan for Air Force leaders to follow. The Moorman Panel issued its report in early January 1993 during the closing days of the Bush administration.[1]

The panel envisioned the future Air Force as a thoroughly integrated air and space force that reflected General McPeak's unprecedented mission statement of June 1992, which declared air and space coequal. Moreover, in the world of the next century, the Air Force would be the linchpin in the nation's strategy of projecting military power rapidly and decisively with expeditionary forces. Space would provide the "global eyes and ears" that would ensure "global reach and global power." In short, space represented the decisive edge for the warfighter. General McPeak's mission statement also flatly asserted that "the Air Force will lead the Defense Department in the acquisition, operation and application of space capabilities to preserve the peace and win in war." The Moorman Panel focused on these three areas in its assessment of the Air Force's future leadership role. In the area of acquisition, the panel examined ways to reduce the costs of acquiring and maintaining space systems, of making operational requirements the driving force in the acquisition process, and of ensuring U.S. space superiority through innovative, sophisticated technological solutions. Operational objectives included establishing space control capabilities equivalent to the air superiority mission, providing responsive space launch and on-orbit control, and leading the armed forces in providing "an integrated aerospace control system—air, missile, and space defense—for combatant commanders." Finally, for space applications, the panel examined how the Air Force could become the "preeminent service for the exploitation of space capabilities" and produce a "space applications mindset" throughout the Air Force.[2]

The Moorman Panel also incorporated the results of a number of space studies that emerged in late 1992 during the closing months of the Bush administration. Three task group reports of the National Space Council addressed America's future in space: "The Future of the U.S. Space Launch Capability," by Edward C. Aldridge, Jr., (November 1992); "The Future of the U.S. Space Industrial Base," by Daniel J. Fink (November 1992); and "A Post Cold War Assessment of U.S. Space Policy," by Laurel Wilkening (December 1992). The Aldridge report, as noted above, called for replacement of the National Launch System with the "Spacelifter." Following publication of the report, Congress directed cancellation of the National Launch System. Senior leaders of the Air Force responded by agreeing among themselves that the Air Force would lead the national effort to develop a responsive launch system. In early 1994 they directed a comprehensive "Space Launch Modernization Study" by a distinguished committee of forty experts from all space sectors and chaired by General Moorman. The Moorman Committee was charged with developing an extensive requirements data base, synthesizing the needs of space launch for the commercial, civil, and national security sectors, then compiling options and "roadmaps." It promised to be the most credible effort to date to solve the launch problem.[3]

The Blue Ribbon Panel also took into account the Fink report, which stressed the importance of coordinated Defense Department-NASA measures to achieve more efficient procurement and lower operating costs while maintaining vital space technologies and facilities within a reduced space industrial base. The Wilkening report advocated more centralization and efficiency across the military and civilian space sectors, as well as increased cooperation among civil, military and commercial space elements to better confront international space competition. Finally, the panel remained well aware of the February 1993 *Triennial Report to Congress on Service Roles, Missions and Functions* by the Chairman of the Joint Chiefs of Staff. The controversial report proposed to eliminate U.S. Space Command and make U.S. Strategic Command responsible for the space mission. Doing so would likely mean the end of the Army and Navy space commands. In the final report, the panel considered their recommendations consistent with the decisions and findings of the space studies that occured during their deliberations.

In its critique of the acquisition area, the panel found widespread "fragmentation and duplication of effort" that resulted in expensive, inefficient, "stove-piped" systems, whereby each agency pursued its own agenda without attempting to support multiple requirements and systems. In a world of declining space budgets, the Defense Department and the Air Force no longer could afford costly duplication and many one-of-a-kind satellite systems. Moreover, operational users continued to lack sufficient voice in the requirements process, which stemmed in large part from ignorance of space capabilities. On one level, the panel called for a "summit" process to spread space knowledge throughout the Air Force. On another, more visible level, it called on the Air Force to "seek designation as the single manager for DoD space acquisition." Although the other services would participate, the Air Force would become the focal point for acquisition.[4]

In the operational area, the panel declared that the "Air Force should be designated as the single manager for DoD space operations." Taking its lead from the January 1993 Joint Chiefs of Staff study on roles and missions, the panel called for an end to the Army and Navy space commands. After all, it argued, the Air Force performed 90 percent of Defense Department space operations, and eliminating the other services's space commands would encouage an end to "stove-piping" and duplication. The panel also recommended development of a new launch capability to replace the unresponsive Eisenhower-era fleet of expendable boosters, production of a space-based antisatellite system to counter the growing space capabilities among potential enemies, and a commitment to producing an effective ballistic missile defense. Finally, the panel called on the Air Force to enhance space support through improved arrangements with allies and commercial space companies, as well as providing the doctrine and capabilities to win the emerging space "information war."[5]

In the third area, applications, the panel found the Air Force woefully behind the Navy and Army in integrating and applying space capabilities on the battlefield. It cited the examples that only five percent of the service's aircraft had Global Positioning System receivers installed, and that little Air Force commitment existed to programs such as Tactical Exploitation of National Capabilities (TENCAP). To right the situation, the panel recommended establishment of a Space Warfare Center devoted to developing new applications for space systems and to educating and training operators on space capabilities and tactical applications. In fact, Air Force Space Command already had begun planning for a Space Warfare Center, which it hoped could attract other service operations personnel as well. Furthermore, the Moorman Panel believed that theater arrangements should find the Air Force component commander formally designated as the focal point for space support. The Air Force also should reexamine all training, education, and personnel policies in order to promote a better understanding of space among the aviation community, as well as of aviation needs among the space community.[6]

The panel also advocated establishing a stronger operational space presence at Air Force headquarters—and throughout the Air Force—one that could provide an operational imperative in place of the budget-and-policy focus that traditionally dominated decisions on space issues. The July 1993 activation of Headquarters Fourteenth Air Force at Vandenberg Air Force Base, California, to manage the nation's military space assets, was one response to this recommendation. The new headquarters became the operational focus under Air Force Space Command with responsibility for "providing ballistic missile warning, space control, space lift, and satellite command and control."[7] Finally, the Moorman report addressed the sensitive issue of the "national" reconnaissance space community's role. Because requirements for national systems were identified in intelligence councils outside the normal defense process, it said, defense needs had to conform to intelligence requirements. It recommended the creation of a more formal system to ensure adequate consideration of service needs in the design of national systems.[8]

The panel concluded by observing that in a world of declining resources, improving support to the warfighter would demand major changes in space acquisition, operations, and application. The Air Force, it declared, found itself "uniquely positioned" to ensure the achievement of these goals. "The Air Force's ability to provide Global Reach and Global Power for America allows us to be the leading edge of military force."[9]

By the spring of 1993, General McPeak had endorsed the Blue Ribbon Panel's findings, had designated various Air Force organizations responsible for implementing the panel's recommendations, and had prepared an implementation plan for Air Force Secretary Sheila Widnall's review. The Blue Ribbon Panel set the stage for a major Air Force effort to maintain its leadership of military space. If this theme seemed overly familiar, Air Force leaders believed that the post-Cold War reality of readiness and power projection amid budget austerity provided an unprecedented opportunity for Air Force action.

The reinvigorated Air Force's assertion of leadership took several forms. One involved proposing to Defense Department officials and congressional members the designatation of the Air Force as the executive agent for space research and development and for acquisition. By the summer of 1994, reports indicated that Deputy Defense Secretary John Deutch had agreed to support the Air Force plan and to argue the case before House and Senate conferees who were preparing to negotiate the fiscal year 1995 defense appropriations bill. At the same time, it became clear that Air Force assertiveness had raised old fears of an Air Force space "takeover." Army and Navy leaders could hardly be expected to stand idly by after the Chairman of the Joint Chiefs of Staff had recommended the elimination of their space commands and the Air Force's Blue Ribbon Panel had endorsed this proposal. The always contentious roles and missions debate among the services seemed about to take center stage once again.[10]

The Air Force proposal drew opposition beyond Army and Navy circles. A General Accounting Office report in the summer of 1994 criticized previous Air Force attempts to become the Defense Department's executive agent for space. The GAO recommended that military space acquisition decisions be centralized within the Office of the Secretary of Defense rather than consolidated under Air Force direction. A House Appropriations Committee report in August noted that "the Air Force dominates the military space budget, yet generates little of the requirement. Nevertheless, its space budget competes with other service-specific Air Force requirements such as aircraft and missiles." The House report questioned the Air Force's ability to handle the varied space needs of the military space community.[11]

Air Force leaders like Chief of Staff General Merrill A. McPeak and Air Force Secretary Sheila E. Widnall sought by means of policy statements and public addresses to allay fears, overcome skepticism, and generate support both within and outside the service. In a speech before the "Spacetalk 94" conference on 16 September 1994, General McPeak squarely faced the controversial issue. Referring to the crucial role of space in the Gulf War, he noted that all the services now worked to make space important to warfighters by ensuring that their requirements for space support were met. Unfortunately, he said, this legitimate concern had become embroiled in the "current Washington debate over the proper allocation of roles and missions among the services." He referred to one headline that asserted the "USAF Aggressively Guns for Roles" and was seeking to completely remove the other services from space operations.[12]

The chief of staff sought to "set the record straight." The Air Force, indeed, should be the lead service for space, he reasoned, because this would be good for the Defense Department and the taxpayer. In an era of steadily declining defense budgets, the military was especially challenged to realize the great potential of space. Cutting costs by reducing overhead and "streamlining" organizations represented one solution. He cited the restructured post-Cold War Air Force as an example of successful adjustment to the new realities. In fact, two days earlier, the chief of staff had given a major address on "reinventing the Air Force" at the Air Force Association Convention in Washington, D.C.[13]

General McPeak proposed a similar consolidated, streamlined approach to the development and acquisition of military space systems. He restated the argument made by the Blue Ribbon Panel that fragmentation in the requirements process too often resulted in one-of-a-kind satellites that drove up costs and produced excessive delays in the space launch schedule. Austere times demanded better management. The Defense Department, he noted, had asked the Air Force to examine ways to improve the development and acquisition process. This made good sense. The Air

Force, after all, managed almost 85 percent of the military space budget, employed more than 90 percent of military space personnel, and owned most of the space infrastructure. He assured his audience that the Air Force proposal was not an attempt to usurp the responsibility of the other services to establish their own space requirements. All requirements would be evaluated by the Joint Requirements Oversight Council, comprised of the service vice chiefs of staff and by a Joint Space Management Board directed by senior officials from the Defense Department and intelligence community. This process would ensure "jointness" and, for the first time, effectively integrate intelligence requirements into the larger military space arena. Hence, the Air Force would not determine the space requirements of others; it would act only as the Defense Department's executive agent with responsibility for developing and acquiring space systems. The Air Force proposal would help lower costs by promoting commonality and standardization and serve to end the barrier between classified and unclassified programs. "If the Air Force becomes the lead service for space development and acquisition," the general asserted, "the other services will come to trust us to meet their requirements in space."[14]

Secretary Widnall also took up the theme of Air Force leadership and tried to alleviate the concerns of critics. Referring to the current roles and missions debate in an October 1994 policy letter, she declared that the chief of staff had been misunderstood when he remarkd that "the Army works on the land, the Navy at sea, and the Air Force in the air, and the Air Force accomplishes the majority of space activities." She flatly stated: "let me state clearly that we are not trying to make the Air Force stronger at the expense of the other services." It simply made good sense financially and organizationally to make the service with the largest space role and the most experience responsible for managing the acquisition of space systems. Consequently, the Air Force had proposed that the Secretary of the Air Force be designated the executive agent for space.[15]

Air Force leaders relied on more than official statements and speeches to spread the word on Air Force space leadership. In the spring of 1993, after the Blue Ribbon Panel had completed its deliberations, General McPeak initiated a comprehensive evaluation of space capabilities and "high-leverage" space technologies for the year 2020 and beyond. *SPACECAST 2020* appeared in the spring of 1994, following a year of analysis by scientists, industrialists, and members of all service space commands under the auspices of Air University. In the tradition of Theodore von Kármán's *Toward New Horizons* and subsequent studies, *SPACECAST 2020* produced eighteen white papers that assessed emerging technologies and described creative space applications that would support the security of the country in the next century. Particularly interesting was the closing address delivered by retired Air Force General Michael P. C. Carns on 10 November 1994 to the National Security Industrial Association, which had provided the forum for the first major *SPACECAST 2020* briefing to industry.[16]

The former vice chief of staff declared that space for thirty years had been shaped not by operators but by functionalists from the national intelligence and the surveillance and warning communities. This prevented widespread appreciation for the opportunities space offered military forces. Only Desert Storm, he asserted, had finally opened the door for the warfighter. But the "operationalization" of space would not occur on its own, because the domain of the specialist continued to promote a testing mindset in the Air Force. After all, despite Air Force Space Command's assumption of the operational space launch mission, "space operations are in the hands of the research, development, test, and evaluation…communities." In Carn's opinion, this had to end. At the same time, he argued, the military should encourage the commercial sector to perform all specialized tasks that did not require particular military involvement. He, too, favored standardization and commonality among the military, civil, and commercial space sectors to promote increased efficiency at lower costs. Above all, General Carns focused on the importance of space operations for the Air Force. He agreed that *SPACECAST 2020* represented a good effort to link space technology, capability, and military operations. Now the Air Force needed to assume the "operational sponsorship of space, a formal commitment…mainstreaming space with all of its aspects into the line Air Force." In short, Air Force leaders would need to institutionalize space operations within the Air Force and the wider military community.[17]

Complementing *SPACECAST 2020* was another important study of Air Force space challenges for the future. At the behest of Secretary of the Air Force Widnall and new Chief of Staff General Ronald R. Fogleman, the Scientific Advisory Board convened a group of experts to address the technological requirements and capabilities facing the Air Force into the 21st century. Titling its study *New World Vistas*, the board pointedly linked its study to its predecessor, *Toward New Horizons*, produced by Theodore von Kármán 50 years earlier. Board chairman Gene H. McCall also noted that his team of specialists worked closely with the *SPACECAST 2020* panel and the Rand Corporation, as well as the Air Force Academy and Air University, in preparing the 15-volume study that appeared late in 1995.[18]

New World Vistas focused on integrated, capability-based technology requirements for long-range planning—more specifically for the next 30 years into the new century. The objective was to apply new technologies to produce affordable capabilities. The board asserted that the emphasis of Air Force technology needed to change given the absence of a known "enemy," the reality of high costs, and the military applicability of commercial technologies. In its assessment of space operations, the board recommended the use of distributed satellite constellations relying on single or dual-purpose satellites. With technologies improving significantly at close to a two-year cycle, the study argued that "time from design to launch should be reduced substantially. A goal of two years is reasonable." Commercial vehicles should be used to launch most military satellites, which could be made compatible with available launchers if the satellites were commercially-produced for distributed systems. Consequently, the Air Force needed to reassess dedicated military satellite communications systems like Milstar and to examine different ways to protect satellite systems in the future. The study proceeded to describe a number of specific technical and procedural measures that would result in cheaper, equally effective satellites and a more responsive launch capability that eliminated the current "cast of thousands" approach to management and operations. Throughout their analysis, the authors emphasized taking advantage of new technologies and the proficiency of the commercial sector. *New World Vistas* declared that unless the Air Force asserted itself to perform its unique mission for the nation, there perhaps should not be a separate air force in the next century. Although the service should expect opposition from the Army and Navy, the Air Force should plan immediately for all air and space activities. *New World Vistas*, the authors argued, would help provide the long-range technology and capability-linked plans to support a clear vision for the Air Force into the 21st century.[19]

An equally forthright call for action appeared in early 1995 in the report prepared by the Air Force Association Advisory Group on Military Roles and Missions. That report attacked the fragmentation, absence of leadership, and divided authority that continued to characterize the nation's space community. Space launch represented the most serious example. After many years and millions of dollars, the lack of consensus on requirements had produced little more than a string of "program corpses"—the Advanced Launch System, the National Launch System, and, most recently, Spacelifter. Similar difficulties had led to elimination of FEWS and, now, threatened the Milstar program. Echoing General McPeak and other Air Force leaders, the report warned that the country would lose its technological advantage and fail to achieve operational space capabilities in the future if it did not confront the organizational dilemma. Large space budgets would not solve the problem; reorganization would.[20]

In order to eliminate duplication, reduce costs, and achieve the great advantages offered by space, the Advisory Group stated, defense leaders needed to turn to the service with the space expertise, capability, and commitment—the only service that included space in its mission statement and operated throughout the full spectrum of space functions. The Air Force should be responsible for research, development, and acquisition of space systems to meet the requirements of all the services. Such restructuring would not represent an Air Force power play but, rather, the most logical solution to an intractable problem. To be effective, however, the Air Force needed to end the perception that the space system requirements of operational commanders-in-chief and the other services could not compete successfully against Air Force demands for new aircraft. One way of minimizing the problem would be to have a more equitable distribution of space costs. While the space portion of the Air Force budget supported all the services as well as the joint forces, the Defense

Department did not recognize the need for balanced apportionment to help the Air Force defray the large investment costs it made on behalf of the entire military space community. Above all, the report declared that space needed to finally become institutionalized in the Air Force, and that the Air Force demonstrate "an unequivocal commitment to exploiting space for all forces."[21]

Would the Air Force's quest for military space leadership prove unequivocally successful? At mid-decade, success seemed doubtful in light of initially strong opposition from Navy and Army leaders. Furthermore, the Defense Department had centered space acquisition in a new Space Architect office within the Pentagon. Many roadblocks from earlier years continued to obstruct the Air Force's progress. Fragmentation and lack of consensus, the very problems identified by Air Force critics, worked against the service's efforts. Responsible for preventing more unified, centralized approaches to space management, a fragmented space community contributed to interservice rivalry over roles and missions and to traditional bureaucratic turf battles.

On the other hand, the world of the 1990s presented a landscape that had been significantly altered. For one thing, cooperative efforts had now become more acceptable to all. Multiuser programs and systems increasingly reflected interest in promoting commonality and "convergence" to end duplication and cut costs. Although few would doubt the continued need for dedicated military space systems like DSCS and DMSP, Air Force leaders had joined the chorus to extol the virtues of cooperative ventures among the military, civil, and commercial sectors. Desert Storm had made them believers, and shrinking budgets for space would continue to foster cooperative efforts.

But what about space launch, the most fundamental element of the space program and the one that stubbornly defied efforts to create a responsive, cost-effective means to reach space? Placing responsibility in the hands of Air Force Space Command had begun the process of making space launch "operational," but much work remained. At mid-decade Air Force leaders looked to the Moorman study on space launch to chart the proper course for the nation. Although space launch represented a national concern, the Air Force provided the leadership to solve the problem. The Achilles heel of the space program might well reinforce the Air Force's argument for space leadership.

Above all, the new world of the 1990s reflected the end of the Cold War and the impact of Desert Storm. Superpower rivalry had given way to regional conflict. The United States needed to be ready to field lean, mobile, highly-trained expeditionary forces capable of decisive action in theater-level contingency operations. This first modern war in which space systems played a vital role confirmed the shift to tactical warfighting, and space systems had shown their ability to apply strategic assets to tactical contingencies. In "reinventing" the Air Force, General McPeak had made space a top priority. He believed that Air Force space systems would provide the critical advantage for the "power projection" strategy of the future.

The altered conditions of the 1990s offered the Air Force a golden opportunity to display its space leadership—in the name of greater operational efficiency and the national interest. The postwar Air Force initiative reflected important institutional thinking about space requirements for the post-Cold War era. It also revealed Air Force thinking about the technical and political means necessary to implement this vision. Air Force reviews represented an impressive, comprehensive internal look at the state of military space, future needs, and integration issues. Unfortunately, the weakness of the effort came from attempting to convert ideas into a roadmap for the whole Defense Department without a full, public review of military space. Alternative proposals focused on Defense Department management and encompassed plans for better integration of "black" and "white" space communities, as well as the evolving "jointness" of military space.

Air Force success in the larger arena, however, had to begin from within. From his vantage point as commander-in-chief of U.S. Space Command, General Charles A. Horner noted that when he assumed command of United States air forces during Desert Storm, "most of us over there were ignorant of the contributions of space assets." A major command-post exercise shortly before the conflict did not integrate space forces into the operation.[22] Although Horner quickly realized the importance of space contributions, his experience suggests the central dilemma facing Air Force space leaders at the dawn of the new century. Much of the Air Force continued to view space as more the province of the technocrats, as something beyond the realm of aviators. To be sure, much had been accomplished over the past decade to "operationalize" space in the Air Force. But much remained to be done before space would become a thoroughly integrated element of all Air Force operations and before air and space would become equal in fact as well as name. Above all, Air Force leaders needed to demonstrate greater commitment to space within the service and institutionalize space as a fundamental element of the Air Force's future. Only by establishing the foundation for space within the service could the Air Force demonstrate its commitment to support the warfighter and maintain its space leadership. As the Air Force approached its 50th anniversary as an independent service, it could look back on a half century of leadership in meeting the challenges of military space. Its space vision aimed to perpetuate that leadership and successfully meet the military space challenges of the new century.

Appendices

Proposed Air Force Astronautics Program, 24 January 1958[*]

I. *609, Ballistic Test and Related Systems*
      1. BRATS, Space Research and Experiments
      2. Aerial Survey and Target Locating System (Recon)

II. *447, Manned Hypersonic Research System*
      3. X-15, Space Research and Experiment
      4. Advanced Hypersonic Research Aircraft
          (Manned Space Flight, Space R&D)

III. *464, Dyna-Soar*
      5. Manned Capsule Test (Manned Space Flight)
      6. Conceptual Test (Manned Space Flight)
      7. Boost Glide Tactical (Weapon Delivery)
      8. Boost Glide Interceptor (Countermeasure)
      9. Satellite Interceptor (Countermeasure)
      10. Global Reconnaissance
      11. Global Bomber (Weapon Delivery)

IV. *WS-117L Satellite System*
      12. Advanced Reconnaissance Satellite
      13. Recoverable Data (Photo Capsule) (Recon)
      14. 24-hour Reconnaissance System
      15. Manned Strategic Station (Weapon Development and Recon)
      16. Strategic Communications Station (Data Transmission)

V. *499, Lunar Base System*
      17. Manned Variable Trajectory and Test Vehicle (Recon and Experiment)
      18. Nuclear Rocket Test (Space Recon and Experiment)
      19. Ion Propulsion Test (Space Recon and Experiment)
      20. Lunar Transport (Manned Space Flight, Recon and Experiment)
      21. Manned Lunar Base (Weapon Development and Recon)

---

[*] Source: Lee Bowen, *The Threshold of Space: The Air Force in the National Space Program,* 1945–1959 (Washington, D.C.: USAF Historical Division Liaison Office, September 1960), p. 23.

APPENDIX 2-2
Projects Transferred to ARPA and Redistributed to the Services*

Projects Transferred to ARPA*:*

1.	Argus (nuclear explosion in exosphere)	4 April 1958
2.	Satellite and Outer Space Programs including Vanguard	1 May 1958
3.	High Performance Solid Propellants	7 June 1958
4.	Minitrack Doppler Fence	20 June 1958
5.	Army and Air Force Ballistic Missile Defense Projects	20 June 1958
6.	Studies of the Effects of Space Weapons Employment on Military Electronic Systems	20 June 1958
7.	Nuclear Bomb-Propelled Space Vehicle	20 June 1958
8.	WS-117L	30 June 1958

Projects Redistributed by ARPA*:*

1.	Sounding Rockets and Ground Instrumentation for Argus	AFSWC/AFCRC
2.	Weapon System to Control Hostile Satellites	ARDC
3.	Nuclear Bomb-Propelled Space Vehicle	ARDC
4.	Effects of Space Weapons on Military Electronic Systems	ARDC
5.	WS-117L	ARDC
6.	Lunar Probes	AFBMD
7.	High Energy Propellants and Liquid hydrogen-Liquid Oxygen Propellants	ARDC
8.	Reentry Studies	ARDC
9.	Project Score	ARDC
10.	1,500,000-pound booster	AOMC
11.	Meteorological Satellite	AOMC
12.	Inflatable Sphere	AOMC

---

* Source: Lee Bowen, *The Threshold of Space: The Air Force in the National Space Program,* 1945–1959 (Washington, D.C.: USAF Historical Division Liaison Office, September 1960), pp. 26-27.

APPENDIX 2-3
ARPA Programs Transferred to NASA*

Program III:
1. Man in Space
2. Special Engines
3. Special Components for Space Systems
4. Project Argus
5. Satellite Tracking and Monitoring Systems
6. Satellite Communications Relay, Meteorological Reporting, Navigational Aid Systems

Program IV:
1. ABMA/JPL Program for Four Scientific Space Vehicles to be Launched in 1958
2. AFBMD Program for Three Lunar Probes
3. NOTS Program, a one-frame television with a mechanical scanner to get "a first look at the other side of the moon"
4. Follow-on Program, vaguely defined as "more of the same"

* Source: Lee Bowen, *The Threshold of Space: The Air Force in the National Space Program, 1945–1959* (Washington, D.C.: USAF Historical Division Liaison Office, September 1960), p. 29.

APPENDIX 2-4
Major Military Uses of Space Identified by the Air Force, Spring 1959**

1. Military Reconnaissance Satellites Utilizing Optical, Infra-red and Electromagnetic Instrumentation*
2. Satellites for Weather Observation*
3. Military Communications Satellites*
4. Satellites for Electronic Counter-measures*
5. Satellites as Aids for Navigation*
6. Manned Maintenance and Resupply Outer Space Vehicles*
7. Manned Defensive Outer Space Vehicles*
    Bombardment Satellites*
8. Manned Lunar Station*
9. Satellite Defense System
10. Manned Detection, Warning and Reconnaissance Space Vehicle
11. Manned Bombardment Space Vehicle or Space Base
12. Target Drone Satellite

---

*indicates those missions listed in NSC 5814/1, "Preliminary U.S. Policy on Outer Space," 18 August 1958.

**Source: Memorandum, AFDAT, subj: Background Information on Current Air Force Position in Space, n.d. [March 1959].

APPENDIX 2-5
Military Space Program Plan, November 1958*

Functions	Projects
Navigation	Transit satellite system; assigned to the Navy on 9 May 1960
Meteorology	TIROS television (RCA) satellite system assigned to NASA; military system proposed, but held to studies while negotiations for a single civil-military system were underway with NASA and the Department of Commerce (Weather Bureau)
Communication	Courier Active (repeater) strategic and tactical communications satellite system; assigned to the Army on 15 September 1960
Missile Detection and Space Defense	Infrared radiometers that detect focused heat sources (Missile Detection and Alarm-MIDAS)
	Satellite inspector
	ROBO/Dyna-Soar (X-20)
	Radar tracking of earth satellites
	Optical tracking of satellites (from IGY Baker-Nunn system)
	Distant Early Warning (DEW) radar net and, by the early 1960s, the Ballistic Missile Early Warning System (BMEWS) radar net
Observation of the Earth	Other automated satellites

---

* Source: R. Cargill Hall, "The Origins of U.S. Space Policy: Eisenhower, Open Skies, and Freedom of Space," *Colloquy* (December 1993): 23.

APPENDIX 3-1
DoD Space and Related Programs, Fiscal Years 1961–1962[*]

Program	FY 61	FY 62 (proposed)
Samos, MIDAS, Discoverer	$461.2	$541.2
Transit	21.6	22.4
Notus (later Advent)	42.0	72.0
Saint	6.1	26.0
Spacetrack and Spasur	11.6	38.3
Blue Scout	5.6	15.0
Westford	3.6	4.3
X-15	14.9	7.0
Dyna-Soar	58.0	106.5
Component development, applied research, other	127.1	169.7
Large solid booster		62.0
Titan launch vehicle		15.0
Total	$751.7	$1,079.4

[*] Source: U.S. House, Committee on Government Operations, *Government Operations in Space (Analysis of Civil-Military Roles and Relationships)*, House Report No. 445, 89th Congress, 1st Session, 4 June 1965, p. 65.

APPENDIX 3-2
U.S. Government Space Activities, Fiscal Years 1969–1984[*]
Historical Budget Summary—Budget Authority (in millions of dollars)

Fiscal Year	NASA Space[a]	NASA % Total Space	Defense Space	Defense % Total Space	Energy	Commerce	Interior	Agriculture	NSF	Total Space
1959	260.9	33.2	489.5	62.3	34.3					784.7
1960	461.5	43.3	560.9	52.6	43.5				0.1	1065.8
1961	926.0	51.2	813.9	45.0	67.7				0.6	1808.2
1962	1796.8	54.5	1298.2	39.4	147.8	50.7			1.3	3294.5
1963	3626.0	66.7	1549.9	28.5	213.9	43.2			1.5	5434.5
1964	5016.3	73.4	1599.3	23.4	210.0	2.8			3.0	6831.4
1965	5137.6	73.8	1573.9	22.6	228.6	12.2			3.2	6955.5
1966	5064.5	72.6	1688.8	24.2	186.8	26.5			3.2	6969.8
1967	4830.2	71.9	1663.6	24.7	183.6	29.3			2.8	6709.5
1968	4430.0	67.7	1921.8	30.3	145.1	28.1	0.2	0.5	3.2	6528.9
1969	3822.0	63.9	2013.0	33.6	118.0	20.0	0.2	0.7	1.9	5975.8
1970	3547.0	66.4	1678.4	31.4	102.8	8.0	1.1	0.8	2.4	5340.5
1971	3101.3	65.4	1512.5	31.8	94.8	27.4	1.9	0.8	2.4	4740.9
1972	3071.0	67.1	1407.0	30.7	55.2	31.3	5.8	1.6	2.8	4574.7
1973	3093.2	64.1	1623.0	33.6	54.2	39.7	10.3	1.9	2.6	4824.9
1974	2758.5	59.4	1766.0	38.0	41.7	60.2	9.0	3.1	1.8	4640.3
1975	2915.3	59.3	1892.4	38.5	29.6	64.4	8.3	2.3	2.0	4914.3
1976	3225.4	60.6	1983.3	37.2	23.3	71.5	10.4	3.6	2.4	5319.9
T.Q.[b]	849.2	63.3	460.4	34.3	4.6	22.2	2.6	0.9	0.6	1340.5
1977	3440.2	57.5	2411.9	40.3	21.7	90.8	9.5	6.3	2.4	5982.8
1978	3622.9	55.6	2728.8	43.2	34.4	102.8	9.7	7.7	2.4	6508.7
1979	4030.4	54.3	3211.3	43.2	58.6	98.4	9.9	8.2	2.4	7419.2
1980	4680.4	53.8	3848.4	44.2	59.6	92.6	11.7	13.7	2.4	8688.8
1981	4992.4	50.0	4827.7	48.0	40.5	87.0	12.3	15.5	2.4	9977.8
1982	5527.6	44.0	6678.7	54.0	60.6	144.5	12.1	15.2	2.0	12440.7
1983	6327.9	41.0	8490.9	54.0	38.9	177.8	4.6	20.4		15588.5
1984 (est)	6590.4	38.0	10590.3	61.0	34.1	234.8	4.7	23.0		17477.3

Notes: a. Excludes amounts for air transportation; b. T.Q.=Transitional Quarter

[*] Source: *Aeronautics and Space Report of the President: 1982 Activities* (Washington, D.C.: NASA, 1983), p. 96, with updated figures from OMB and DoD, included in Paul B. Stares, *The Militarization of Space: U.S. Policy, 1945–1984* (Ithica: Cornell University Press, 1985), p. 255.

APPENDIX 3-3

Funding of U.S. Defense Department Space Programs by Mission, Fiscal Years 1961–1984*

Mission[a]	FY 1961[f]	1962	1963	1964	1965	1966	1967	1968	1969	1970	1971	1972	1973
Manned Space Flight	58.0	100.0	131.8	89.7	47.0	151.4	237.1	431.0	515.0	121.8			
Communications	55.2	104.6	35.4	80.2	25.7	61.1	58.6	60.6	62.0	126.8	84.6	54.2	182.8
Navigation	23.6	22.0	42.1	27.9	27.6	15.5	22.0	24.7	11.5	5.1	7.4	6.5	10.8
Early Warning	122.5	180.8	102.4	61.5	50.4	58.6	45.6	68.0	(159.0)[H]	(211.8)	(266.5)	(228.4)	233.2
Weather													21.7
Geodesy									12.1	8.6	5.8	5.8	4.9
Space Defense[b]	8.2	33.0	40.9	66.9	39.1	8.0	16.6	15.4	16.5	9.8	3.9	3.4	3.2
Vehicle and Engine Develop.	3.7	68.3	286.1	389.8	274.4	190.0	98.9	72.5	72.3	74.6	46.0	36.2	32.9
Space Ground Support[c]	57.8	102.6	167.7	171.9	235.1	240.1	316.9	260.2	183.9	169.3	150.5	152.3	143.6
Research and Development[d]	74.2	155.7	174.3	157.9	167.1	156.9	131.1	124.1	79.9	82.6	84.6	118.0	116.4
General Support[e]	420.7	531.2	569.2	553.5	713.0	807.2	732.8	864.5	900.9	868.0	863.0	801.7	873.0
Total	813.9	1298.2	1549.9	1599.3	1579.4[G]	1688.8	1663.6	1921.8	2013.1	1678.4	1512.3	1406.5	1622.5

* Source: Paul B. Stares, *The Militarization of Space: U.S. Policy, 1945–1984* (Ithica, NY: Cornell University Press, 1985), pp. 256–257.

APPENDIX 3-3 (cont.)

Mission[a]	1974	1975	1976	1976(TQ)[j]	1977	1978	1979	1980	1981	1982	1983	1984
Communications	275.3	361.5	361.4	57.29	720.9	574.2	458.6	506.2	687.4	986.1	1329.6	1406.0
Navigation	38.1	47.6	104.8	23.1	104.9	93.8	117.7	185.6	167.0	431.2	295.4	460.0
Early Warning	103.7	136.5	88.7	16.0	87.9	150.0	214.3	207.3	267.3	565.7	707.2	756.9
Weather	24.5	29.1	54.7	8.1	67.9	78.7	61.2	67.9	86.5	109.7	232.4	110.1
Geodesy	6.6	7.7	46.3	2.2	7.7	7.3	8.5	10.3	11.6	28.1	60.8	79.9
Space Defense[b]	(3,5)[i]	(3.2)										
Vehicle and Engine Develop.	26.0	36.8	54.7	18.5	106.3	289.6	509.6	661.0	758.5	842.3	1072.5	1214.2
Space Ground Support[c]	189.9	91.9	111.7	24.9	123.6	173.7	210.9	242.3	337.8	472.0	614.4	806.9
Research and Development[d]	133.8	137.1	159.3	50.1	209.5	296.0	434.0	427.7	573.7	759.6	848.7	1098.7
General Support[e]	968.1	1044.2	1047.9	260.3	981.2	1065.5	1196.5	1196.5	1540.1	2586.7	3329.3	4657.6
Total	1766.0	1892.4	1983.3	460.4	2411.1	2728.8	3211.3	3211.3	3848.4	6681.1	8490.9	10590.3

Notes:

a. The missions listed exclude satellite reconnaissance due to the March 1962 DoD Directive that ended all official reference to reconnaissance activities. However as the official totals for FY 1959 and FY 1960 include satellite reconnaissance projects it is likely that subsequent funding for this activity has been hidden in the above totals. Funding for CIA satellite intelligence gathering activities is excluded from this table. These figures are not publicly available. The mission elements have also varied in subsequent years as reflected in the table.

b. Space Defense includes such projects as SAINT; Program 505, 437.

c. Defined as including range support, instrumentation, satellite detection, tracking and control.

d. Defined as including basic and applied research and component development.

e. Defined as including laboratory and research center in house programs, development support organizations, general operational support, and space related military construction not otherwise charged to specific space projects.

f. Figures for FY 1961 are the first available using this mission break down.

g. This total is different from Appendix 3-4 where the amount is given as $1573.9.

h. For FY 1969–72 there are no separate figures for the Early Warning Mission. However the entry 'miscellaneous' which is in parenthesis includes Early Warning.

i. Space Defense funding for FY 1974 and FY 1975 is added in parenthesis but it is excluded from subsequent listings for these years.

j. TQ=Transitional Quarter.

APPENDIX 3-4
U.S. Space Activities: Historical Budget Summary*

Historical Summary and 1970 Recommendations January 1969—New Obligation Authority (in millions of dollars)

	NASA		DoD	AEC	Commerce	Interior	Agriculture	NSF	Total
	Total	Space[1]							
1955	56.9	56.9	3.0						59.9
1956	72.7	72.7	30.3	7.0				7.3	117.3
1957	78.2	78.2	71.0	21.3				8.4	178.9
1958	117.3	117.3	205.6	21.3				3.3	347.5
1959	305.4	235.4	489.5	34.3					759.2
1960	523.6	461.5	560.9	43.3				0.1	1065.8
1961	964.0	926.0	813.9	67.7				0.6	1808.2
1962	1824.9	1,796.8	1,298.2	147.8	50.7			1.3	3294.8
1963	3673.0	3,626.0	1,549.9	213.9	43.2			1.5	5434.5
1964	5099.7	5,046.3	1,599.3	210.0	2.8			3.0	6861.4
1965	5249.7	5,167.6	1,573.9	228.6	12.2			3.2	6985.5
1966	5174.9	5,094.5	1,688.8	186.8	26.5	4.1		3.2	7003.9
1967	4967.6	4,862.2	1,663.6	183.6	29.3	3.0		2.8	6744.5
1968	4588.8	4,452.5	1,921.8	145.1	28.1	2.0	0.5	3.2	6553.2
1970 budget:									
1969	3994.9	3,844.8	2,082.5	117.2	20.2	2.2	0.7	3.3	6070.9
1970	3760.5	3,599.0	2,218.7	105.5	9.7	6.0	3.6	3.7	5946.2

[1] Excludes amounts for aircraft technology in 1959 and succeeding years. Amounts for NASA-NACA aircraft and space activities not separately identifiable prior to 1959.

Note: Details may not add to totals because of rounding.

* Source: Report to Congress, U.S. Aeronautics and Space Activities 1968, p. 110.

APPENDIX 3-4 (cont.)
U.S. Space Budget: New Obligation Authority*

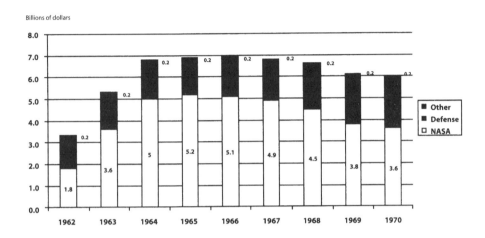

Billions of dollars

* Source: Report to Congress, *U.S. Aeronautics and Space Activities 1968*, p. 110.

295

APPENDIX 6-1
Strategic Air Command to Space Command Transfers,
15 November 1982–1 May 1983*

Project, Sensor, Unit, Site:

Overseas Ground Station (OGS) Availability Improvments	15 November 1982
DSP-1 Upgrade	
Sensor Evolutionary Development	
BMEWS Missile Impact Predictor	1 December 1982
BMEWS Radar Upgrades	
Large Processing Station Upgrade	31 January 1983
Operational Support Module	
DSP Peripheral Upgrade	
PAVE PAWS Expansion	1 February 1983
Tactical Warning Sensor Tech Support Center	
ARPA Maui Optical Station (AMOS) Compensated Imaging	
Ground-based Electro-optical Deep Space Surveillance (GEODSS) System, Sites 4 and 5	
Mobile Ground System	1 March 1983
Improved Radar Calibration System	
C-Band Upgrades	
Space Based Surveillance System	
Retrograde Sensors	
Pacific Barrier (PACBAR) III radar	
Diyarbakir Extended Range Modification	1 April 1983
Peterson AFB	
Detachment 1, 20th Missile Warning Squadron (FSS-7)	1 May 1983
20th Missile Warning Squadron (FPS-85)	
Perimeter Acquisition Radar Characterization System	
COBRA DANE	
FPS-79, Pirinclik, Turkey	
PAVE PAWS	
Defense Support Program (DSP)	
Simplified Processing Station	

APPENDIX 6-1 (cont.)

Ballistic Missile Early Warning System (BMEWS)
Defense Meteorological Satellite Program (DMSP)
GEODSS Sites 1, 2, and 3
Maui Optical Tracking and Identification Facility (MOTIF)
Western Space and Missile Center Sensor Support
Eastern Space and Missile Center Sensor Support
Baker-Nunn cameras
ARPA Lincoln C-Band Observables Radar (ALCOR)
SACLOG
4614th Contracting Squadron
BMEWS Radome Replacement
4602d Computer Services Squadron and Detachment 1
FPS-85 System Programming Agency
Operating Location BE, 3900th Computer Services Squadron
Activities of Lincoln Laboratories
PACBAR I radar
PACBAR II radar
Signal Analysis System Acquisition
Large Processing Station Facility Upgrade
Thule AB, Greenland
Sondrestrom AB, Greenland
Clear AFS, Alaska

* Source: History, Space Command ADCOM, January–December 1983, pp. 3-4.

APPENDIX 6-2
USAF Space Plan Tasking Summary, November 1984 *

1a	Space Weapons Plan	HQ USAF/XOS	October 1985
2a	ALMV Summary Plan	SPACECOM/XPSD	Done
3a	SDI Operational Requirements	SPACECOM/XPSD	January 1985
3b	SDI Strategic Implications	HQ USAF/XOS	January 1985
4a	Space Systems Architecture Study	SPACECOM/XPSS	January 1985
5a	Force Structure Operational Concepts	SPACECOM/XPSS	+
5b	Force Structure Support	SPACECOM/LGXP	June 1985
5c	Satellite Replacement	SPACECOM/XPSS	June 1985
6a	Space Transportation Master Plan	SAF/ALS	March 1985
7a	Satellite Autonomy Plan	Space Division/XRP	January 1985
7b	Satellite Control Network Documentation	Space Division/XRP	January 1985
7c	Space Station R&D Requirements	Space Division/XRP	June 1985
7d	Space Station Operational Requirements	SPACECOM/XPSF	February 1985
8a	Staffing Decisions	HQ USAF/MPMP	Done
8b	Training	HQ USAF/MPPTS	Done
8c	Education Assessment	HQ USAF/MPPE	May 1985
8d	Career Development and Retention	HQ USAF/MPPTS	May 1985
9a	Military Crews in Space	SPACECOM/XPSF	March 1985
10a	Military Use of Commercial Systems	SPACECOM/KRQS	March 1985
10b	Commercial Systems Survivability	SPACECOM/KRQS	March 1984
11a	Logistics Support	AFLC/XRXO	May 1985
11b	Depot Support	AFLC/XRXO	May 1985
11c	Logistics Integration	AFLC/XRXO	May 1985
12a	New Actions	HQ USAF/XOS	May 1985
13a	Air Force Manual 2-XK	SPACECOM/XPXX	February 1985
13b	Basic Doctrine	HQ USAF/XOXIS	May 1985
14a	Radio Frequency Interference	SPACECOM/DOCE	June 1985
14b	Space Debris	Space Division/XRP	June 1985
15a	Planning Gap	HQ USAF/XOS	January 1985

+ Upon publication of Air Force Regulation 55-24

---

* Source: Space Command/XPXX to HQ USAF/XOS *et al*, subj: 15-16 Nov 84 Space Plan Meeting Minutes, 19 December 1984, attachment 3.

Notes

Introduction. The Dawn of the Space Age

1. Project RAND, *Preliminary Design of an Experimental World-Circling Spaceship* (Santa Monica, CA: Douglas Aircraft Co, 2 May 1946), p. 1.

2. In recognizing this dichotomy, one must not overlook the Navy's contribution to Project Vanguard and both Army and Navy missile development work throughout this period.

3. Theodore von Kármán with Lee Edson, *The Wind and Beyond: Theodore von Kármán, Pioneer in Aviation and Pathfinder in Space* (Boston: Little, Brown and Company, 1967), pp. 225-227; Michael H. Gorn, *The Universal Man. Theodore von Kármán's Life in Aeronautics*, Smithsonian History of Aviation Series, ed. Von Hardesty. (Washington, D.C.: Smithsonian Institution Press, 1992), p. 82.

4. Gorn, *The Universal Man*, pp. 73-92; von Kármán, *Wind and Beyond*, pp. 234-248; Robert Frank Futrell, *Ideas, Concepts, Doctrine: Basic Thinking in the United States Air Force, 1907–1960.* Vol. I. Maxwell Air Force Base, AL: Air University Press, 1989, p. 543, (hereafter cited as Futrell, *Ideas, 1907–1960*); Frank J. Malina, "Origins and First Decade of the Jet Propulsion Laboratory," in Eugene M. Emme, ed. *The History of Rocket Technology: Essays on Research, Development, and Utility* (Detroit: Wayne State University Press, 1964), pp. 46-52.

5. Roger E. Bilstein, *Orders of Magnitude. A History of the NACA and NASA, 1915–1990.* The NASA History Series. NASA SP-4406, (Washington, D.C.: National Aeronautics and Space Administration, 1989), pp. 1-14; Loyd S. Swenson Jr., James M. Grimwood, and Charles C. Alexander, *This New Ocean: A History of Project Mercury.* The NASA Historical Series. NASA SP-4201 (Washing-

ton, D.C.: National Aeronautics and Space Administration, 1966), pp. 6-9.

6. Quoted in, U.S. House, Committee on Government Operations, *Government Operations in Space (Analysis of Civil-Military Roles and Relationships)*, House Report No. 445, 89th Congress, 1st Session, June 4, 1965, p. 23.

7. See Bilstein, *Orders of Magnitude*, pp. 15-42; Swenson, *This New Ocean*, pp. 9-13; Futrell, *Ideas, 1907–1960*, pp. 541-543.

8. Space generally came to mean the region above 50 miles in altitude, beyond which aerodynamic vehicles had insufficient oxygen. Orbiting spacecraft needed approximately 100 miles altitude to remain in Earth orbit. See Robert L. Perry, *Origins of the USAF Space Program, 1945–1956. V, History of DCAS 1961.* AFSC Historical Publications Series 62-24-10 (Los Angeles, CA: Air Force Systems Division, Space Systems Division, 1961), p 1; Swenson, *This New Ocean*, p. 13; Rip Bulkeley, *The Sputniks Crisis an Early United States Space Policy: A Critique of the Historiography of Space* (Bloomington, ID: Indiana University Press, 1991), pp. 46-48.

9. Discussion of the space pioneers is based on the following: Perry, *Origins of the USAF Space Program*, pp. 2-6; Swenson, *This New Ocean*, pp. 13-18; Walter A. McDougall, *…the Heavens and the Earth: A Political History of the Space Age* (New York: Basic Books, 1985), pp. 20-21, 76-78; Bilstein, *Orders of Magnitude*, pp. 12-14.

10. Oberth had returned to Rumania in 1930 to teach and, because of his difficult personality and Rumanian citizenship, did not join the Peenemuende project until 1941. Although the von Braun group did not acknowledge its debt to Oberth, Oberth's patents covered their innovations, which

were accomplished independently of Goddard's work. For German rocket developments, see Michael J. Neufeld, *The Rocket and the Reich: Peenemuende and the Coming of the Ballistic Missile Era* (New York: The Free Press, 1995).

11. Malina, "Origins and First Decade of the Jet Propulsion Laboratory," pp. 46-52; von Kármán, *Wind and Beyond*, pp. 240-242.

12. Malina, "Origins and First Decade of the Jet Propulsion Laboratory," pp. 46-52.

13. von Kármán, *Wind and Beyond*, p. 243; Gorn, *Universal Man*, pp. 82-92; Malina, "Origins and First Decade of the Jet Propulsion Laboratory," pp. 48-60.

14. Gorn, *Universal Man*, pp. 82-92; Malina, "Origins and First Decade of the Jet Propulsion Laboratory," pp. 48-60.

15. Malina, "Origins and First Decade of the Jet Propulsion Laboratory," pp. 48-66. In the words of Frank Malina, JPL became the country's initial "center for space and long-range missile development."

16. H. H. Arnold, *Global Mission* (New York: Harper & Brothers, Publishers, 1949), p. 532.

17. Theodore von Kármán, *Toward New Horizons: Science, the Key to Air Supremacy. Commemorative Edition, 1950–1992* (Headquarters Air Force Systems Command, 1992), pp. vii-ix; Gorn, *Universal Man*, pp. 108-117.

18. von Kármán, *Toward New Horizons: Science, the Key to Air Supremacy*; Gorn, *Universal Man*, pp. 108-117; von Kármán, *Wind and Beyond*, pp. 289-294; Thomas A. Sturm, *The USAF Scientific Advisory Board: Its First Twenty Years, 1944–1964* (Washington, D.C.: USAF Historical Division Liaison Office, 1967), pp. 2-15.

19. Bruce L. R. Smith, *The RAND Corporation* (Cambridge, MA: 1966), pp. 30-65; R. Cargill Hall, "The Origins of U.S. Space Policy: Eisenhower, Open Skies, and Freedom of Space" *Colloquy* (December 1993): 5; Futrell, *Ideas, 1907–1960*, pp. 209; McDougall, *...the Heavens and the Earth*, p. 89; von Kármán, *Wind and Beyond*, p. 302.

20. Hall, "Origins of U.S. Space Policy," p. 5; Futrell, *Ideas, 1907–1960*, pp. 209; Edmund Beard, *Developing the ICBM: A Study in Bureaucratic Politics* (New York: Columbia University Press, 1976), pp. 24-26, 112.

21. Perry, "Origins of the USAF Space Program," p. 9; Hall, "Origins of U.S. Space Policy," pp. 5-6; Beard, *Developing the ICBM*, pp. 71-73.

22. Perry, "Origins of the USAF Space Program," p. 9; Hall, "Origins of U.S. Space Policy," pp. 5-6; Beard, *Developing the ICBM*, pp. 71-73.

23. von Kármán, *Toward New Horizons: Science, the Key to Air Supremacy*; Gorn, *Universal Man*, pp. 108-117; von Kármán, *Wind and Beyond*, pp. 289-294; Thomas A. Sturm, *The USAF Scientific Advisory Board: Its First Twenty Years, 1944–1964* (Washington, D.C.: USAF Historical Division Liaison Office, 1967), pp. 2-15.

24. von Kármán, *Wind and Beyond*, p. 293.

25. Bulkeley, *The Sputniks Crisis*, p. 40; Jacob Neufeld, *The Development of Ballistic Missiles in the United States Air Force 1945–1960. General Histories* (Washington, D.C.: Office of Air Force History, 1990), p. 24.

26. Gorn, *Universal Man*, p. 117.

27. Futrell, *Ideas, 1907–1960*, pp. 219-220; Hall, "Early U.S. Proposals," p. 68; Beard, *Developing the ICBM*, pp. 69-72. Bush was slow to change his views. In his 1949 book, *Modern Arms and Free Men*, he conceded that long range missiles could be built, but at too high a cost. Only after Sputnik did he revise his earlier views.

28. von Kármán, *Wind and Beyond*, p. 230.

29. Futrell, *Ideas, 1907–1960*, pp. 480-481; Bulkeley, *The Sputniks Crisis*, pp. 37-44.

Chapter 1. Before Sputnik

1. Hall, "Early U.S. Proposals," pp. 69-74; Perry, "Origins of the USAF Space Program," pp. 8-13; Hall, "Origins of U.S. Space Policy," pp. 5-6; see also the review of postwar rocket and satellite developments in Constance McLauglin Green and Milton Lomask, *Vanguard: A History* (Washington, D.C.: Smithsonian Institution Press, 1971).

2. The following discussion of the Rand proposal is based on: Project RAND, *Preliminary Design of an Experimental World-Circling Spaceship* (Santa Monica, CA: Douglas Aircraft Co., 2 May 1946); Merton E. Davies and William R. Harris, *RAND's Role in the Evolution of Balloon and Satellite Observation Systems and Related U.S. Space Technology* (The RAND Corporation, September 1988. R-3692-RC), pp. 3-9; Hall, "Early U.S. Proposals," pp. 74-80; Perry, "Origins of the USAF Space Program," pp. 12-17.

3. It should be noted that Arthur C. Clarke, intrepid British member of the British Planetary Society, suggested such a course using geosynchronous satellite positioning the previous year. He comments on this in his, *The Promise of Space* (New York: Harper & Row, Publishers,1968), pp. 97-101.

4. Project RAND, *Preliminary Design of an Experimental World-Circling Spaceship* (Santa Monica, CA: Douglas Aircraft Co, 2 May 1946), p. 19.

5. *Ibid.*, pp. 1-2.

6. It also should be noted that the Key West Agreement worked out by the Joint Chiefs of Staff at Key West, Florida, in March of 1948 accorded the Air Force operations in air space, but did not address activities in outer space. Logically, the Air Force thus devoted its attention and budget priorities to air-breathing cruise missiles. Futrell, *Ideas, 1907–1960,* p. 198.

7. For postwar V-2 experiments, see especially David H. DeVorkin, *Science with a Vengeance: How the Military Created the US Space Sciences after World War II* (New York: Springer-Verlag, 1992); Neufeld, *Development of Ballistic Missiles,* p. 36; Malina, "Origins and First Decade of the Jet Propulsion Laboratory," pp. 48-66; U.S. House, Committee on Government Operations, *Government Operations in Space (Analysis of Civil-Military Roles and Relationships),* House Report No. 445, 89th Congress, 1st Session, June 4, 1965, p. 24-26.

8. Malina, "Origins and First Decade of the Jet Propulsion Laboratory," p. 65; Green and Lomask, *Vanguard,* p. 10.

9. DeVorkin, *Science with a Vengeance,* pp. 168-182; Green and Lomask, *Vanguard,* pp. 10-11. Despite development by the Office of Naval Research, the Vanguard generally is regarded as a largely civilian program in contrast with its competitors for America's scientific satellite entry in the International Geophysical Year program.

10. Neufeld, *Development of Ballistic Missiles,* pp. 24-27, 44-50; Beard, *Developing the ICBM,* pp. 43-82.

11. Beard, *Developing the ICBM,* pp. 17-29; Neufeld, *Development of Ballistic Missiles,* pp. 13-23.

12. Beard, *Developing the ICBM,* pp. 17-29; Neufeld, *Development of Ballistic Missiles,* pp. 13-23; Donald R. Baucom, *The Origins of SDI, 1944–1983* (Lawrence, KS: University Press of Kansas, 1992), pp. 1-15. As Baucom points out, following the 1958 decision, the Army dominated the missile defense arena until the Strategic Defense Initiative Organization began operations in 1984. He argues that the Air Force continued to lag in the missile defense activities, and at present ranks behind both the Army and the Navy in terms of size and scope of missile defense programs. Donald R. Baucom, "Manuscript Review Comments," 8 December 1995.

13. Neufeld, *Development of Ballistic Missiles,* pp. 50-56.

14. Beard, *Developing the ICBM*, pp. 111-112.

15. *Ibid.*, pp. 52-55.

16. Beard, *Developing the ICBM*, p. 61.

17. Bulkeley, *The Sputniks Crisis*, p. 71.

18. Beard, *Developing the ICBM*, pp. 107-113; Futrell, *Ideas, 1907–1960*, pp. 275-278.

19. Doolittle's role is recounted by Ivan Getting in, Gen Bernard A. Schriever, *et al*, *Reflections on Research and Development in the United States Air Force*, Jacob Neufeld, ed. (Washington, D.C.: Center for Air Force History, 1993), p. 40.

20. Bulkeley, *The Sputniks Crisis*, pp. 74-77.

21. Futrell, *Ideas, 1907–1960*, pp. 488-489; Beard, *Developing the ICBM*, pp. 129-130.

22. Beard, *Developing the ICBM*, pp. 130-145. Ongoing doubts about the project's technical feasibility rather than roles and missions concerns apparently prompted the Air Staff to refer ARDC's request to the Guided Missiles Committee. Since spring 1950, the Air Force had been authorized exclusive development of long-range strategic missiles, although the Army and Navy continued to contest both development and operational responsibility for missiles. Neufeld, *Development of Ballistic Missiles*, p. 56.

23. Beard, *Developing the ICBM*, pp. 129-151.

24. The following analysis is based on: Davies, *RAND's Role...Satellite Observation Systems*, pp. 9-19; Hall, "Early U.S. Proposals," pp. 80-84.

25. Perry, "Origins of the USAF Space Program," p. 23.

26. Bulkeley, *The Sputniks Crisis*, pp. 78-82; Hall, "Early U.S. Proposals," pp. 85-88.

27. Davies, *RAND's Role...Satellite Observation Systems*, pp. 18-19; Hall, "Early U.S. Proposals," pp. 88-91.

28. Perry, "Origins of the USAF Space Program," pp. 29-37; Davies, *RAND's Role ...Satellite Observation Systems*, pp. 23-35.

29. Davies, *RAND's Role...Satellite Observation Systems*, pp. 35-39; Hall, "Origins of U.S. Space Policy," pp. 6, 19-20.

30. Robert A. Divine, *The Sputnik Challenge* (New York: Oxford University Press, 1993), p. 18; Futrell, *Ideas, 1907–1960*, pp. 424-428; Fred I. Greenstein, *The Hidden-Hand Presidency: Eisenhower as Leader* (New York: Basic Books, 1982), p. 70.

31. James R. Killian, Jr., *Sputnik, Scientists, and Eisenhower: A Memoir of the First Special Assistant to the President for Science and Technology* (Cambridge: The MIT Press, 1977), p. 68; Hall, "Origins of U.S. Space Policy," pp. 19-20.

32. Davies, *RAND's Role...Satellite Observation Systems*, pp. 48-56.

33. Killian, *Sputnik, Scientists, and Eisenhower*, p. 68; Hall, "Origins of U.S. Space Policy," pp. 19-20.

34. Divine, *Sputnik Challenge*, pp. 22-23; Beard, *Developing the ICBM*, 145-151.

35. For the most comprehensive study of the missile program and Gardner's role, see Beard, *Developing the ICBM*, pp. 143-194, and Neufeld, *Development of Ballistic Missiles*, pp. 95-151; John T. Greenwood, "The Air Force Ballistic Missile and Space Program (1954 –1974)," *Aerospace Historian* 21 Winter (1974): 190-197. The von Neumann Committee is often referred to as the Teapot Committee. According to Dr. Simon Ramo, however, this designation applies only to a second committee, which Gardner formed at the same time, to examine non-strategic missile programs. The latter received the name Teapot Committee when Gardner objected to Ramo's first suggestion, Tea Garden, because he believed the association with his own name was too close. By contrast, the von Neumann Committee should receive no other designation than Strategic Missile Evaluation Committee (SMEC). Dr. Simon Ramo, telephone conversation with Mr. George W. Bradley, Director of History, Air Force Space Command, 7 March 1997.

36. Bruno W. Augenstein, *Evolution of the U.S. Military Space Program, 1945–1960: Some*

Key Events in Study, Planning, and Program Management (Rand Corporation, September 1982), pp. 6-7; Davies, *RAND's Role...Satellite Observation Systems*, pp. 48-56.

37. Divine, *Sputnik Challenge*, p. 22.

38. Beard, *Developing the ICBM*, pp. 160-161.

39. Beard, *Developing the ICBM*, p. 184; Neufeld, *Development of Ballistic Missiles*, pp. 122-123.

40. Schriever, *et al*, *Reflections on Research and Development*, pp. 53-58; Greenwood, "The Air Force Ballistic Missile and Space Program," pp. 190-197.

41. Beard, *Developing the ICBM*, pp. 185-194; Divine, *Sputnik Challenge*, pp. 22-23.

42. Schriever, *et al*, *Reflections on Research and Development*, pp. 53-58; Greenwood, "The Air Force Ballistic Missile and Space Program," pp. 190-197; Beard, *Developing the ICBM*, pp. 185-194; Divine, *Sputnik Challenge*, pp. 22-23.

43. Perry, "Origins of the USAF Space Program," pp. 40-45; McDougall, *...the Heavens and the Earth*, pp. 107-111.

44. Perry, "Origins of the USAF Space Program," pp. 40-45; McDougall, *...the Heavens and the Earth*, pp. 107-111.

45. Davies, *RAND's Role ...Satellite Observation Systems*, pp. 53-55; Hall, "Origins of U.S. Space Policy," pp. 18-21.

46. Perry, "Origins of the USAF Space Program," pp. 40-45; Augenstein, *Evolution of the U.S. Military Space Program*, pp. 6-8.

47. Perry, "Origins of the USAF Space Program," pp. 42-43.

48. *Ibid.*

49. *Ibid.*

50. *Ibid.*, pp. 43-45; Davies, *RAND's Role ...Satellite Observation Systems*, p. 63.

51. Hall, "Origins of U.S. Space Policy," pp. 19-21; Killian, *Sputnik, Scientists, and Eisenhower*, pp. 67-86; Bulkeley, *Sputniks Crisis*, pp. 147-148.

52. Hall, "Origins of U.S. Space Policy," pp. 19-21; Killian, *Sputnik, Scientists, and Eisenhower*, pp. 67-86; Bulkeley, *Sputniks Crisis*, pp. 147-148. It should be noted that the Killian Panel's deliberations on the U-2 were too secret even to appear in their formal report.

53. Hall, "Origins of U.S. Space Policy," pp. 19-21; Killian, *Sputnik, Scientists, and Eisenhower*, pp. 67-86; Bulkeley, *Sputniks Crisis*, pp. 147-148; Divine, *Sputnik Challenge*, pp. 23-25.

54. For a comprehensive treatment of the IGY issue, see Bulkeley, *Sputniks Crisis*, pp. 89-122; Hall, "Origins of U.S. Space Policy," pp. 20-21, notes 38 and 39.

55. Perry, "Origins of the USAF Space Program," pp. 45-47; McDougall, *...the Heavens and the Earth*, pp. 118-119; Green and Lomask, *Vanguard*, pp. 16-18.

56. Hall, "Origins of U.S. Space Policy," pp. 20-21; Bulkeley, *Sputniks Crisis*, pp. 157-158, 213; R. Cargill Hall, "The Eisenhower Administration and the Cold War: Framing American Astronautics to Serve National Security" *Prologue* (Spring 1995) 27 no.1: 63.

57. Hall, "Origins of U.S. Space Policy," pp. 20-24; Bulkeley, *Sputniks Crisis*, pp. 179-181; McDougall, *...the Heavens and the Earth*, pp. 118-123.

58. House Report, *Government Operations in Space*, pp. 27-29; Paul B. Stares, *The Militarization of Space: U.S. Policy, 1945–1984* (Ithica, NY: Cornell University Press, 1985), pp. 33-35; Perry, "Origins of the USAF Space Program," p. 48; Futrell, *Ideas, 1907–1960*, pp. 547-548; Bulkeley, *Sputniks Crisis*, pp. 127-128. Bulkeley points out that during the summer of 1955 Project Orbiter added the JPL's radio link, then under development and later named Microlock, to its specifications in order to enhance its competitive position. Like Minilock, which would become NASA's primary tracking and telemetry system, Microlock was a radio interferometer tracking system. Less accurate than Minilock, the JPL system produced a peak

accuracy of 3 to 4 minutes of arc, in contrast to Minitrack's 20 to 30 seconds of arc. See Bulkeley, *Sputniks Crisis*, pp. 128, 206, 244. Minitrack operational characteristics are also described in Green and Lomask, *Vanguard*, pp. 146-148.

59. House Report, *Government Operations in Space*, pp. 27-29; Stares, *The Militarization of Space*, pp. 33-35; Perry, "Origins of the USAF Space Program," p. 48; Futrell, *Ideas, 1907–1960*, pp. 547-548; Bulkeley, *Sputniks Crisis*, pp. 127-128.

60. National Security Council, "U.S. Scientific Satellite Program," NSC 5520, drafted 20 May, approved 26 May 1955, quoted in Bulkeley, *Sputniks Crisis*, p. 135.

61. Bulkeley, *Sputniks Crisis*, pp. 132-136; Vernon Van Dyke, *Pride and Power: The Rationale of the Space Program* (Urbana, IL: University of Illinois Press, 1964), p. 14; Green and Lomask, Vanguard, pp. 41-56; Roger D. Launius, *NASA: A History of the U.S. Civil Space Program* (Malabar, FL: Krieger Publishing Company, 1994), p. 22.

62. Perry, "Origins of the USAF Space Program," pp. 49-55.

63. *Ibid.*

64. *Ibid.*

65. *Ibid.*

66. Perry, "Origins of the USAF Space Program," pp. 42-43; Davies, *RAND's Role*

...*Satellite Observation Systems*, p. 64.

67. Hall, "Origins of U.S. Space Policy," pp. 21-22; Perry, "Origins of the USAF Space Program," pp. 55-57; Stares, *Militarization of Space*, pp. 30-31; *Space and Missile Systems Organization: A Chronology 1954–1976*, AFSC Historical Publications (SAMSO, Chief of Staff, History Office), p. 36; Kenneth E. Greer, "Corona," *Studies in Intelligence*, Supplement, 17 (Spring 1973): 6, reprinted in Kevin C. Ruffner, ed., *Corona: America's First Satellite Program* (CIA History Staff, Center for the Study of Intelligence, Washington, D.C., 1995).

68. Davies, *RAND's Role...Satellite Observation Systems*, pp. 23-94; Bilstein, *Orders of Magnitude*, pp. 42-44; Swenson, *This New Ocean*, pp. 55-74.

69. Bulkeley, *Sputniks Crisis*, pp. 136-142.

70. *Ibid*; Green and Lomask, *Vanguard*, pp. 130-131.

71. Divine, *Sputnik Challenge*, pp. 29-30.

72. Hall, "Origins of U.S. Space Policy," p. 22; Perry, "Origins of the USAF Space Program," pp. 55-57.

73. Quoted in Stares, *Militarization of Space*, p. 32.

74. Hall, "Origins of U.S. Space Policy," p. 22; Futrell, *Ideas, 1907–1960*, p. 545.

75. Bulkeley, *Sputniks Crisis*, p. 42.

Chapter 2. From Eisenhower to Kennedy

1. Robert A. Divine, *The Sputnik Challenge* (New York: Oxford University Press, 1993), pp. xiii-xiv.

2. *Ibid.*, pp. xiii-xviii; Walter A. McDougall, *...the Heavens and the Earth: A Political History of the Space Age* (New York: Basic Books, 1985), pp. 141-156; Lee Bowen, *The Threshold of Space: The Air Force in the National Space Program, 1945–1959* (Washington, D.C.: USAF Historical Division Liaison Office, September 1960), pp. 8-15; Paul B. Stares, *The Militarization of Space:*

U.S. Policy, 1945–1984 (Ithica, NY: Cornell University Press, 1985), pp. 38-40.

3. Particularly unfortunate was the comment of presidential chief of staff Sherman Adams, who said America had no plans to enter "an outer space basketball game" with the Soviets. Divine, *Sputnik Challenge*, p. xv.

4. Divine, *Sputnik Challenge*, p. 6; R. Cargill Hall, "The Origins of U.S. Space Policy: Eisenhower, Open Skies, and Freedom of Space," *Colloquy* (December 1993): 23.

5. Divine, *Sputnik Challenge*, pp. 47-52;

James R. Killian, Jr., *Sputnik, Scientists and Eisenhower: A Memoir of the First Special Assistant to the President for Science and Technology* (Cambridge: The MIT Press, 1977, pp. 26-27.

6. Maj John B. Hungerford, Jr., *Organization for Military Space: A Historical Perspective, Report No. 82-1235* (Maxwell AFB, AL: Air Command and Staff College, 1982), pp. 22-23; Maj Gen John B. Medaris, *Countdown for Decision* (New York: G. P. Putnam's Sons, 1960), pp. 155-167; McDougall, *...the Heavens and the Earth*, pp. 131, 150, 168. The intrepid von Braun initially promised a launch within 60 days, but Medaris countered with 90, while the official Army estimate became 120 days. The first attempt to launch Vanguard on 6 December 1957 proved embarrassingly unsuccessful. See Divine, *Sputnik Challenge*, pp. 26-27.

7. U.S. House, Committee on Government Operations, *Government Operations in Space (Analysis of Civil-Military Roles and Relationships)*, House Report No. 445, 89th Congress, 1st Session, June 4, 1965, p. 31.

8. Robert Kipp, *Space Detection and Tracking: A Chronology* (Peterson Air Force Base, CO: Air Force Space Command Office of History), March 1990; Bruno W. Augenstein, *Evolution of the U.S. Military Space Program, 1945–1960: Some Key Events* in *Study, Planning, and Program Management* (Rand Corporation, September 1982), p. 11; Thomas A. Sturm, *The USAF Scientific Advisory Board: Its First Twenty Years, Office of Air Force History Special Studies* (Washington, D.C.: Government Printing Office, 1986; reprint of 1967 edition), pp. 80-82.

9. Memorandum, Col V. A. Adduci, Assistant Director, Legislative Liaison to Assistant Deputy Chief of Staff, Plans and Programs, subj: Policy Coordination Section Activities, 7 November 1957. Col Adduci and his staff seemed most alarmed about high congressional interest in the Army's

NIKE-ZEUS ballistic missile defense system.

10. General Thomas D. White, "At the Dawn of the Space Age," *The Air Power Historian* v, no. 1 (January 1958): 15-19.

11. Robert Frank Futrell, *Ideas, Concepts, Doctrine: Basic Thinking in the United States Air Force, 1907–1960*. Vol. I. (Maxwell Air Force Base, AL: Air University Press, 1989), pp. 553-554 (hereafter cited as Futrell, *Ideas, 1907–1960*).

12. Augenstein, *Evolution of the U.S. Military Space Program*, p. 11; Bowen, *The Threshold of Space*, pp. 20-22.

13. Bowen, *The Threshold of Space*, pp. 21-22.

14. Hall, "Origins of U.S. Space Policy," p. 22.

15. Bowen, *The Threshold of Space*, p. 22; USAF Scientific Advisory Board, *Report of the Scientific Advisory Board Ad Hoc Committee on Space Technology*, 6 December 1957. Committee members included David T. Griggs, Clark B. Milligan, Mark M. Mills, W. H. Radford, H. Guyford Stever, Edward Teller, and C. S. White.

16. Bowen, *The Threshold of Space*, pp. 22-24; Futrell, *Ideas, 1907–1960*, p. 590; Divine, *Sputnik Challenge*, p. 98. The grandstanding by Air Force generals could not convince their audience that lunar bases could play a strategic role, an issue disposed of earlier by the Killian Committee. See Killian, *Sputnik, Scientists, and Eisenhower*, p. 129.

17. Memorandum, SAFGC to Assistant Secretary of Defense (International Security Affairs), subj: Proposal for a National Policy on Outer Space, 18 March 1958.

18. The following discussion of ARPA is based on: Bowen, *The Threshold of Space*, pp. 24-34; Futrell, *Ideas, 1907–1960*, pp. 590-594; House Report, *Government Operations in Space*, pp. 31-55; Hungerford, *Organization for Military Space*, pp. 25-28; History, Air Research and Development Command, Vol I, 1 July–31 December 1959, Chapter 1.

19. This discussion of NASA is based on the following: Bowen, *The Threshold of*

Space, pp. 24-34; House Report, *Government Operations in Space*, pp. 31-55; Futrell, *Ideas, 1907–1960*, pp. 594-606; McDougall, *...the Heavens and the Earth*, pp. 157-176; Divine, *Sputnik Challenge*, pp.104-105, 111-112, 145-154.

20. Roger E. Bilstein, *Orders of Magnitude. A History of the NACA and NASA, 1915–1990*. The NASA History Series. NASA SP-4406 (Washington, D.C.: National Aeronautics and Space Administration, 1989), p. 47.

21. See Loyd S. Swenson Jr., James M. Grimwood, and Charles C. Alexander, *This New Ocean. A History of Project Mercury*. The NASA Historical Series. NASA SP-4201 (Washington, D.C.: National Aeronautics and Space Administration, 1966), pp. 75-77.

22. Stares, *Militarization of Space*, pp. 41-43.

23. Quoted in Futrell, *Ideas, 1907–1960*, pp. 594-595.

24. Divine, *Sputnik Challenge*, pp. 81, 104.

25. Deputy Secretary of Defense Quarles to Director, Bob Stans, April 1, 1958.

26. House Report, *Government Operations in Space*, p. 39; Killian, *Sputnik, Scientists, and Eisenhower*, p. 134.

27. See especially, House Report, *Government Operations in Space*, pp. 39, 46.

28. Divine, *Sputnik Challenge*, pp. 147-148.

29. *Ibid.*, pp. 105-107; Stares, *Militarization of Space*, p. 47; Killian, *Sputnik, Scientists, and Eisenhower*, pp. 123-124.

30. Stares, *Militarization of Space*, pp. 47-57; Hall, "Origins of U.S. Space Policy," p. 23-24.

31. For example, see Gen Bernard A. Schriever, "Does the Military Have a Role in Space?" in *Space: Its Impact on Man and Society*, ed. Lillian Levy (New York: W. W. Norton & Co., 1965), pp. 59-68.

32. J. D. Hunley, ed., *The Birth of NASA: The Diary of T. Keith Glennan*. Vol. NASA SP-4105. The NASA History Series (Washington, D.C.: NASA History Office, 1993), pp. 9-15; House Report, *Government Operations in Space*, pp. 49-53.

33. ARPA's FY 1959 budget totaled $331,726,000 and NASA's $384,073,532. The remaining funds appeared against the military service budgets and those of the AEC and DoD agencies. The Air Force maintained no space budget as such. See Bowen, *The Threshold of Space*, pp. 38-43.

34. Memorandum, Under Secretary of the Air Force to Secretary of Defense, subj: Space Programs, 17 September 1958; Memorandum, Director ARPA to Under Secretary of the Air Force, subj: Space Programs, 31 October 1958.

35. House Report, *Government Operations in Space*, pp. 51-52.

36. On broader aspects of the Defense Reorganization Act of 1958, which created the DDR&E, see Futrell, *Ideas, 1907–1960*, pp. 573-586. See also House Report, *Government Operations in Space*, pp. 53-55.

37. Memorandum, Col C. R. Roderick, Chief, Committee Liaison Division, Office of Legislative Liaison to Assistant Director, Legislative Liaison, September 1958. Although NASA received the meteorological satellite mission in 1958, and would achieve success with its TIROS I satellite, the Air Force continued to seek responsibility for its own weather satellite. In the 1960s, NASA relinquished its meteorological satellite mission to the Air Force, which developed the Defense Meteorological Satellite Program as a classified project in support of the Strategic Air Command and the NRO. Gen Thomas S. Power, CINCSAC, to Gen Thomas D. White, CSAF, 1 December 1960, Thomas D. White Papers, Library of Congress, Manuscript Division, Box 34, "2-15 SAC."

38. Memorandum, AF/DCS Plans to AF/DCS Development, subj: Air Force Objectives in Space, 5 February 1959.

39. *Ibid.*; Memorandum, AFDAT, subj: Background Information on Current Air Force Position in Space, n.d. [March 1959].

40. Less sympathetic observers might point out that Symington's charges were

without merit, and he as Secretary of the Air Force a decade earlier did considerable damage to military space activities.

41. *Ibid.*

42. *Ibid.* Interestingly, the AFDAT analysis noted that the problem with ARPA would remain if DDR&E were to follow the same procedures.

43. AFBMD, *Relationship between Ballistic Missile Programs and Space*, January 1959. On 25 April 1959, shortly after the Symington Committee hearings, Maj Gen Schriever relinquished his Ballistic Missile Division command for command of ARDC and promotion to Lt Gen.

44. *Ibid.*

45. U.S. Senate, *Investigation of Governmental Organization for Space Activities*, Hearings Before the Subcommittee on Governmental Organization for Space Activities of the Committee on Aeronautical and Space Sciences, 86th Congress, 1st Session, March 24-May 7, 1959, pp. 413-418. MacIntyre and Schriever testimony also appears on pp. 352, 354, 360-361, 396-403, 405, 409, 426-428, 448-49, and 461-62.

46. The following descriptions of systems and programs are based on: Department of Defense, *Annual Report of the Secretary of Defense and the Reports of the Secretary of the Army, Secretary of the Navy, Secretary of the Air Force, July 1, 1958 to June 30, 1959* (Washington, D.C.: U.S. Government Printing Office, 1960), pp. 20-25, 334-335; James Baar and William E. Howard, *Spacecraft and Missiles of the World, 1962* (New York: Harcourt, Brace & World, Inc., 1962), Chapter 3, Chapter 4; Horace Jacobs and Eunice Engelke Whitney, *Missile and Space Projects Guide 1962* (New York: Plenum Press, 1962); Bowen, *The Threshold of Space*, pp. 43-48; Max Rosenberg, *The Air Force in Space, 1959–1960*, Vol. SHO-S-62/112 (Washington, D.C.: USAF Historical Division Liaison Office, June 1962), pp. 31-49.

47. Samos had long been considered an acronym for Satellite and Missile Observation System. It seems that WS 117L Project Director Col Fritz Oder adopted the name only to signify the island home of Midas. For background on the reconnaissance program, see Jeffrey T. Richelson, *America's Secret Eyes in Space: The U.S. Keyhole Spy Satellite Program* (New York: Harper Business, 1990), pp. 44-64.

48. Bowen, *The Threshold of Space*, pp. 46-47. For a discussion of the recently declassified Project CORONA, see Kenneth E. Greer, "Corona," *Studies in Intelligence*, Supplement, 17 (Spring 1973): 6, reprinted in Kevin C. Ruffner, ed., *Corona: America's First Satellite Program* (CIA History Staff, Center for the Study of Intelligence, Washington, D.C., 1995); Robert A. MacDonald, "CORONA: Success for Space Reconnaissance, A Look into the Cold War, and Revolution for Intelligence," *PE&RS*, Vol. LXI, No. 6, June 2, 1995, pp. 689-720; Message from Roger Bossart to Bob Peterson, "CORONA Program Profile," Lockheed Press Release, May 24, 1995; Jeffrey T. Richelson, *America's Secret Eyes in Space: The U.S. Keyhole Spy Satellite Program* (New York: Harper Business, 1990), Chapters 1 and 2; Dwayne A. Day, "CORONA: America's First Spy Satellite," *Quest* (Summer 1995): 4-21; Jonathan McDowell, "US Reconnaissance Satellite Programs. Part I: Photoreconnaissance," *Quest* (Summer 1995): 22-33.

49. *Ibid.*, pp. 36-38. Initially the Air Force and Army shared responsibility for Notus, a two-part satellite communications program consisting of 24-hour synchronous satellites in equatorial orbit (Decree), and the Courier delayed repeater satellite. By May 1959, Air Force requests for a system that would fulfill SAC's long range communications needs in polar regions resulted in the addition of Steer, a one-channel communications satel-

lite, and Tackle, a multichannel system. Later in the fall, DDR&E, citing technical, funding, and schedule problems, would eliminate Steer and Tackle, leaving only Decree, which was transferred to the Army under the name Advent. The Air Force continued to lobby for a feasible polar system.

50. Kipp, *Space Detection and Tracking: A Chronology*.

51. The following discussion is based on: Space Systems Division (AFSC), *Chronology of Early Air Force Man-In-Space Activity*, AFSC Historical Publications Series 65-21-1, January 1965; Swenson, *This New Ocean*, pp. 33-97; Jean Evans, "History of the School of Aviation Medicine," *The Air Power Historian*, October 1958, pp. 245-261; Bowen, *The Threshold of Space*, pp. 31-49; Rosenberg, *The Air Force in Space, 1959–1960*, pp. 46-49; Clarence J. Geiger, *History of the* X-20A *Dyna-Soar*. Vol. 1. AFSC Historical Publications Series 63-50-1. (Wright-Patterson AFB, OH: Air Force Systems Command, Aeronautical Systems Division, October 1963).

52. Less ambitious space medicine programs took place at the Navy School of Aviation Medicine at Pensacola, Florida, and NACA's Lewis Flight Propulsion Laboratory in Cleveland.

53. Bowen, *The Threshold of Space*, p. 31.

54. See Senate, *Investigation of Governmental Organization for Space Activities*, pp. 484-493, 271-272B.

55. For the Navy proposal, see "General Proposal for Organization for Command and Control of Military Operations in Space," [n.d.], Eisenhower Library, White House Office, Office of the Spacial Assistant for Science and Technology, Records (James R. Killian and George B. Kistiakowsky, 1957–1961). Box 15 "Space [July–December 1959] (7)." Rosenberg, *The Air Force in Space, 1959–1960*, pp. 18-21.

56. Lt Gen B. A. Schriever, Commander, ARDC to Lt Gen R. C. Wilson, Deputy Chief

of Staff, Development, 18 May 1959, w/atch, draft letter to Secretary of Defense Neil H. McElroy.

57. Rosenberg, *The Air Force in Space 1959–1960*, pp. 18-21; Carl Berger, *The Air Force in Space, Fiscal Year 1961*, Vol. SHO-S-66/142 (Washington: USAF Historical Division Liaison Office, April 1966), pp. 29-30.

58. *Ibid.*; Futrell, *Ideas, 1907–1960*, pp. 593, 601.

59. AFBMD, *Impact of the OSD 'Space Operations' Memorandum on the AFBMD Space Program*, October 1959.

60. Memorandum, AF/CS to Secretary of the Air Force, subj: Proposed Assignment of ABMA to Department of the Air Force, 29 September 1959.

61. Rosenberg, *The Air Force in Space, 1959–1960*, pp. 15-18; Derek W. Elliott, "Finding an Appropriate Commitment: Space Policy Development Under Eisenhower and Kennedy, 1954–1963," unpublished dissertation (Washington, D.C.: The George Washington University, 10 May 1992), pp. 95-97; Hunley, *The Birth of NASA*, pp. 18-24; Medaris, *Countdown to Decision*, pp. 265-300.

62. Memorandum, AFDAT to AFCCS, subj: Statement of Critical Problem Concerning SATURN, DYNA SOAR, and Air Force Space Responsibilities, 30 October 1959. General Wilson's failure to have General Boushey's focal point office for Air Staff space activities redesignated from Director of Advanced Technology to Assistant for Astronautic Systems served as a reminder that the administration remained sensitive to the term "astronautics" and any suggestion of a higher military profile that might call attention to the nation's military space program.

63. Rosenberg, *The Air Force in Space, 1959–1960*, pp. 27-31; House Report, *Government Operations in Space*, pp. 53-55.

64. Rosenberg, *The Air Force in Space, 1959–1960*, pp. 27-31; House Report, *Government Operations in Space*, pp. 53-55.

65. Bowen, *The Threshold of Space,* pp. 38-43.

66. Rosenberg, *The Air Force in Space, 1959–1960,* pp. 1-3.

67. *Ibid.,* pp. 3-4; Stares, *Militarization of Space,* pp. 47-57; Hall, "Origins of U.S. Space Policy," p. 24; Elliott, "Finding and Appropriate Commitment," pp. 100-101.

68. White House to Congress, subj: Proposed Amendments to the National Aeronautics and Space Act of 1958, 14 January 1960.

69. House Report, *Government Operations in Space,* pp. 55-58; Rosenberg, *The Air Force in Space, 1959–1960,* pp. 4-10.

70. House Report, *Government Operations in Space,* pp. 55-58; Rosenberg, *The Air Force in Space, 1959–1960,* pp. 4-10.

71. U.S. Senate, *NASA Authorization for Fiscal Year 1961,* Hearings Before the NASA Authorization Subcommittee of the Committee on Aeronautical and Space Sciences, Pt 1, 86th Congress, 2nd Session, 28-30 March 1960, p. 505.

72. Rosenberg, *The Air Force in Space, 1959–1960,* pp. 21-26. Begun in October 1959, the 120-page document provided the detailed planning necessary to achieve the objectives and capabilities prescribed in the the two unapproved papers.

73. Berger, *The Air Force in Space,* FY 61, pp. 26-29. This monograph has been declassified except for a number of pages in the section on Samos. Part of the Air Force's dilemma is apparent in the terminology used at this time. Air Force leaders and government historians who examined this era often refer to Air Force space policy and policy guidance, even though space policy is established by the President and the National Security Council and the service is charged with implementing that policy.

74. Rosenberg, *The Air Force in Space, 1959–1960,* p. 13.

75. *Ibid.,* pp. 20-21. Gates replaced Neil McElroy in December 1959.

76. The following discussion is based on: Berger, *The Air Force in Space,* FY 61, pp. 34-35, 41-43; R. Cargill Hall, "The Air Force in Space," Unpublished classified draft chapter, May 1984, information used is unclassified; Richelson, *America's Secret Eyes in Space,* Chapter 2; Stares, *Militarization of Space,* pp. 44-46. Despite the passage of time, Berger remains one of the most convincing sources for Samos developments.

77. Albert D. Wheelon, the directed Project CORONA from 1963 to 1966, has noted that critics of Samos also doubted whether the camera's small focal length and the microwave downlink's narrow bandwidth would be capable of achieving the 60-foot resolution forecast by the planners. Although the Air Force continued to focus on film readout for the first phase of Samos (Project 101A-E2), it pursued a recovery capability for the second phase (Project 101B-E5). The third phase (Program 201-E6), also involved film recovery. The Samos project concluded, unsuccessfully, in 1962, and it would be another fifteen years before digital technology made possible effective film readout from high magnification satellite camera systems. Albert D. Wheelon, "CORONA: A Triumph of American Technology," Presentation given at Piercing the Curtain, a joint symposium held by the CIA's Center for Study of Intelligence and the George Washington University's Space Policy Institute at the George Washington University on May 23-24, 1995; Message from Roger Bossart to Bob Peterson, "CORONA Program Profile," Lockheed Press Release, May 24, 1995; R. Cargill Hall, "The Eisenhower Administration and the Cold War: Framing American Astronautics to Serve National Security" *Prologue* (Spring 1995), Vol. 27, no.1, pp. 67-68; McDowell, "US Reconnaissance Satellite Programs," pp. 28-29.

78. General Greer's west coast Samos office included both the national Project CORONA and the Air Force Samos satellite programs. The term National Reconnaissance Office (NRO) remained classified. Memorandum, Secretary of the Air Force Dudley C. Sharp to Chief of Staff Thomas D. White, 13 September 1960, w/atch, "Organization and Functions of the Office of Missile and Satellite Systems and Satellite Reconnaissance Advisory Council," Box 36, Thomas D. White Papers, Manuscript Division, Library of Congress.

79. The Aerospace Corporation, *The Aerospace Corporation. Its Work: 1960–1980* (Los Angeles: Times Mirror Press, 1980), pp. 15-25.

80. Rosenberg, *The Air Force in Space, 1959–1960*, p. 30.

81. Berger, *The Air Force in Space, FY 61*, p. 2.

82. Lt Gen B. A. Schriever, ARDC/CC, to Trevor Gardner, Chairman and President, Hycon Manufacturing Co, subj: [committee on space development program], 11 October 1960.

83. Berger, *The Air Force in Space, FY 61*, pp. 3-5; Stares, *Militarization of Space*, pp. 60-62; Memorandum, Department of the Air Force, Memorandum for Chief of Staff, USAF, subj: Report to the President-Elect of the Ad Hoc Committee on Space, January 1961.

84. *Ibid.*, pp. 29-31.

85. Berger, *The Air Force in Space, FY 61*, pp. 5-6, 30-31. In his 15 December 1959 letter, General White said: "I would like every member of the Air Force to do everything within his power to maintain the same degree of harmony and cooperation with NASA [as had existed with the National Advisory Committee for Aeronautics]." Quoted in Rosenberg, *The Air Force in Space, 1959–1960*, pp. 13-14.

86. Gen Thomas D. White, CSAF to Overton Brooks, Chairman, Science and Astronautics Committee, House of Representatives, 19 January 1961; Overton Brooks to Dr. T. Keith Glennan, President, Case Institute of Technology, 14 February 1961, Box 47, Thomas D. White Papers, Manuscript Division, Library of Congress.

87. *Ibid.*; Futrell, *Ideas, 1907–1960*, pp. 604-606; Space Systems Division (AFSC), *Chronology of Early Air Force Man-In-Space Activity*, p. 41.

88. Berger, *The Air Force in Space, FY 61*, pp. 5-6.

89. DoD Directive 5160.32, "Development of Space Systems," 6 March 1961.

90. For the background of the reorganization see, Berger, *The Air Force in Space, FY 61*, pp. 6-10; History, Air Research and Development Command, Vol I, 1 January–31 March 1961, History of Air Force Systems 1 April–30 June 1961, I-22 to I-52; U.S. Air Force Oral History Interview, Gen Bernard A. Schriever, (USAF, Ret) with Maj Lyn R. Officer and Dr. James C. Hasdorff, Albert F. Simpson Historical Research Center, Air University, Montgomery, Alabama, No. K239.0512-676, 20 June 1973, pp. 23-25; Rick W. Sturdevant, "The United States Air Force Organizes for Space: The Operational Quest", in *Organizing for the Use of Space: Historical Perspectives on a Persistent Issue*, ed. Roger D. Launius (San Diego, CA: American Astronautical Society, 1995).

91. Berger, *The Air Force in Space, FY 61*, pp. 10-14; Memorandum, Trevor Gardner, *et al* to Secretary of the Air Force Eugene M. Zuckert, 23 April 1961.

92. Berger, *The Air Force in Space, FY 61*, pp. 15-25.

93. *Ibid.*, pp. 23-24.

94. *Ibid.*, pp. 24-25, 59-66. These enormous boosters represented an important commitment to large-scale space operations. While NASA's liquid propellant Saturn clustered rocket and single-chamber F-1 Nova rocket motor would be capable of

producing 1.5-million pounds of thrust, the Air Force's large, segmented solid rocket motor development plan called for a 3-million pound thrust capability. Later, a Webb-McNamara agreement of November 1961 would eliminate the large solid motors as backup to the Nova and Saturn, and the Air Force would focus on using the solid-propellant motors in its Minuteman ICBM project. See Carl Berger, *The Air Force in Space, Fiscal Year 1962*, Vol. SHO-6-66/198 (Washington, D.C.: USAF Historical Division Liaison Office, June 1966), pp. 54-60; James Baar and William E. Howard, *Spacecraft and Missiles of the World, 1962* (New York: Harcourt, Brace & World, 1962), pp. 30-31.

95. Interview, Secretary of the Air Force Eugene Zuckert. Oral History Interview for the Kennedy Memorial Library by Lawrence McQuade, File No. 168.7050-1, Albert F. Simpson Historical Research Center (Maxwell AFB, AL: Air University, May-June 1964), part 6, pp. 5-6, 14-15.

Chapter 3. The Air Force in the Era of Apollo

1. Message from the President of the United States, *United States Aeronautics and Space Activities, 1961* (Washington, D.C.: Government Printing Office), p. 1.

2. *Ibid.*, p. iii.

3. U.S. House, *Committee on Government Operations, Government Operations in Space (Analysis of Civil-Military Roles and Relationships)*, House Report No. 445, 89th Congress, 1st Session, 4 June 1965, p. 61. Total funding for Project Apollo amounted to just over $16 billion. See Courtney G. Brooks, James M. Grimwood, and Loyd S. Swenson, *Chariots for Apollo: A History of Manned Lunar Spacecraft*, The NASA History Series, NASA SP-4205 (Washington, D.C.: NASA, 1979), pp. 409-411.

4. Carl Berger, *The Air Force in Space, Fiscal Year 1961*, Vol. SHO-S-66/142 (Washington, D.C.: USAF Historical Division Liaison Office, April 1966), pp. 24-25.

5. Air Force leaders seldom discussed their views on space leadership with uniformity and precision during the Kennedy and Johnson years. Clearly they always advocated greater Air Force space responsibilities, but at times their pretensions embraced leadership not only of the military space effort, but the national space program—this despite the large, recognized mission responsibilities of NASA and the NRO. Indeed, despite NASA's manned lunar mission, the Air Force continued to pursue a mission for military man-in-space, while service leaders refused to forego proposals for an offensive military space capability despite national policy to the contrary. Air Force space pretensions did not normally find widespread support among civilian Air Force officials. Indeed, uniformed service leaders and their civilian counterparts often differed over a greater Air Force space role in the national program.

6. *United States Aeronautics and Space Activities, 1961.*

7. DoD Directive 5160.32, "Development of Space Systems," 6 March 1961; Department of Defense, "Annual Report for Fiscal Year 1961," p. 20. House Report, *Government Operations in Space*, pp. 70-71. The so-called Wiesner Report asserted that 90 percent of space resources within the military belonged to the Air Force. Memorandum, Secretary of the Air Force to Chief of Staff, USAF, subj: Report to the President-Elect of the Ad Hoc Committee on Space, January 1961. During congressional testimony on 16 January 1961, Director of Defense Research and Engineering Herbert F. York stated that "the total amount of money spent in the Department of Defense on space programs, either directly by the Air Force or through the Air Force,

311

is 91 percent." U.S. House, *Hearings before the Committee on Science and Astronautics*, 37th Congress, 1st Session, February 16, 1961 (Washington, D.C.: Government Printing Office), p. 9.

8. As noted in the previous chapter, Air Force space advocates like General Schriever favored a larger military space program and one that would include offensive weapons in space regardless of national policy to the contrary. More accurately described as a policy of "freedom of space" and the "peaceful uses of space," the Eisenhower administration encouraged passive military space missions but not offensive platforms and weapons that could threaten the valuable national space reconnaissance mission. Although service leaders were aware of the administration's concern, they nevertheless pursued the characteristic approach to defense that called for preparing for potential threats regardless of the venue. They would likely argue that the only way to ensure "peaceful overflight" and the reconnaissance space flight would be to deploy weapons in space that would deny the Soviets space supremacy. Continued Air Force pressure would lead the Kennedy administration to reaffirm the Eisenhower policy and stymie service efforts to achieve "space supremacy."

9. Carl Berger, *The Air Force in Space, Fiscal Year 1962*, Vol. SHO-S-66/198 (Washington, D.C.: USAF Historical Division Liaison Office, June 1966), p. 4.

10. Berger, *The Air Force in Space, FY 62*, pp. 4-5.

11. *Ibid.*, pp. 5-6.

12. *Ibid.*, p. 6.

13. Quoted in *Ibid.*, p. 8. For Schriever's speech, see the *Washington Post*, 13 October 1961.

14. Quoted in Berger, *The Air Force in Space, FY 62*, pp. 7-8. General LeMay became Chief of Staff in June 1961 following Gen Thomas D. White's retirement.

15. *Ibid.*, p. 8.

16. *Ibid.*, p. 18; Air Force Space Plan (Draft), Headquarters USAF, September 1961.

17. Air Force Space Plan (Draft), September 1961, p. 7.

18. *Ibid.*, p. 83.

19. *Ibid.*, "Summary of Conclusions and Recommendations," pp. 83-88.

20. *Ibid*, pp. 20-21, 83-88.

21. Berger, *The Air Force in Space, FY 62*, pp. 19-20; Col Harry L. Evans, Deputy for Satellite Systems, SSD to Col Appold, SSD/DCL, subj: USAF Space Program FY 63-64, 2 January 1962. Later, on 8 September 1962, Air Force headquarters designated Ferguson's office, DCS/R&D, the focal point for Air Force space issues.

22. Berger, *The Air Force in Space, FY 62*, pp. 20-21; Presentation, Lt Gen James Ferguson, Deputy Chief of Staff, Research & Technology, USAF. Statement before the House Committee on Armed Services, February 1962, SAMSO Archives, Space, General, 1962. The Ferguson address is also found in *Aviation Week and Space Technology*, 5 March 1962, pp. 75, 77, 79, 83, 87, 89, 91-92, 94, 96.

23. Presentation, Statement by Lt Gen James Ferguson, Deputy Chief of Staff, Research & Technology, USAF, before the House Committee on Armed Services, February 1962.

24. *Ibid.* The eleven areas included improved boosters, in-space propulsion, power supplies required, communications technology, re-entry and recovery, and, most importantly, man's space role.

25. *Ibid*; AFXPD-PA-LRP, *Air Force Space Program FY 1963*, Panel One, Ferguson Task Force, 22 December 1961.

26. Presentation, Statement by Lt Gen James Ferguson, Deputy Chief of Staff, Research & Technology, USAF, before the House Committee on Armed Services, February 1962.

27. Memorandum, Secretary of Defense to Secretary of the Air Force, "The Air Force Manned Military Space Program," 23 February 1962.

28. Berger, *The Air Force in Space, FY 62*, pp. 22-23; Presentation, Address by Gen Curtis E. LeMay, Chief of Staff, United States Air Force, Assumption College, Worcester, Massachusetts, 28 March 1962, SAMSO Archives, Space, General, 1962.

29. Berger, *The Air Force in Space, FY 62*, pp. 21-25. The McNamara-LeMay meeting is also described in, Presentation, Address to Space Technical Objectives Group, by Lt Gen Howell M. Estes, Jr., DCAS, Los Angeles California, 16 April 1962, p. 2.

30. The other panels are: Unmanned Spacecraft Panel, Space Flight Ground Environment Panel, Supporting Space Research and Technology Panel, Aeronautics Panel. U.S. House, Subcommittee on NASA Oversight of the Committee on Science and Astronautics, *The NASA-DoD Relationship*, 88th Congress, 2nd Session, 1964, pp. 11-15; House Report, *Government Operations in Space*, p. 65.

31. House Report, *Government Operations in Space*, pp. 65-66, 126-127; Berger, *The Air Force in Space*, FY 62, pp. 54-60; Arnold S. Levine, *Managing NASA in the Apollo Era*. The NASA History Series. NASA SP-4102 (Washington, D.C.: National Aeronautics and Space Administration, 1982), pp. 225-228.

32. Berger, *The Air Force in Space, FY 62*, pp. 10-12; House Report, *Government Operations in Space*, p. 66.

33. Berger, *The Air Force in Space, FY 62*, pp. 12-14.

34. DoD Directive 5030.18, "Department of Defense Support of National Aeronautics and Space Administration (NASA)," 24 February 1962.

35. House Report, *Government Operations in Space*, pp. 65-67; Report, "Air Force/NASA Space Program Management Panel 3 Report (Ferguson Task Force) USAF Space Program FY 63-64," n.d. [December 1961–January 1962]; Levine, *Managing NASA*, pp. 211-214; House Report, *NASA-DoD Relationship*, pp. 3-6. Despite NASA's interest in the Navy's Transit system, the agency never developed a navigation satellite.

36. House Report, *Government Operations in Space*, pp. 80-83; Paul B. Stares, *The Militarization of Space: U.S. Policy, 1945–1984* (Ithica, NY: Cornell University Press, 1985), pp. 77-78.

37. House Report, *Government Operations in Space*, pp. 80-83; Stares, *Militarization of Space*, pp. 77-78.

38. House Report, *Government Operations in Space*, pp. 80-84; Stares, *Militarization of Space*, pp. 77-78, 65-66, 69-70. Although these projects were removed from public scrutiny, the government officially announced all launches and the trade press continued to discuss DoD space issues and activities.

39. The most comprehensive treatment of the 156 Committee deliberations is: Raymond L. Garthoff, "Banning the Bomb in Outer Space," *International Security*, Winter 1980/81 (Vol. 5, No. 3), pp. 25-40. Although Garthoff describes persistent JCS opposition to a unilateral declaration without verification strictures, he at no point refers to Air Force actions as precipitating Kennedy's decision to establish the committee. For a discussion of disarmament space developments during the 1960s, see Stares, *Militarization of Space*, pp. 59-105.

40. House Report, *Government Operations in Space*, p. 73; Roger E. Bilstein, *Orders of Magnitude. A History of the NACA and NASA, 1915–1990*. The NASA History Series. NASA SP-4406 (Washington, D.C.: National Aeronautics and Space Administration, 1989), p. 63.

41. Berger, *The Air Force in Space, FY 62*, p. 9.

42. *Ibid.*, pp. 19-20.

43. *Ibid.*, pp. 23-25.

44. *Ibid.*, pp. 14-17; Levine, *Managing NASA*, pp. 217-225. The two sides settled the range dispute through an agreement of 14 January 1963, but the reimbursement issue dragged on for six years.

45. For data on the Samos program, see Jonathan McDowell, "US Reconnaissance Satellite Programs. Part I: Photorecon-naissance," *Quest* (Summer 1995): 22-33. Despite continued Air Force concerns in the 1960s about a possible Soviet orbiting wea-pon threat, détente became the order of the day and a sense of urgency about Soviet space threats diminished.

46. Testimony of DoD Comptroller Charles J. Hitch, quoted in Levine, *Managing NASA*, p. 217.

47. DoD Directive 5160.32, "Development of Space Systems," 6 March 1961.

48. House Report, *Government Operations in Space*, pp. 70-71.

49. *Ibid.*

50. For coverage of Titan III development, see Berger, *The Air Force in Space, FY 62*, pp. 43-53, and Robert F. Piper, *History of the Titan III, 1961–1963*. Vol.1, AFSC Historical Publications Series 64-22-1 (Washington, D.C.: Air Force Systems Command, Space Systems Division, June 1964). The new, rigorous review process in the McNamara Pentagon was part of the comprehensive Planning, Programming, Budgeting System management procedures adopted by DoD.

51. Berger, *The Air Force in Space, FY 62*, p. 45. Although involved in a variety of important projects as Assistant Secretary of the Air Force, Brockway McMillan's most important responsibility involved his posi-tion as Director of the National Reconnais-sance Office.

52. *Ibid.*

53. *Ibid.*, pp. 45-53. The Titan IIIA com-prised a basic Titan II with transtage capable of launching a 5,8000-pound payload into a 100-mile orbit, while the Titan III consisted of a Titan A booster with two strap-on solid rockets designed to place a 25,000-pound payload into low earth orbit.

54. *Ibid.*, pp. 50-51.

55. Report to the Congress from the President of the United States, *United States Aeronautics and Space Activities, 1962*, (Washington, D.C.: National Aeronautics and Space Council, 1963), p. 33.

56. Cantwell, *The Air Force in Space, Fiscal Year 1964*, pp. 21-23; Gen Bernard A. Schriever, *et al*, *Reflections on Research and Development in the United States Air Force*, Jacob Neufeld, ed. (Washington, D.C.: Center for Air Force History, 1993), pp. 72-73.

57. Berger, *The Air Force in Space, FY 62*, pp. 77-78; Stares, *Militarization of Space*, p. 79.

58. The most comprehensive analysis of the three initiatives is found in the critique by Launor F. Carter, Chief Scientist of the Air Force. See Launor F. Carter, *An Inter-pretive Study of the Formulation of the Air Force Space Plan*, (Washington, D.C.: Head-quarters USAF, 4 February 1963), SAMSO Archives.

59. Memorandum, Secretary of Defense to Secretary of the Air Force, subj: The Air Force Manned Military Space Program, 23 February 1962.

60. Presentation, Address to Space Tech-nical Objectives Group by Lt Gen Howell M. Estes, Jr., DCAS, Los Angeles California, 16 April 1962. Nineteen separate Air Force or Air Force related organizations were represented in the Estes-directed effort.

61. Briefing, SSD, "Briefing on Task Group Capabilities vs Space Plan Objectives," 1 June 1962, SAMSO Archives; Carter, *Formu-lation of the Air Force Space Plan*, pp. 3-5.

62. The remaining programs listed are: Lifting Body Re-entry Vehicle, Manned Hypersonic Test Vehicle, Manned Maneu-verable Spacecraft, Strategic Earth Orbiting Base, Space Logistic Support System,

Manned Satellite Interceptor, Required Reconnaissance System, Advanced Interceptor System, Orion, and Nuclear Rocket. Moreover, only MODS and Blue Gemini represented new programs, and neither received funding approval. See Carter, *Formulation of the Air Force Space Plan*, pp. 5-12; Memorandum, Secretary of the Air Force to Secretary of Defense, subj: Five Year Space Program, 9 November 1962; Col N. C. Appold, AFSC/STP to Col K. W. Schultz, DCS/DevPlanning, subj: Data for Section I of USAF Five-Year Space Program, 27 August 1962; Briefing, AFSC Task Group, "Status Briefing, USAF 5 Year Space Program," 20 August 1962; Briefing, SSD/SSEH, "Briefing for Dr. Launor F. Carter. Evolution of the 5 Year Space Program," 27 November 1962; Briefing, SSD, "USAF Five Year Space Program Briefing," September 1962; Briefing, SSD, "USAF Five Year Space Program Briefing, AFSC Recommended Program," September 1962, SAMSO Archives, AF Space Program, 1962–63.

63. USAF Space Plan (Draft Working Paper), Revised July 1962.

64. The stated requirements were: strategic earth orbital base, strategic aerospace vehicle (aerospace plane), unmanned space reconnaissance system, manned space reconnaissance system, ballistic missile interceptor, space surveillance-missile detection, aerospace surveillance and warning system, space surveillance-satellite inspection, satellite interceptor system, command control, earth based, communications satellites, space logistic support system, astro tug. *Ibid.*, pp. 30-34; Carter, *Formulation of the Air Force Space Plan*, pp. 2-3; Paper, SSD, "Preliminary Remarks on Military Technical Objectives in Space for use by the Systems Requirements Panel of the Technical Objectives Task Group (AFSC)," 2 May 1962.

65. Carter, *Formulation of the Air Force Space Plan*, pp. 13-15.

66. The National Aeronautics and Space Act of 1958 (Public Law 58-568), Section 102(c)(6).

67. Levine, *Managing NASA*, pp. 211-214.

68. *NASA Data Book*, table 3-26.

69. House Report, *Government Operations in Space*, pp. 65-67; Levine, *Managing NASA*, pp. 217-221. The extensive AF/DoD-NASA organizational interfaces are described in House Report, *NASA-DoD Relationship*. Descriptions of formal agreements are found in House Report, *Government Operations in Space*, pp. 123-133.

70. Increased cooperation between the Air Force and NASA is described in Gerald T. Cantwell, *The Air Force in Space, Fiscal Year 1964*, Vol. SHO-S-67/52 (Washington, D.C.: USAF Historical Division Liaison Office, June 1967), pp. 8-14. Early in 1964 at NASA's request, Air Force Brig Gen Samuel C. Phillips received responsibility for managing Project Apollo as NASA's Deputy Director of the Apollo Program.

71. Berger, *The Air Force in Space, FY 62*, pp. 36-41.

72. Air Force Space Plan (Draft), September 1961, pp. 36-37.

73. Presentation, Statement by Lt Gen James Ferguson, Deputy Chief of Staff, Research & Technology, USAF, before the House Committee on Armed Services, February 1962, SAMSO Archives; Berger, *The Air Force in Space, FY 62*, p. 40.

74. Maj Timothy D. Killebrew, *Military Man in Space: A History of Air Force Efforts to Find a Manned Space Mission*. Air Command and Staff College Report No. 87-1425 (Maxwell AFB, AL: Air University Press, 1987), pp. 25-26; Stares, *Militarization of Space*, p. 79; Berger, *The Air Force in Space, FY 62*, pp. 41-42; Cantwell, *The Air Force in Space, Fiscal Year 1964*, pp. 15-16; Jeffrey T. Richelson, *America's Secret Eyes in Space* (New York: Harper Collins, 1990), pp. 83-84; Barton C. Hacker and James M. Grimwood, *On the Shoulders of Titans: A History of*

Project Gemini, The NASA History Series, NASA SP-4203 (Washington, D.C.: NASA, 1977), pp. 117-122.

75. Levine, *Managing NASA,* pp. 230-231; W. Fred Boone, NASA Office of Defense Affairs, *The First Five Years, December 1, 1962, to January 1, 1968,* Historical Division, Office of Policy (Washington, D.C.: NASA, December 1970), pp. 83-87; Killebrew, *Military Man in Space,* pp. 25-26.

76. Boone, *The First Five Years,* pp. 83-87; House Report, *Government Operations in Space,* pp. 84-85, 129; Cantwell, *The Air Force in Space, Fiscal Year 1964,* pp. 15-16.

77. Boone, *The First Five Years,* pp. 83-87.

78. House Report, *Government Operations in Space,* pp. 86-87; U.S. Senate, Hearing before the Subcommittee of the Committee on Appropriations, *Department of Defense Appropriations for 1964,* 88th Congress, 1st Session, 24 April 1963.

79. House Report, *Government Operations in Space,* pp. 85-88.

80. Maj Barkley G. Sprague, USAF, *Evolution of the Missile Defense Alarm System (MIDAS) 1955–1982,* Report No. 85-2580 (Maxwell AFB, AL: Air Command and Staff College, 1985), pp. 15-19. For discussion of the MIDAS program, see Chapter 4.

81. *Ibid.,* pp. 78-80; Stares, *Militarization of Space,* pp. 71-82; Report to the Congress from the President of the United States, *United States Aeronautics and Space Activities, 1963,* pp. 39-60; Report to the Congress from the President of the United States, *United States Aeronautics and Space Activities, 1964,* pp. 41-67.

82. U.S. House, Committee on Armed Services, *Hearings on Military Posture,* 88th Congress, 1st Session, January–February 1963, pp. 467, 471.

83. Cantwell, *The Air Force in Space, Fiscal Year 1964,* p. 24.

84. Levine, *Managing NASA,* pp. 231-232.

85. Cantwell, *The Air Force in Space, Fiscal Year 1964,* pp. 24-25.

86. *Ibid.,* pp. 16-18; Boone, *The First Five Years,* pp. 88-93.

87. *Ibid.*

88. Boone, *The First Five Years,* pp. 88-93.

89. Cantwell, *The Air Force in Space, Fiscal Year 1964,* pp. 16-18.

90. *Ibid.;* Boone, *The First Five Years,* pp. 94-96.

91. Cantwell, *The Air Force in Space, Fiscal Year 1964,* p. 18. ASSET provided the acronym for the Aerothermodynamic/Elastic Structural Systems Environmental Test program. Launched by single-stage THOR boosters, ASSET non-orbiting, unmanned, glide-reentry vehicles conducted a broad range of reentry experiments.

92. Cantwell, *The Air Force in Space, Fiscal Year 1964,* pp. 18-20.

93. *Ibid.,* pp. 25-28; Levine, *Managing NASA,* p. 232; Killebrew, *Military Man in Space,* pp. 22-24; House Report, *Government Operations in Space,* pp. 87-88. Dyna-Soar received the X-20 designation on 26 June 1962.

94. Cantwell, *The Air Force in Space, Fiscal Year 1964,* pp. 28-30.

95. House Report, *Government Operations in Space,* pp. 88-89; Report to the Congress from the President of the United States, *United States Aeronautics and Space Activities, 1963,* pp. 41-42; Killebrew, *Military Man in Space,* pp. 27-28.

96. Quoted in House Report, *Government Operations in Space,* pp. 88-89.

97. *Ibid.* Because the space station could not change orbit, its reconnaissance and surveillance capability was limited to objects when in view.

98. Boone, *The First Five Years,* pp. 97-99; House Report, *Government Operations in Space,* pp. 88-89.

99. Levine, *Managing NASA,* pp. 232-233; Cantwell, *The Air Force in Space, Fiscal Year 1964,* p. 21.

100. Cantwell, *The Air Force in Space, Fiscal Year 1964*, pp. 21-23.

101. *Ibid.*

102. House Report, *Government Operations in Space*, pp. 88-89; Killebrew, *Military Man in Space*, pp. 27-28; Richelson, *America's Secret Eyes in Space*, pp. 84-86, 90-92.

103. Richelson, *America's Secret Eyes in Space*, pp. 84-86, 90-92. For a listing of the specific experiments, see Killebrew, *Military Man in Space*, pp. 31-33.

104. Richelson, *America's Secret Eyes in Space*, pp. 90-92; Killebrew, *Military Man in Space*, pp. 31-33; Robert Frank Futrell, *Ideas, Concepts, Doctrine: Basic Thinking in the United States Air Force, 1961–1984*, Vol. II (Maxwell Air Force Base, AL: Air University Press, December 1989), p. 681 (hereafter cited as Futrell, *Ideas, 1961–1984*). See also The Aerospace Corporation, *The Aerospace Corporation: Its Work: 1960–1980* (Los Angeles: Times Mirror Press, 1980), pp. 87-89.

105. Cantwell, *The Air Force in Space, Fiscal Year 1964*, pp. 1-5.

106. *Ibid.*, p. 5.

107. *Ibid.*, pp. 5-6. Schriever's comments are found in Gen Bernard A. Schriever, *et al*, *Reflections on Research and Development in the United States Air Force*, Jacob Neufeld, ed. (Washington, D.C.: Center for Air Force History, 1993), pp. 72-73. For the general impact of *Project Forecast* on Air Force scientific development, see Michael H. Gorn, *Harnessing the Genie: Science and Technology Forecasting for the Air Force, 1944–1986*. Air Staff Historical Study (Washington, D.C.: Office of Air Force History, 1988), pp. 87-130.

108. Boone, *The First Five Years*, pp. 97-103.

109. Futrell, *Ideas, 1961–1984*, pp. 681-682; Richelson, *America's Secret Eyes in Space*, pp. 88-92; Killebrew, *Military Man in Space*, pp. 31-33; Stares, *Militarization of Space*, p. 98. Harold Brown moved from DDR&E to become Secretary of the Air Force in October 1965.

110. Futrell, *Ideas, 1961–1984*, pp. 682-683; Richelson, *America's Secret Eyes in Space*, pp. 101-103; Killebrew, *Military Man in Space*, pp. 34-39; Stares, *Militarization of Space*, pp. 98-99.

111. Levine, *Managing NASA*, pp. 234-235; House Report, *Government Operations in Space*, pp. 88-91; Futrell, *Ideas, 1961–1984*, pp. 682-683; Richelson, *America's Secret Eyes in Space*, pp. 101-103; Killebrew, *Military Man in Space*, pp. 34-39; Stares, *Militarization of Space*, pp. 98-99. NASA began with an ambitious post-Apollo space station concept similar to DDR&E's choice in late 1963, but ended with Skylab, a mini orbiting space station similar to the Air Force's MOL that would involve experiments with three astronauts in 1973-74.

112. Futrell, *Ideas, 1961–1984*, pp. 682-683; Richelson, *America's Secret Eyes in Space*, pp. 101-103; Killebrew, *Military Man in Space*, pp. 34-39; Stares, *Militarization of Space*, pp. 98-99.

113. Quoted in Stares, *Militarization of Space*, pp. 159-160.

114. Quoted in Futrell, *Ideas, 1961–1984*, p. 683. According to Richelson, Laird did not take part in the initial decision to cancel the MOL. Rather, that decision involved President Nixon, National Security Advisor Henry Kissinger, and Robert Mayo, Bureau of the Budget director—with the support of CIA Director Richard Helms, whose agency considered manned reconnaissance platforms potentially intolerable to the Soviets. Laird opposed canceling the MOL. See Richelson, *America's Secret Eyes in Space*, pp. 102-103.

115. Futrell, *Ideas, 1961–1984*, p. 683; Richelson, *America's Secret Eyes in Space*, p. 103. After spending nearly $8 billion over a twenty year period, the Vandenberg space shuttle and launch complex, known as SLC-6, was "abandoned in place." For discussion of the Vandenberg project, see Roger G.

Guillemette, "Vandenberg: Space Shuttle Launch and Landing Site. Part 1—Construction of Shuttle Launch Facilities" *Spaceflight*, Vol 36 (October 1994), pp. 354-357; Roger G. Guillemette, "Vandenberg: Space Shuttle Launch and Landing Site: Part 2—Abandoned in Place" *Spaceflight*, Vol 36 (November 1994), pp. 378-381.

116. Killebrew, *Military Man in Space*, p. 36.

117. Futrell, *Ideas, 1961–1984*, p. 683.

118. Maj John B. Hungerford, Jr., *Organization for Military Space: A Historical Perspective.* Air Command and Staff College Report 82-1235 (Maxwell AFB, AL.: Air University Press, 1982), pp. 47-49; John T. Greenwood, "The Air Force Ballistic Missile and Space Program (1954–1974)," *Aerospace Historian 21* (Winter 1974):199. Although the Air Staff on 16 January 1967 created a new agency to provide a focus for developing space policy and plans, its responsibilities proved far less inclusive that General Schriever had desired. Outside its purview remained the sensitive national reconnaissance program, while other Air Force agencies either retained or acquired a portion of space systems responsibilities. See Gerald T. Cantwell, *The Air Force in Space, Fiscal Year 1967*, Part I (Washington, D.C.: USAF Historical Division Liaison Office, May 1970), pp. 5-6. See also Rick W. Sturdevant, "The United States Air Force Organizes for Space: The Operational Quest," in *Organizing for the Use of Space: Historical Perspectives on a Persistent Issue*, ed. Roger D. Launius (San Diego, CA: American Astronautical Society, 1995).

119. Futrell, *Ideas, 1961–1984*, pp. 684-686. The role of the Space Shuttle in Air Force space developments is examined in Chapter 5.

120. Futrell, *Ideas, 1961–1984*, p. 683.

Chapter 4. From the Ground Up

1. The Aerospace Corporation, *The Aerospace Corporation: Its Work: 1960–1980* (Los Angeles: Times Mirror Press, 1980), p. 46 (hereafter cited as *AerospaceCorp*). See also David Spires and Rick W. Sturdevant, "From Advent to Milstar: The United States Air Force and the Challenges of Military Satellite Communications," in *Proceedings of the NASA Conference. Beyond the Ionosphere: The Development of Satellite Communications* (Washington, D.C.: NASA, 1996).

2. *Ibid.*; Maj Robert E. Lee, *History of the Defense Satellite Communications System (1964–1986).* Air Command and Staff College Report No. 87-1545 (Maxwell AFB, AL: Air University Press, 1987), pp. 1-4 (hereafter cited as Lee, *History of DSCS*); David K. van Keuren, "Moon in Their Eyes: Moon Communication Relay (MCR) at the Naval Research Laboratory, 1951–1962," in *Proceedings of the NASA Conference. Beyond the Ionosphere: The Development of Satellite Commu-* nications (Washington, D.C.: NASA, 1996).

3. Carl Berger, *The Air Force in Space, Fiscal Year 1961*, Vol. SHO-S-66/142 (Washington, D.C.: USAF Historical Division Liaison Office, April 1966), pp. 84-93; *AerospaceCorp*, pp. 47-49; Lee, *History of DSCS*, pp. 5-8.

4. *AerospaceCorp*, pp. 47-49; Lee, *History of DSCS*, pp. 5-8, Berger, *The Air Force in Space, FY61*, pp. 84-93.

5. *AerospaceCorp*, p. 48.

6. Carl Berger, *The Air Force in Space, Fiscal Year 1962*, Vol. SHO-S-66/198 (Washington, D.C.: USAF Historical Division Liaison Office, June 1966), pp. 61-70.

7. Berger, *The Air Force in Space, FY62*, pp. 65-68; *AerospaceCorp*, pp. 48-52.

8. Military and commercial capabilities would not share the same satellite until the Navy contracted with COMSAT in 1973 for "Gapfiller" service pending completion of its FLTSATCOM system. See Michael Kinsley,

Outer Space and Inner Sanctums: Government, Business, and Satellite Communication (New York: John Wiley & Sons, 1976), pp. 199-200; Anthony Michael Tedeschi, *Live Via Satellite: The Story of COMSAT and the Technology that Changed World Communication* (Washington, D.C.: Acropolis Books, 1989), p. 150; Martin, *Communication Satellites, 1958–1992*, pp. 104-105, 183-186.

9. *AerospaceCorp*, pp. 48-52; Lee, *History of DSCS*, pp. 8-10; U.S. House, Committee on Government Operations, *Government Operations in Space (Analysis of Civil-Military Roles and Relationships)*, House Report No. 445, 89th Congress, 1st Session, 4 June 1965, pp. 78-79.

10. Gerald T.Cantwell, *The Air Force in Space, Fiscal Year 1964*, Vol. SHO-S-67/52 (Washington, D.C.: USAF Historical Division Liaison Office, June 1967), pp. 69-76; Gerald T.Cantwell, *The Air Force in Space, Fiscal Year 1965*, Vol. SHO-S-68/186 (Washington, D.C.: USAF Historical Division Liaison Office, April 1968), pp. 42-51; *AerospaceCorp*, pp. 48-52; Thomas Karas, *The New High Ground* (New York: Simon and Schuster, 1983), p. 73.

11. Lee, *History of DSCS*, pp. 10-13; *AerospaceCorp*, p. 51.

12. Lee, *History of DSCS*, pp. 10-13; *AerospaceCorp*, pp. 50-52; Chronology, AFSC/ Space Division, *Space and Missile Systems Organization: A Chronology, 1954–1979* (Los Angeles, CA: Space Systems Division, n.d.), pp. 173-174, 176, 182, 193 (hereafter cited as *SAMSO Chronology*).

13. Lee, *History of DSCS*, pp. 13-14; *AerospaceCorp*, pp. 50-52; Horace Jacobs and Eunice Engelke Whitney, *Missile and Space Projects Guide 1962* (New York: Plenum Press, 1962), pp. 190-191.

14. Lee, *History of DSCS*, pp. 11-13.

15. Lee, *History of DSCS*, pp. 13-14; *AerospaceCorp*, pp. 52-55; Thomas Karas, *The New High Ground: Strategies and Weapons of Space-Age War* (New York: Simon and Schuster, 1983), pp. 73-75.

16. Lee, *History of DSCS*, pp. 15-21; *AerospaceCorp*, pp. 55-57; Karas, *The New High Ground*, pp. 75-76; Jacob Neufeld, *The Air Force in Space, 1970–1974* (Washington, D.C.: Office of Air Force History, August 1976), pp. 15-19.

17. Lee, *History of DSCS*, pp. 15-21; *AerospaceCorp*, pp. 55-57; Karas, *The New High Ground*, pp. 75-76; Neufeld, *The Air Force in Space, 1970–1974*, pp. 15-19.

18. Lee, *History of DSCS*, pp. 15-28; *AerospaceCorp*, pp. 55-57; Karas, *The New High Ground*, pp. 75-76; Neufeld, *The Air Force in Space, 1970–1974*, pp. 15-19; Donald H. Martin, *Communication Satellites, 1958–1992* (El Segundo, CA: Aerospace Corporation, 1991), pp. 100-102.

19. *AerospaceCorp*, p. 59; Neufeld, *The Air Force in Space, 1970–1974*, pp. 19-23.

20. *AerospaceCorp*, p. 59; Neufeld, *The Air Force in Space, 1970–1974*, pp. 19-23; *SAMSO Chronology*, pp. 154, 187-188, 194.

21. *AerospaceCorp*, p. 59; Neufeld, *The Air Force in Space, 1970–1974*, pp. 19-23.

22. *AerospaceCorp*, p. 59; Neufeld, *The Air Force in Space, 1970–1974*, pp. 19-23; *SAMSO Chronology*, p. 196.

23. *AerospaceCorp*, pp. 59-61; Neufeld, *The Air Force in Space, 1970–1974*, pp. 19-23; Karas, *The New High Ground*, pp. 76-77.

24. Project RAND, *Preliminary Design of an Experimental World-Circling Spaceship* (Santa Monica, CA: Douglas Aircraft Co, 2 May 1946), Introduction, p. 11.

25. A good, concise background sketch is found in *AerospaceCorp*, pp. 73-74.

26. Maj Michael D. Abel, *History of the Defense Meteorological Satellite Program: Origin Through 1982*. Air Command and Staff College Report No. 87-0020 (Maxwell AFB, AL: Air University Press, 1987), pp. 3-5, (hereafter cited as Abel, *History of DMSP*). The Nimbus satellite weighed 650 pounds.

Along with its cloud-mapping RCA video cameras, its infrared radiometer provided night pictures for the first time. Between 1964 and 1978, the space agency launched seven Nimbus satellites, which provided a 24-hour observation capability. See Jacobs and Whitney, *Missile and Space Projects Guide 1962*, p. 121; Roger D. Launius, *NASA: A History of the U.S. Civil Space Program* (Malabar, FL: Krieger Publishing Company, 1994), p. 130.

27. Abel, *History of DMSP*, pp. 3-5, 19-20; *AerospaceCorp*, pp. 73-74. Although an additional declassification decision in 1975 released more information, much of DMSP's early development remained classified.

28. Neufeld, *The Air Force in Space, 1970–1974*, pp. 27-30.

29. *Ibid.*; *AerospaceCorp*, p. 74; Abel, *History of DMSP*, pp. 31-38.

30. Neufeld, *The Air Force in Space, 1970–1974*, pp. 27-30; *AerospaceCorp*, pp. 74-77; Abel, *History of DMSP*, pp. 31-38. Meanwhile the Air Force continued to support NASA's Nimbus ("Observer") satellites with booster and range support.

31. Maj Dennis L. Alford, *History of the Navstar Global Positioning System (1963–1985)*, Air Command and Staff College Report No. 86-0050 (Maxwell AFB, AL: Air University Press, 1986), pp. 1-3, (hereafter cited as Alford, *History of Navstar/GPS*); *AerospaceCorp*, pp. 63-64; The Aerospace Corporation, *The Global Positioning System. A Record of Achievement* (Los Angeles, CA: The Aerospace Corporation, 1994), pp. 3-4 (hereafter cited as *AeroCorp/GPS*); Ivan A. Getting, "The Global Positioning System," *IEEE Spectrum*, December 1993, pp. 36-37. Also see Karas, *The New High Ground*, pp. 124-134.

32. Alford, *History of Navstar/GPS*, pp. 2-4; *AerospaceCorp*, pp. 63-64; *AeroCorp/GPS*, pp. 5-6; Getting, "The Global Positioning System," pp. 37-38.

33. Getting, "The Global Positioning System," p. 37.

34. Alford, *History of Navstar/GPS*, pp. 4-8; *AerospaceCorp*, pp. 64-65; *AeroCorp/GPS*, pp. 6-8; Getting, "The Global Positioning System," pp. 38-39; Neufeld, *The Air Force in Space, 1970–1974*, pp. 24-27. Planners also referred to the project as the Defense Navigation Satellite System (DNSS).

35. Alford, *History of Navstar/GPS*, pp. 4-8; *AerospaceCorp*, pp. 64-65; *AeroCorp/GPS*, pp. 6-8; Getting, "The Global Positioning System," pp. 38-44; Neufeld, *The Air Force in Space, 1970–1974*, pp. 24-27.

36. Alford, *History of Navstar/GPS*, pp. 4-8; *AerospaceCorp*, pp. 64-67; *AeroCorp/GPS*, pp. 6-9; Getting, "The Global Positioning System," pp. 38-44; Neufeld, *The Air Force in Space, 1970–1974*, pp. 24-27.

37. I am indebted to R. Cargill Hall for use of his unpublished article, "The Air Force in Space," May 1984.

38. As Hall points out, the reconnaissance satellites initially embraced photographic requirements through low-resolution area cameras, which developed film on board and radioed the results to read-out stations on earth, and high resolution or so-called "close-look" cameras, whose film capsules were sent earthward by means of reentry capsules for ocean recovery. Hall, "The Air Force in Space," pp. 6-7. For an unclassified account of the nation's space reconnaissance effort, see Jeffrey T. Richelson, *America's Secret Eyes in Space* (New York: Harper Collins, 1990); CORONA Program Profile, Lockheed Press Release, 24 May 1995.

39. It should be noted that the Air Force also conducted operations involving electronic surveillance with so-called ferret satellites. These spacecraft would record radio and radar frequencies and transmit to earth from 300-mile altitude polar orbits. They proved especially valuable for determining the operational status of various

weapons and developing appropriate electronic countermeasures. See Hall, "The Air Force in Space," pp. 8-9.

40. *AerospaceCorp*, p. 69.

41. *AerospaceCorp*, pp. 69-70. Vela is Spanish for "watchman" and Hotel signified the participating agencies: ARPA, AEC, USAF. See Hall, "The Air Force in Space," pp. 9-10.

42. Berger, *The Air Force in Space, FY61*, pp. 102-103; Berger, *The Air Force in Space, FY62*, pp. 106-107; Cantwell, *The Air Force in Space, FY64*, pp. 87-88; *AerospaceCorp*, pp. 70-71.

43. Cantwell, *The Air Force in Space, FY65*, pp. 81-83; *AerospaceCorp*, pp. 70-71.

44. *AerospaceCorp*, pp. 70-71; Neufeld, *The Air Force in Space, 1970–1974*, pp. 41-42; Hall, "The Air Force in Space," pp. 9-10.

45. Cantwell, *The Air Force in Space, FY64*, p. 51; Alec Galloway, "A Decade of US Reconnaissance Satellites," *International Defense Review*, June 1972, pp. 249-253.

46. Berger, *The Air Force in Space, FY61*, pp. 44-48.

47. *Ibid*. With the March 1961 reorganization, the Space Division assumed planning responsibilities for MIDAS.

48. Berger, *The Air Force in Space, FY62*, pp. 72-82.

49. *Ibid*.

50. *Ibid*.

51. *Ibid*. The "other" available early warning system was the Over-the-Horizon-Backscatter (OTH-B) radar that, despite its potential, would experience considerable deployment delays because of technical challenges.

52. Cantwell, *The Air Force in Space, FY64*, pp. 51-59.

53. *Ibid*.

54. *Ibid*.

55. *Ibid*.

56. Cantwell, *The Air Force in Space, FY65*, pp. 36-39; Curtis Peebles, *High Frontier: The United States Air Force and the Military Space*

Program (Washington, D.C.: Government Printing Office, 1997), pp. 34-36.

57. Neufeld, *The Air Force in Space, 1970–1974*, pp. 31-40.

58. *Ibid*. The politically sensitive Woomera site soon prompted officials to consider alternatives elsewhere.

59. *Ibid*.; Galloway, "A Decade of US Reconnaissance Satellites," pp. 249-253.

60. Neufeld, *The Air Force in Space, 1970–1974*, pp. 31-40; Galloway, "A Decade of US Reconnaissance Satellites," pp. 249-253.

61. Chronology, AFSPACECOM/HO, *Space Detection and Tracking: A Chronology, 1957–1983*; Berger, *The Air Force in Space, FY61*, pp. 76-83; Rick W. Sturdevant, "The United States Air Force Organizes for Space: The Operational Quest," in *Organizing for the Use of Space: Historical Perspectives on a Persistent Issue*, ed. Roger D. Launius (San Diego, CA: American Astronautical Society, 1995), pp. 172-174. The NORAD Combat Operations Center became operational in April 1966.

62. Chronology, AFSPACECOM/HO, *Space Detection and Tracking: A Chronology, 1957–1983*; Berger, *The Air Force in Space, FY61*, pp. 76-83.

63. Cantwell, *The Air Force in Space, FY65*, pp. 71-75; *Aerospace Defense: A Chronology of Key Events, 1945–1990* (Peterson AFB, CO: AFSPACECOM/HO, 1991);

64. Cantwell, *The Air Force in Space, FY65*, pp. 71-75; *Aerospace Defense: A Chronology of Key Events*; Chronology, AFSPACECOM/HO, *Space Detection and Tracking: A Chronology, 1957–1983*.

65. Cantwell, *The Air Force in Space, FY65*, pp. 71-75; *Aerospace Defense: A Chronology of Key Events*; Chronology, AFSPACECOM/HO, *Space Detection and Tracking. A Chronology, 1957–1983*; Peebles, *High Frontier: The United States Air Force and the Military Space Program*, pp. 32-34.

66. *AerospaceCorp*, p. 95.

67. *Ibid.*, pp. 95-96, 104-113; Hall, "The Air Force in Space," pp. 15-16; Air Command and Staff College, *Space Handbook* (Maxwell AFB, AL: Air University, 1977), Chapter 6; Air Force Space Command, "Titan" (Peterson AFB, CO: AFSPC/HO, n.d.).

68. *AerospaceCorp*, pp. 95-96, 104-113; Hall, "The Air Force in Space," pp. 16-21; Air Command and Staff College, *Space Handbook* (Maxwell AFB, AL: Air University, 1977), Chapter 6; James Baar and William E. Howard, *Spacecraft and Missiles of the World, 1962* (New York: Harcourt, Brace & World, Inc., 1962), p. 33.

69. *AerospaceCorp*, pp. 104-113.

70. *Ibid.*

71. *Ibid*; Air Force Space Command, "Titan" (Peterson AFB, CO: AFSPC/HO, n.d.).

72. *Ibid.*

73. *Ibid.*, pp. 117-119; Memorandum, Patrick AFB, FL/OI, subj: The Eastern Test Range, September, 1969.

74. *AerospaceCorp*, pp. 117-119; Memorandum, Patrick AFB, FL/OI, subj: The Eastern Test Range, September, 1969; Hall, "The Air Force in Space," pp. 21-22.

75. *AerospaceCorp*, pp. 120-121; Hall, "The Air Force in Space," pp. 21-22.

76. *AerospaceCorp*, pp. 123-127; Hall, "The Air Force in Space," pp. 22-27; AFSC/SCFHO, *Air Force Satellite Control Facility: Historical Brief and Chronology, 1954–Present* (Sunnyvale, CA: AFSCF/HO, n.d.) (hereafter cited as *SCF Chronology*).

77. *AerospaceCorp*, pp. 123-127; Hall, "The Air Force in Space," pp. 22-27; *SCF Chronology*.

78. *AerospaceCorp*, pp. 123-127; Hall, "The Air Force in Space," pp. 22-27; *SCF Chronology*.

79. *AerospaceCorp*, pp. 123-127; Hall, "The Air Force in Space," pp. 22-27; *SCF Chronology*.

80. *AerospaceCorp*, pp. 123-127; Hall, "The Air Force in Space," pp. 22-27; *SCF Chronology*.

81. Rick W. Sturdevant, "The United States Air Force Organizes for Space: The Operational Quest," in *Organizing for the Use of Space: Historical Perspectives on a Persistent Issue*, ed. Roger D. Launius (San Diego, CA: American Astronautical Society, 1995), pp. 172-175; Maj John B. Hungerford., Jr., *Organization for Military Space: A Historical Perspective*, Report No. 82-1235 (Maxwell AFB, AL: Air Command and Staff College, 1982), pp. 46-47 (hereafter cited as Hungerford, *Organization for Military Space*). In the next decade the launch sites themselves were placed directly under SSD's successor SAMSO.

82. Rick W. Sturdevant, "The United States Air Force Organizes for Space: The Operational Quest," in *Organizing for the Use of Space: Historical Perspectives on a Persistent Issue*, ed. Roger D. Launius (San Diego, CA: American Astronautical Society, 1995), pp. 172-175; Hungerford, *Organization for Military Space*, pp. 46-49; *SCF Chronology*.

83. Maj Ernie R. Dash and Maj Walter D. Meyer, "The Meteorological Satellite: An Invaluable Tool for the Military Decision-Maker," *Air University Review*, Vol. xxix, No. 3 (March–April 1978), p. 13.

84. Henry W. Brandli, "The Use of Meteorological Satellites in Southeast Asia Operations," *Aerospace Historian*, Vol. 29, No. 3 (September 1982), pp. 172-175; Gen William W. Momyer, *Air Power in Three Wars* (Washington, D.C.: Office of Air Force History, 1978), pp. 228-231; Neufeld, *The Air Force in Space, 1970–1974*, pp. 27-28.

85. Lt Col John J. Lane, Jr., *Command and Control and Communications Structures in Southeast Asia* (Maxwell AFB, AL: Air University, 1981), pp. 113-114. The DoD quest for inexpensive SATCOM during the Vietnam War gave rise to the "30 circuits" episode. When several carriers learned in summer 1966 that COMSAT Corporation intended to lease 30 circuits directly to DoD, in direct violation of the FCC's recently promulgated Authorized Users decision, they formally protested. In February 1967,

the FCC ordered a "composite" rate of $7,100 per half-circuit and split the traffic evenly three ways among ITT, RCA, and WUI. The FCC required COMSAT Corporation to sell the circuits to the carriers at $3,800. See Michael Kinsley, *Outer Space and Inner Sanctums: Government, Business, and Satellite Communication* (New York: John

Wiley & Sons, 1976), pp. 60-62.

86. Lane, *Command and Control and Communications Structures in Southeast Asia*, pp. 113-114; Kinsley, *Outer Space and Inner Sanctums*, pp. 60-62.

87. DoD Directive 5160.32, "Development of Space Systems," 8 September 1970.

Chapter 5. Organizing for Space

1. Paul B. Stares, *The Militarization of Space: U.S. Policy, 1945–1984* (Ithica, NY: Cornell University Press, 1985), p. 176; Walter A. McDougall, *...the Heavens and the Earth: A Political History of the Space Age* (New York: Basic Books, 1985), pp. 421-422.

2. Robert Kipp, AFSPACECOM/HO, "Trends in Military Space: Organizational Growth and Maturity," 30–31 May 1990.

3. DoD Directive 5160.32, "Development of Space Systems," 8 September 1970.

4. Gen Jacob E. Smart, "Strategic Implications of Space Activities," *Strategic Review* (Fall 1974), pp. 19-21.

5. Working Paper, AF/XO, "Outline for 'White Paper' on Space Management and Planning Initiatives," n.d. [September 1978].

6. For a detailed discussion of the Soviet buildup and comparison of American and Soviet nuclear arsenals, see the yearly issues of *The Military Balance*, especially *The Military Balance, 1975–1976* (London: The International Institute of Strategic Studies, 1975), pp. 3-4, 73, and Lawrence Freedman, *US Intelligence and the Soviet Strategic Threat* (Boulder, CO: Westview Press, 1977). The number of American launchers had remained constant since 1967 at 1,054 ICBMs and 656 SLBMs. By contrast, in 1967 the Soviet inventory consisted of 460 ICBMs and 130 SLBMs. These figures, to be sure, do not reflect evolving capabilities in terms of missile size, range, throw-weight, and MIRV technology. Characteristic of concern by American officials about the Soviet build-up

is Secretary of Defense Melvin Laird's comment in early 1970 warning of the "continuing rapid expansion of Soviet strategic offensive forces." Quoted in Freedman, *US Intelligence and the Soviet Strategic Threat*, p. 153. Such worries continued unabated through the decade.

7. Freedman, *US Intelligence and the Soviet Strategic Threat*, pp. 129-168.

8. Report of Secretary of Defense James R. Schlesinger to the Congress on the FY 1976 and FY 1977T (transition) Budgets, 5 February 1975 (Washington: Government Printing Office, 1975), pp. I-13, 14; II-1 to II-11; II-18.

9. John Newhouse, *War and Peace in the Nuclear Age* (New York: Alfred A. Knopf, 1989), p. 233.

10. For discussion of SALT I and the ABM treaty, see John Newhouse, *Cold Dawn: The Story of SALT* (New York: Holt, Rinehart and Winston, 1973). Both agreements are reproduced in full in the appendix, pp. 273-281. Also see Newhouse, *War and Peace in the Nuclear Age*, pp. 209-265; Freedman, *US Intelligence and the Soviet Strategic Threat*, pp. 166-168.

11. Owen E. Jensen, "The Years of Decline: Air Defense from 1960 to 1980," in Stephen J. Cimbala, ed., *Strategic Air Defense* (Wilmington, DE: Scholarly Books, 1989), pp. 40-41.

12. Summary Report, *Joint DoD/NASA Study of Space Transportation Systems*, 16 June 1969; Jacob Neufeld, *The Air Force in*

Space, 1970–1974 (Washington, D.C.: Office of Air Force History, August 1976), pp. 1-14; McDougall, *…the Heavens and the Earth*, p. 421.

13. Neufeld, *The Air Force in Space, 1970–1974*, pp. 1-15; Roger D. Launius, "Toward an Understanding of the Space Shuttle: A Historiographical Essay," *Air Power History* (Winter 1992), pp. 3-10. As noted in Chapter 3, the Dyna-Soar, despite its multi-purpose mission profile and reusable launch capability, fell victim in the early 1960s to the triple threat of high costs, technical uncertainty, and low priority. For a detailed treatment of Shuttle development, see Dennis R. Jenkins, *Space Shuttle: The History of Developing the National Space Transportation System* (Marceline, Missouri: Walsworth Publishing Company, 1993).

14. Quoted in Robert Frank Futrell, *Ideas, Concepts, Doctrine: Basic Thinking in the United States Air Force, 1961–1984*, Vol. II (Maxwell Air Force Base, AL: Air University Press, December 1989), p. 685 (hereafter cited as Futrell, *Ideas, 1961–1984*). Later, when officials canceled the payload retrieval mission, the rationale for NASA's ambitious annual flight schedule no longer applied.

15. Roger D. Launius, "Toward an Understanding of the Space Shuttle: A Historiographical Essay," *Air Power History* (Winter 1992), pp. 3-10; As Neufeld points out, the Air Force supplied development costs in the "token amount" of $4 million or less per year from 1971 to 1974. See Neufeld, *The Air Force in Space, 1970–1974*, p. 3.

16. History, Space and Missile Systems Organization, 1 July 1972–30 June 1973, pp. 201-206; Working Papers, DoD/DRE, "Space and Space-Related Program Data, Fiscal Years 1979–1986," March 1981, pp. 169-174; Neufeld, *The Air Force in Space, 1970–1974*, pp. 4-5. Lockheed's designer of the Space Tug was Saunders B. Kramer.

17. Neufeld, *The Air Force in Space, 1970–*1974, pp. 4-8; see also John Logsdon, "The Decision to Develop the Space Shuttle," *Space Policy* (May 1986), pp. 103-118. In the wake of the *Challenger* disaster, Logsdon argues that NASA's focus on costs from the beginning compromised crucial technology and allowed budget priorities to degrade the project.

18. Logsdon, "The Decision to Develop the Space Shuttle," pp. 103-118; Working Papers, DoD/DRE, "Space and Space-Related Program Data, Fiscal Years 1979–1986," March 1981, pp. 169-174; Neufeld, *The Air Force in Space, 1970–1974*, pp. 8-9. In the mid-1970s, Shuttle planners predicted that each Shuttle flight would cost $15.4 million, based on use of five orbiters and an average of 60 flights yearly. By contrast, Titan IIIC flight costs were estimated to average $35 million per flight. See GAO Report, "Issues Concerning the Future of the Space Transportation System," GAO/MASAD-83-6, 28 December 1982, pp. i-iii; Edgar Ulsamer, "Space. High-Flying Yankee Ingenuity," *Air Force Magazine* (September 1976), pp. 98-104.

19. Neufeld, *The Air Force in Space, 1970–1974*, pp. 9-10; History, Space and Missile Systems Organization, 1 July 1972–30 June 1973, pp. 201-206. For discussion of the Vandenberg project, see Roger G. Guillemette, "Vandenberg. Space Shuttle Launch and Landing Site: Part 1—Construction of Shuttle Launch Facilities" *Spaceflight*, Vol 36 (October 1994), pp. 354-357; Roger G. Guillemette, "Vandenberg. Space Shuttle Launch and Landing Site: Part 2—Abandoned in Place" *Spaceflight*, Vol 36 (November 1994), pp. 378-381.

20. McDougall, *…the Heavens and the Earth*, p. 423.

21. Neufeld, *The Air Force in Space, 1970–1974*, pp. 11-13; History, Space and Missile Systems Organization, 1 July 1973–30 June 1975, pp. 160-192.

22. History, Space and Missile Systems

Organization, 1 July 1973–30 June 1975, pp. 160-192.

23. CINCNORAD to CSAF, subj: Operational Responsibility for Space Transportation System, 25 November 1974, w/atch, Background Paper, "Operation of the Space Transportation System"; Talking Paper, ADCOM/XPDQ, "The Space Transportation System (STS)," 26 April 1978; Neufeld, *The Air Force in Space, 1970–1974*, pp. 13-14.

24. Col Morgan W. Sanborn, "National Military Space Doctrine," *Air University Review*, Vol. xxvii, No. 2, (January–February 1977), pp. 75-79.

25. Working Papers, DoD/DRE, "Space and Space-Related Program Data, Fiscal Years 1979–1986," March 1981, pp. 29-33; Maj Robert E. Lee, *History of the Defense Satellite Communications System (1964–1986)*. Air Command and Staff College Report No. 87-1545 (Maxwell AFB, AL: Air University Press, 1987); The Aerospace Corporation, *The Aerospace Corporation: Its Work: 1960–1980* (Los Angeles: Times Mirror Press, 1980), pp. 56-57; *Aeronautics and Space Report of the President. 1980 Activities* (Washington, D.C.: NASA, 1981), pp. 37-38.

26. Working Papers, DoD/DRE, "Space and Space-Related Program Data, Fiscal Years 1979–1986," March 1981, pp. 145-149; Maj Dennis L. Alford, *History of the Navstar Global Positioning System (1963–1985)*. Air Command and Staff College Report No. 86-0050 (Maxwell AFB, AL: Air University Press, 1986); Lt Col John F. Scheerer and Maj Joseph Gassmann, "Navstar GPS: Past, Present, and Future," *The Navigator* (Winter 1983).

27. Maj Barkley G. Sprague, *Evolution of the Missile Defense Alarm System (MIDAS) 1955–1982*. Air Command and Staff College Report No. 85-2580 (Maxwell AFB, AL: Air University Press, 1985), pp. 27-33.

28. Maj Michael D. Abel, *History of the Defense Meteorological Satellite Program: Origin Through 1982*. Air Command and Staff College Report No. 87-0020 (Maxwell AFB, AL: Air University Press, 1987), pp. 31-60. Abel describes specific problems encountered by each of the series 5D-1 satellites.

29. Robert Kipp, "Background Paper on Space Detection and Tracking System (SPADATS)," 16 March 1988; AFSPACECOM/HO, "Space Detection and Tracking: A Chronology, 1957–1983," March 1990; Col James E. Strub and Col Thomas S. Moorman, Jr., "Space Surveillance and Detection," Presentation before the American Institute of Aeronautics and Astronautics, 20–21 May 1982.

30. Maj John B. Hungerford, Jr., *Organization for Military Space: A Historical Perspective*. Report No. 82-1235 (Maxwell AFB, AL: Air Command and Staff College, 1982), pp. 57-64.

31. For details on Project 437, see Wayne R. Austerman, *Program 437: The Air Force's First Antisatellite System*, AFSPACECOM/HO, April 1991; Stares, *Militarization of Space*, pp. 120-128; 201-209.

32. Quoted in Stares, *Militarization of Space*, p. 173.

33. Memorandum, DDR&E to Asst SECAF/R&E, subj: Space Surveillance Program Planning, 20 April 1976; Talking Paper, NORAD/XPDQ, "Space Defense," 10 June 1976; Memorandum, SECAF/R&D to ODDR&E, subj: Plan for the Evolution of Space Surveillance Capabilities, 1 June 1976, w/atch, Initial Report. The most important studies included: *A Study of Future Air Force Space Policy and Objectives* (Washington, D.C.: Headquarters USAF, July 1977); *Space Missions Organizational Planning Study (SMOPS)* (Washington, D.C.: Headquarters USAF, February 1979); USAF Scientific Advisory Board (SAB), *The Air Force in Space*, Summer Study (Washington, D.C.: SAB, August 1980); and *Space Policy and Requirements Study (SPARS)* (Washington, D.C.: Headquarters USAF, May 1981).

34. Memorandum, Secretary of Defense to Secretaries of the Military Departments, *et al*, subj: Space Defense Working Group, 4 November 1976.

35. On the Carter administration and defense, see Newhouse, *War and Peace in the Nuclear Age*, pp. 292-332; Gregg Herken, *Counsels of War* (New York: Alfred A. Knopf, 1985), pp. 279-302; Thomas Powers, "Choosing a Strategy for World War III," *The Atlantic Monthly* (November 1982), pp. 82-110.

36. On the "countervailing strategy," see Department of Defense, *Report of Defense Secretary Harold Brown to the Congress on the* FY *1981 Budget,* FY *1982 Authorization Request and* FY *1981–1985 Defense Programs, January 29, 1980* (Washington, D.C.: Government Printing Office, 1980), pp. 65-68.

37. Stares, *Militarization of Space*, pp. 180-182.

38. *Ibid.*, pp. 185-186; Futrell, *Ideas, 1961–1984*, p. 688.

39. One study asserts that the real problem lay not in the directives themselves, but in their enforcement by the always hesitant bureaucracy. While the civilian side moved effectively to implement the elements of PD 42, those on the military side did not. Other institutional issues affecting DoD and the Air Force are also discussed in LTC Neal E. Lamping and LTC Richard P. MacLeod, *Space—A National Security Dilemma: Key Years of Decision* (Washington, D.C.: National Defense University, June 1979).

40. Briefing, AF/XO, "Air Force Space Policy," November 1976.

41. AF/XO to AF/CV, subj: Air Force Space Policy, 29 January 1977.

42. AF/CC to ALMAJCOM/CC, subj: Air Force Space Policy, 9 May 1977; Futrell, *Ideas, 1961–1984*, p. 688. As noted in Chapters 3 and 4, Air Force leaders referred to service policy despite the fact that policy is established by the President and, technically, DoD sets objectives to achieve that policy.

43. *Future Air Force Space Policy and Objectives*, July 1977.

44. Members of the Space Operations Steering Group included several elements within the Plans and Operations Office: the Director of Operations, Director of Concepts, as well as the Director of Space from the Research and Development office. Although Air Force leaders would continue in future to pursue the "sole agent" objective, by the end of the Carter administration in 1981 enthusiasm had waned considerably. Ultimately, the Air Force would witness the establishment of the a unified command for military space, which it had opposed since the Eisenhower administration.

45. Rick W. Sturdevant, "The United States Organizes for Space: The Operational Quest," in *Organizing for the Use of Space: Historical Perspectives on a Persistent Issue*, ed. Roger D. Launius (San Diego, CA: American Astronautical Society,1995), pp. 174-175.

46. Robert Kipp, AFSPACECOM/HO, "The Reorganization of 1979 and the Space Organization Issue," 8 March 1988.

47. Under Secretary of the Air Force to CINCNORAD, 9 May 1978.

48. Background Paper, ADCOM/SPDQ, "Need for a USAF Space Operations Command," 8 June 1978.

49. ADCOM Staff Meeting Notes, 11 October 1978.

50. Interview Transcript, Maj Gen Thomas S. Moorman, Jr., with Robert Kipp and Thomas Fuller, 27 July 1988. The nine general officers were Lt Gen A.B. Anderson, Jr., USAF DCS/Operations, Plans and Readiness, Maj Gen H.S. Vandenberg, Jr., USAF DCS/XO, Maj Gen J.R. Bickle, USAF DCS/RD, Maj Gen S.H. Sherman, Jr., USAF DCS/MPM, Maj Gen B.K. Brown, ADCOM DCS/DO, Maj Gen J.E. Kulpa, Jr., SAFSP, Maj Gen L.A. Skantze, AFSC DCS/SD, Maj Gen

D.L. Gray, SAC DCS/XP, and Brig Gen J.R. McCarthy, AFLC DCS/XP.

51. Interview Transcript, Moorman, pp. 3-5; Futrell, *Ideas, 1961–1984*, pp. 689-690; LTC Neal E. Lamping and LTC Richard P. MacLeod, *Space—A National Security Dilemma: Key Years of Decision* (Washington: National Defense University, June 1979), pp. 79-81.

52. Interview, Gen James E. Hill, USAF (Ret.), USAF Oral History Interview by James C. Hasdorff, No. K239.0512-1324, Albert F. Simpson Historical Research Center (Maxwell AFB, AL: Air University, 3-4 May 1982), p. 106.

53. Hungerford, *Organization for Military Space*, pp. 52-55; Kipp, "Trends in Military Space;" History, Space Division, 1 October 1981–30 September 1982, pp. 1-15; Newhouse, *War and Peace in the Nuclear Age*, pp. 304, 318-323.

54. Hungerford, *Organization for Military Space*, pp. 52-55; Kipp, "Trends in Military Space;" History, Space Division, 1 October 1981–30 September 1982, pp. 1-15. See also Chapter 4, pp. 167-168, 170-171, this study.

55. Talking Paper, AF/XOXFD, Space Related Initiatives, 22 October 1979; Hungerford, Organization for Military Space, pp. 52-55; Kipp, "Trends in Military Space;" History, Space Division, October 1981–September 1982, pp. 1-15.

56. History, Space Division, 1 October 1981–30 September 1982, pp. 110-115.

57. Paper, AF/XOXFD, Space Related Initiatives, 22 October 1979; History, ADCOM/ADC, 1 January–31 December 1979, pp. 111-115.

58. Hungerford, *Organization for Military Space*, p. 63.

59. Point Paper, AF/XPDS, USAF/XO Corona Fall Briefing, 21 August 1981; Point Paper, AF/XOXFD, Organization for Space, 26 April 1981.

60. John L. McLucas to Dr. Raymond Bisplinghoff, Chairman, Scientific Advisory Board, 13 August 1980, w/atch, Conclusions and Recommendations.

61. *Ibid.*

62. AF/XO to AF/RD, subj: USAF Scientific Advisory Board Summer Study on Space, 2 January 1981; Interview Transcript, Brig Gen Earl S. Van Inwegen, with Rick W. Sturdevant, 1 November 1995. Van Inwegen provides an insider's account of the evolution of the Air Force's Space Command from his position on General Chain's staff in the Directorate of Operations and Readiness and, from September 1981, the Directorate of Space. According to Van Inwegen, Chain provided much of the momentum behind O'Malley's efforts for space operational reform. Shortly after O'Malley became Deputy Chief of Staff for Operations, Plans and Readiness, Chain had his staff prepare what he called the "Fester" Briefing, which recommended how the Air Force should organize for space and would sit on O'Malley's desk and fester until the latter took action on it.

63. Interview (tape recording), Secretary of the Air Force John L. McLucas, by the author, Washington, D.C., 30 March 1995.

64. Gen Lew Allen, Jr., CSAF, to Dr. Raymond E. Bisplinghoff, Chairman, USAF Scientific Advisory Board, 15 April 1981. General Allen, a physicist, also served in the National Reconnaissance Office. For General Allen's cautious approach to space organizational initiatives, see Interview, General Lew Allen, Jr., USAF (Ret.), USAF Oral History Interview by Dr. James C. Hasdorff, Albert F. Simpson Historical Research Center, No. K239.0512-1694 (Maxwell AFB, AL: Air University, 8-10 January 1986), pp. 163-165.

65. Interview, Lt Gen Richard C. Henry, in *Air Force Magazine*, Vol. 65, No. 6 (June 1982), p. 41.

66. Hungerford, *Organization for Military Space*, pp. 67-68; Futrell, *Ideas, 1961–1984*, pp. 690-691.

67. HQ USAF/XO, *Space Policy and*

Requirements Study (SPARS), 18 May 1981.

68. *Ibid.*; Interview (tape recording), Col Samuel Beamer, ADCOM/XO, USAF (Ret), by the author, Colorado Springs, CO, 10 July 1991.

69. Point Paper, AF/XPDS, USAF/XO Corona Fall Briefing, 21 August 1981; Point Paper, AF/XOXFD, "Organization for Space," 26 April 1981.

70. Staff Summary Sheet, AF/J5SX, CSAF Tasking on Space Policy and Organization, 15 January 1982; History, Space Command, ADCOM, ADC, 1982, pp. 6-7; Thomas Karas, *The New High Ground: Strategies and Weapons of Space-Age War* (New York: Simon and Schuster, 1983), p. 19; Edward C. Aldridge, Jr., "Defense in the Fourth Dimension," *Defense 83* (January 1983), pp. 3-10.

71. Staff Summary Sheet, AF/J5SX, CSAF Tasking on Space Policy and Organization, 15 January 1982; History, Space Command, ADCOM, ADC, 1982, pp. 6-7; Thomas Karas, *The New High Ground: Strategies and Weapons of Space-Age War* (New York: Simon and Schuster, 1983), p. 19; Edward C. Aldridge, Jr., "Defense in the Fourth Dimension," *Defense 83* (January 1983), pp. 3-10; Interview Transcript, Van Inwegen. Although Van Inwegen served as one of the staff officers directed to prepared the "Navajo Blanket," he mistakenly places the event in 1977 during the controversy over ADCOM's demise.

72. Karas, *New High Ground*, p. 19; Futrell, *Ideas, 1961–1984*, pp. 695-696.

73. Henry M. Narducci, *Strategic Air Command and the Space Mission, 1977–1984*, (Offutt AFB, NE: SAC, 10 October 1985), pp. 24-33; History, Space Command, ADCOM, ADC, 1982, pp. 6-11; Aldridge, "Defense in the Fourth Medium," pp. 3-10; Kipp, "Trends in Military Space." In later years, General Hartinger would often cite the West Point connection and friendship with Marsh and Henry as responsible for important personal initiatives and a relatively smooth decision-making process. See Interview, Gen James V. Hartinger, USAF (Ret.), USAF Oral History Interview by Capt Barry J. Anderson, No. K239.0512-1673, Albert F. Simpson Historical Research Center (Maxwell AFB, AL: Air University, 5-6 September 1985).

74. Narducci, *Strategic Air Command and the Space Mission, 1977–1984*, pp. 24-33; History, Space Command, ADCOM, ADC, 1982, pp. 6-11; Aldridge, "Defense in the Fourth Medium," pp. 3-10; Kipp, "Trends in Military Space." The ad hoc working group found particularly difficult the issue of how to integrate the Consolidated Space Operations Center (CSOC) into the organization structure.

75. Narducci, *Strategic Air Command and the Space Mission, 1977–1984*, pp. 24-33; History, Space Command, ADCOM, ADC, 1982, pp. 6-11; Aldridge, "Defense in the Fourth Medium," pp. 3-10; Kipp, "Trends in Military Space." General O'Malley became Vice Chief of Staff on 1 June 1982.

76. Interview Transcript, Allen, p. 164.

77. Narducci, *Strategic Air Command and the Space Mission, 1977–1984*, pp. 24-33; History, Space Command, ADCOM, ADC, 1982, pp. 6-11; Aldridge, "Defense in the Fourth Medium," pp. 3-10; Kipp, "Trends in Military Space."

78. Narducci, *Strategic Air Command and the Space Mission, 1977–1984*, pp. 24-33; History, Space Command, ADCOM, ADC, 1982, pp. 6-11; Aldridge, "Defense in the Fourth Medium," pp. 3-10; Kipp, "Trends in Military Space."

79. The official news release stated the following: "Creation of Space Command will further consolidate USAF operational space activities, provide a link between the space-related research and development process and operational users, and retain North American Aerospace Defense Com-

mand authority and responsibilities as currently organized. It is the Air Force's hope and belief that Space Command will develop quickly into a unified command." Quoted in History, Space Command, ADCOM, ADC, 1982, pp. 9-10. For details of the O'Malley-Nagler meeting, see Interview Transcript, Van Inwegen. A member of the Space Operations Steering Committee, Van Inwegen noted that, after the proposed public announcement had been taken to General O'Malley for his final review, he was seen coming out of his office "with his arm around Admiral Nagler's shoulder."

80. Futrell, *Ideas, 1961–1984*, p. 696; Stares, *Militarization of Space*, pp. 180-182.

81. AFM 1-6, Aerospace Basic Doctrine: Military Space Doctrine, 15 October 1982; Futrell, *Ideas, 1961–1984*, p. 696; Stares, *Militarization of Space*, pp. 180-182.

82. History, Space Command, ADCOM, 1983, pp. 51-54; *Space Surveillance Architecture Study*, (Washington, D.C.: Headquarters USAF, 10 June 1983).

Chapter 6. From Star Wars to the Gulf War

1. "Normalize" and "operationalize" become frequently used "buzz" words in the late 1980s and early 1990s to buttress the case for creating an operational focus for military space. For a useful discussion of the space "normalization" movement, see History, Air Force Space Command, January–December 1990, pp. 107-114.

2. History, Space Command, ADCOM, January–December 1983, pp. 2-12; History, Air Force Space Command, January–December 1987, pp. 1-2; Rick W. Sturdevant, "The United States Air Force Organizes for Space: The Operational Quest," in *Organizing for the Use of Space: Historical Perspectives on a Persistent Issue*, ed. Roger D. Launius (San Diego, CA: American Astronautical Society, 1995), pp. 180-181.

3. At this time, DSP was being used operationally but had not been formally turned over from Air Force Systems Command. History, Space Command, ADCOM, January–December 1983, pp. 2-12; History, Air Force Space Command, January–December 1987, pp. 1-2; Sturdevant, "The United States Air Force Organizes for Space," pp. 180-183.

4. History, Space Division, 1 October 1983–30 September 1984, pp. 242-262; *Aeronautics and Space Report to the President. 1983 Activities* (Washington, D.C.: NASA, 1984), pp. 29-30; *Aeronautics and Space Report to the President. 1984 Activities* (Washington, D.C.: NASA, 1985), pp. 39-40.

5. History, Space Division, 1 October 1983–30 September 1984, pp. 211-238.

6. *Aeronautics and Space Report to the President. 1983 Activities* (Washington, D.C.: NASA, 1984), pp. 27-29; *Aeronautics and Space Report to the President. 1984 Activities* (Washington, D.C.: NASA, 1985), pp. 37-39; United States General Accounting Office Report, DoD Acquisition. *Case Study of the MILSTAR Satellite Communications System*, 31 July 1986; Point Paper, AFSPACECOM/KRQS, "MILSTAR," 27 March 1984; Background Paper, AFSPACECOM/SIMRS, "MILSTAR Issues," 18 November 1985; "Milstar Satellite Project Under Watchful Eye," *Defense Electronics* (June 1986), p. 64.

7. History, Space Division, 1 October 1983–30 September 1984, pp. 179-205; *Aeronautics and Space Report to the President. 1983 Activities* (Washington, D.C.: NASA, 1984), p. 29; *Aeronautics and Space Report to the President. 1984 Activities* (Washington, D.C.: NASA, 1985), p. 39. NASA's PAM-DII comprised one of three solid-fueled upper stage vehicles under development at this time. NASA built the PAM for relatively light payloads designed for flight aboard the Shuttle or Delta booster. The other two were

the Air Force's Inertial Upper Stage (IUS) for heavier payloads, and the jointly procured modified Centaur upper stage for use with the heaviest payloads and interplanetary missions. History, Space Division, 1 October 1983–30 September 1984, pp. 98-99; History, Space Division, October 1985–September 1986, p. 73.

8. Following a reassessment of space system requirements after the *Challenger* tragedy, on 17 February 1988, Secretary of Defense Frank Carlucci authorized an expansion of the GPS space segment to 24 satellites, consisting of 21 active and three on-orbit spares. History, Air Force Space Command, January–December 1988, p. 148.

9. Commander, Space Command to Commander-in-Chief, SAC, 20 January 1983; Memorandum, AFSPACECOM, Director, Space to DCS/Plans, subj: Resource Management, 22 March 1984; Commander-in-Chief, SAC to Commander, Space Command, 26 April 1984, w/Point Paper, "Command Missions and Resource Management."

10. Commander, Space Command to Commander-in-Chief, SAC, 18 June 1984.

11. History, Space Division, October 1984–September 198, pp. 11-12; Sturdevant, "The United States Air Force Organizes for Space," p. 180.

12. History, Air Force Space Command, January–December 1987, pp. 144-148; AFSPACECOM/HO, *Aerospace Defense: A Chronology of Key Events, 1945–90*, 1 October 1991, p. 68.

13. In 1987 DoD issued a new policy on space launch that designated expendable launch vehicles rather than the Shuttle the primary military launch vehicles. As a result, the Shuttle Operations and Planning Complex in the CSOC was no longer needed. In December 1986, Air Force headquarters had decided to delete funding for the CSOC, and in early 1987 directed complete phaseout of the development program by October of

that year. History, Air Force Space Command, January–December 1987, pp. 144-148, 162-163; History, Air Force Space Command, January–December 1990, pp. 181-84.

14. History, Space Command, ADCOM, ADC, 1982, p. 10.

15. History, United States Space Command, September 1985 to December 1986, pp. 1-5. See Chapter 5, this study, pp. 196-207 for discussion of events and issues leading to establishment of Space Command.

16. *Ibid.*; Paper, USSPACECOM/HO, "United States Space Command," 17 August 1987; Sturdevant, "The United States Air Force Organizes for Space," p. 184.

17. For the political background to SDI see B. Bruce-Briggs, *The Shield of Faith* (New York, 1988); Sanford Lakoff and Herbert F. York, *A Shield in Space?* (Berkeley, 1989); Stephen J. Cimbala, ed., *Strategic Air Defense* (Wilmington, DE, 1989); Philip M. Boffey and others, *Claiming the Heavens* (New York, 1988); Robert C. Richardson III, USAF (Ret), "High Frontier: 'The Only Game in Town,'" *Journal of Social, Political and Economic Studies* VII, nos. 1 and 2 (1982): 55-66; Speech, Brig Gen Robert R. Rankine, Jr., USAF, Special Assistant for Strategic Defense Initiative, Address to the National Defense Institute, Lisbon, Portugal, 3 June 1985.

18. History, United States Space Command, September 1985 to December 1986, pp. 1-5; Paper, USSPACECOM/HO, "United States Space Command," 17 August 1987; Sturdevant, "The United States Air Force Organizes for Space," p. 184. Air Force Space Command gained responsibility for the SDI Test Facility when DoD in 1985 decided to locate the National Test Facility in the CSOC at Falcon Air Station. History, Air Force Space Command, January–December 1986, pp. 91-92.

19. History, United States Space Command, September 1985 to December 1986, pp. 1-5; Background Paper, USSPACECOM/

HO, "United States Space Command," 17 August 1987.

20. Gen Robert T. Herres replaced the retiring Hartinger on 30 July 1984. With the formation of U.S. Space Command, Herres briefly wore four hats, as Space Command commander and Aerospace Defense Command, NORAD, and U.S. Space Command commander-in-chief. Shortly thereafter, on 1 October 1986, Air Force Space Command received a separate commander, Maj Gen Maurice C. Padden, while on 19 December 1986, Aerospace Defense Command was inactivated. Air Force Space Command and U.S. Space Command would remain separated until 23 March 1992. AFSPACECOM/ HO, *Aerospace Defense: A Chronology of Key Events, 1945–90,* 1 October 1991, pp. 63-66; History, Air Force Space Command, January–December 1993, pp. 14-15.

21. History, Air Force Space Command, January–December 1989, pp. 15-16.

22. History, Space Command, ADCOM, January–December 1983, pp. 52-54; History, Space Command, ADCOM, January– December 1984, pp. 87-88; Point Paper, AFSPACECOM/XPXX, "USAF Space Plan," 20 January 1988.

23. For a comprehensive discussion of the Space Plan before and after the *Challenger* accident, see History, Air Force Space Command, January–December 1986, pp. 79-91.

24. *Ibid.*

25. *Ibid.* A revised mission statement did not receive approval until 14 May 1990.

26. History, Air Force Space Command, January–December 1986, pp. 128-132; History, Space Division, October 1985–September 1986, pp. xxxiii, 55-56, 60, 132. See also John M. Logsdon and Ray A. Williamson, "U.S. Access to Space," *Scientific American,* Vol. 260, No. 3 (March 1989), pp. 34-40.

27. History, Space Division, October 1985–September 1986, pp. 60-67; "*Challenger* Panel is Seen Rebuking NASA Officials," *Wall Street Journal,* 12 May 1986, p. 2. For literature focusing on the Shuttle tragedy, see Roger D. Launius, "Toward an Understanding of the Space Shuttle: A Historiographical Essay," *Air Power History* (Winter 1992), pp. 15-18.

28. History, Space Division, October 1985–September 1986, pp. 60-67; History, Air Force Space Command, January–December 1986, pp. 128-132; Logsdon and Williamson, "U.S. Access to Space," pp. 34-36.

29. GAO Report, Report to the Chairman, Committee on Government Operations, House of Representatives, *Implications of Joint NASA/DoD Participation in Space Shuttle Operations,* GAO/NSIAD-84-13, 7 November 1983, pp. 3-21; Edgar Ulsamer, "Space. High-Flying Yankee Ingenuity," *Air Force Magazine* (September 1976), pp. 98-104.

30. GAO Report, *Implications of Joint NASA/DoD Participation in Space Shuttle Operations,* pp. 3-21; History, Space Division, 1 October 1982–30 September 1983, pp. 68-72.

31. GAO Report, *Implications of Joint NASA/DoD Participation in Space Shuttle Operations,* pp. 3-21; Statement by Dr. William J. Perry, Under Secretary of Defense for Research and Engineering before the Senate Subcommittee on Science, Technology, and Space, 4 June 1979; Edgar Ulsamer, "Space Shuttle Mired in Bureaucratic Feud," *Air Force Magazine* (September 1980), pp. 72-77; GAO Report, *Issues Concerning the Future Operation of the Space Transportation System,* GAO/MASAD-83-6, 28 December 1982, pp. i, iii, 3, 12.

32. Interest Paper, USSPACECOM/J5SX, "Mixed Fleet," 7 October 1987.

33. History, Space Division, 1 October 1982–30 September 1983, pp. 68-72; Ulsamer, "Space Shuttle Mired in Bureaucratic Feud," pp. 72-77; Interest Paper, ADCOM/NORAD, "Liquid Upper Stage (Centaur)," 7 June 1982.

34. GAO Report, *Issues Concerning the Future Operation of the Space Transportation*

System, pp. i, iii, 3, 12.

35. Interest Paper, USSPACECOM/J5SX, "Mixed Fleet," 7 October 1987; Presentation, Edward C. Aldridge, Jr., Under Secretary of the Air Force, Address to the National Space Club, 18 November 1981; "NASA Shouldn't Operate Shuttle, AF Under Secretary Says," *Aerospace Daily*, Vol. 112, No. 14, 20 November 1981, p. 105.

36. History, Space Division, 1 October 1983–30 September 1984, pp. 88-95; Point Paper, AFSPACECOM/XPSS, "Space Transportation System, 13 February 1984." The Atlas II would have the capability of launching 14,5000 pounds to a 100-nautical mile easterly orbit, and 6,100 pounds into a geosynchronous transfer orbit. GAO Fact Sheet, "Military Space Programs: An Unclassified Overview of Defense Satellite Programs and Launch Activities," GAO/NSIAD-90-154FS, p. 54.

37. Memorandum, Secretary of Defense for the Secretaries of the Military Departments, *et al*, subj: Defense Space Launch Strategy, 7 February 1984.

38. History, Space Division, 1 October 1983–30 September 1984, pp. 88-95; Draft Memorandum, Glynn Lunney to General Beer, 2 December 1983. On the issue of "militarization," see for example, "A New Image for the Space Shuttle," *Science*, 18 January 1985, pp. 276-277; William J. Broad, "As Shuttle Orbits, a Debate Grows Over Military's Role," *The New York Times*, 6 October 1985.

39. Logsdon and Williamson, "U.S. Access to Space," pp. 34-40; Tom Dworetzky, "The Launch Gap," *Discover* (July 1988), pp. 54-62; William J. Broad, "Pentagon Leaves the Shuttle Program," *The New York Times*, 7 August 1989, pp. 5-6.

40. History, Space Division, 1 October 1983–30 September 1984, pp. 88-95; Memorandum, Secretary of the Air Force to Secretary of Defense, subj: DoD Space

Launch Vehicles—Decision Memorandum," May 1994.

41. U.S. House, *Assured Access to Space During the 1990s*, Joint Hearings Before the Subcommittee on Space Science and Applications of the Committee on Armed Services, 99th Congress, 1st Session, 23, 24, 25 July 1985, pp. 43-46, 52-54, 91-98, 143-145.

42. History, Space Division, October 1986–September 1987, pp. 69-75.

43. For example, see the articles under the general heading, "National Security and the U.S. Space Program After the *Challenger* Tragedy," by Albert D. Wheelon, "Pete" Aldridge, Richard R. Colino, Richard L. Garwin, Hans Mark, Bruce Murray, and James A. Van Allen in *International Security*, Vol. 11, No. 4 (Spring 1987), pp. 141-186.

44. The booster companies found it difficult to retool quickly for production. History, Space Division, October 1985–September 1986, pp. 60-67; Dworetzky, "The Launch Gap," pp. 54-62.

45. History, Space Division, October 1985–September 1986, pp. 60-67; Dworetzky, "The Launch Gap," pp. 54-62; Logsdon and Williamson, "U.S. Access to Space," pp. 34-40; Air Staff Transition Papers, HQ USAF/DOSS, "DoD Space Launch Systems," 16 December 1988; William Welling, "Questions Remain for Space Command," *Advanced Propulsion* (February/March 1986), pp. 26-30.

46. History, Space Division, October 1987–September 1988, pp. 49-51; History, Space Division, October 1985–September 1986, pp. 60-67; History, Space Division, October 1986–September 1987, pp. xli-xliii; History, Air Force Space Command, January–December 1986, pp. 128-132; Dworetzky, "The Launch Gap," pp. 54-62; Logsdon and Williamson, "U.S. Access to Space," pp. 34-40; Air Staff Transition Papers, "DoD Space Launch Systems," 16 December 1988; Welling, "Questions

Remain for Space Command," pp. 26-30. The Air Force awarded a Delta II contract to McDonnell Douglas in January 1987. The first of 13 Titan IIs were expected to be operational in 1988. The Delta II, with PAM D upper stage, would prove capable of launching nearly 1900 pounds to Navstar GPS's 10,900-nautical mile altitude transfer orbit, and 11,100 pounds to an 100-nautical mile easterly orbit. GAO Fact Sheet, "Military Space Programs, p. 58; Edward H. Kolcum, "First USAF/McDonnell Douglas Delta 2 Launch Begins New Military Space Era." *Aviation Week & Space Technology*, 20 February 1989.

47. History, Space Division, October 1987–September 1988, pp. 49-51; History, Space Division, October 1985–September 1986, pp. 60-67; History, Space Division, October 1986–September 1987, pp. xli-xliii; History, Air Force Space Command, January–December 1986, pp. 128-132; Dworetzky, "The Launch Gap," pp. 54-62; Logsdon and Williamson, "U.S. Access to Space," pp. 34-40; Air Staff Transition Papers, "DoD Space Launch Systems," 16 December 1988; Welling, "Questions Remain for Space Command," pp. 26-30.

48. History, Air Force Space Command, January–December 1987, pp. 98-100; Memorandum, HQ USAF/Chief Space C3 & EC Division, Directorate of Programs & Eval, subj: Air Force Space Policy White Paper, 6 July 1987, w/atch, "White Paper on Air Force Space Policy." It should be noted that other important studies appeared at this time to help generate support for space in the Air Force. For example, the Chief of Staff in 1985 established an Aerospace Forum consisting of major command plans chiefs to examine whether the "aerospace force should perform its roles and missions with space fully integrated." After an initial report in May 1986, the Aerospace Forum seemed to succumb to later, more important

initiatives. HQ USAF/XO, Aerospace Forum Report, 22 May 1986.

49. History, Air Force Space Command, January–December 1987, pp. 98-100; Memorandum, HQ USAF/Chief Space C3 & EC Division, Directorate of Programs & Eval, subj: Air Force Space Policy White Paper, 6 July 1987, w/atch, "White Paper on Air Force Space Policy."

50. History, Air Force Space Command, January–December 1987, pp. 98-100; Memorandum, HQ USAF/Chief Space C3 & EC Division, Directorate of Programs & Eval, subj: Air Force Space Policy White Paper, 6 July 1987, w/atch, "White Paper on Air Force Space Policy." By the fall of 1987 the Air Force had a draft policy letter well underway, and on 19 October 1987, Air Force Space Command received a new, three-star commander, Lt Gen Donald J. Kutyna.

51. History, United States Space Command, January 1987 to December 1988, p. xvii.

52. *Ibid.*, pp. 56-60.

53. *Ibid.*, p. 245.

54. *Ibid.*, pp. 242-249; History, Air Force Space Command, January–December 1987, pp. 130-135. To assist the process, Air Force Space Command's operations staff created STOPLIGHT, a satellite systems analysis model that would indicate coverage gaps by projecting the on-orbit satellite constellation status through 1994.

55. History, Air Force Space Command, January–December 1987, pp. 135-136; History, United States Space Command, January 1987 to December 1988, pp. 242-249; Point Paper, AFSPCACECOM/XPSS, "Operational Space Launch," 16 October 1987. Also see the discussion of launch strategy in U.S. Senate, *Air Force Space Launch Policy and Plans*, Hearing Before the Subcommittee on Strategic Forces and Nuclear Disarmament of the Committee on Armed Services, 100th Congress, 1st Session, 6 October 1987.

56. Paper, AFSPACECOM, "Point Paper on

Transfer of Space Launch Operations to AFSPACECOM," 8 October 1987; History, Air Force Space Command, January–December 1990, pp. 81-84.

57. Background Paper, AFSPACECOM/ XPSS, "Military Man in Space (MMIS)," 28 September 1989, w/Briefing, same subject, 25 September 1989; History, Air Force Space Command, January–December 1986, pp. 126-128; History, Space Division, October 1987–September 1988, pp. 617-620; History, Air Force Space Command, January–December 1988, pp. 136-140; History, Air Force Space Command, January–December 1989, p. 161.

58. History, Air Force Space Command, January–December 1990, pp. 81-84.

59. History, Space Division, October 1987–September 1988, pp. 547-554.

60. *Ibid.*; History, Space Division, October 1988–September 1989, pp. 588-591, 632-635. See also Gen John L. Piotrowski, "C3I for Space Control," *Signal* (June 1987), pp. 23-33; Gen John L. Piotrowski, "The Right Space Tools," *Military Forum* (March 1989), pp. 46-48.

61. History, Air Force Space Command, January–December 1988, pp. 91-97; Point Paper, AFSPACECOM/XPXX, "CSAF Blue Ribbon Panel on Space Status Briefing," 26 November 1990, w/Briefing, "Blue Ribbon Panel Status," 27 November 1990.

62. History, Air Force Space Command, January–December 1988, pp. 91-97; Point Paper, AFSPACECOM/XPXX, "CSAF Blue Ribbon Panel on Space Status Briefing," 26 November 1990, w/Briefing, "Blue Ribbon Panel Status," 27 November 1990.

63. The "roadmap" format was a common approach taken by major Air Force commands to establish priorities and actions. History, Air Force Space Command, January–December 1988, pp. 91-97; Point Paper, AFSPACECOM/XPXX, "CSAF Blue Ribbon Panel on Space Status Briefing,"

26 November 1990, w/Briefing, "Blue Ribbon Panel Status," 27 November 1990; History, Air Force Space Command, January– December 1989, pp. 91-100; History, Space Division, October 1988–September 1989, pp. 573-574.

64. History, Air Force Space Command, January–December 1989, pp. 91-100.

65. Background Paper, AFSPACECOM/ XPSS, "Transfer of Functions [From 'discussion book' senior executive offsite at USAF Academy, 9 January 1990]," 12 December 1989; History, United States Space Command, January 1987 to December 1988, pp. 207-210; History, Air Force Space Command, January–December 1988, pp. 125-129; History, Air Force Space Command, January–December 1989, pp. 15-18. Although Air Force Space Command initiated planning for the transfer of the Cheyenne Mountain operations centers, the transfer itself never took place, the action having been disapproved by U.S. Space Command.

66. For discussion of the ALS, see History, Air Force Space Command, January– December 1987, pp. 133-135; History, Air Force Space Command, January–December 1988, pp. 140-144; History, Air Force Space Command, January–December 1989, pp. 153-159; History, Air Force Space Command, January–December 1990, pp. 156-159; History, Space Division, October 1986– September 1987, pp. 69-75; History, Space Division, October 1987–September 1988, pp. 115-134; History, Space Division, October 1989–September 1990, pp. 754-805; Dworetzky, "The Launch Gap," pp. 54-62; Logsdon and Williamson, "U.S. Access to Space," pp. 34-40; Richard DeMeis, "Sweetening the orbital bottom line," *Aerospace America* (August 1988), pp. 26-30; Commander-in-Chief, USSPACECOM to Vice Chairman, JCS, [Mission Need Statement for ALS] , 6 September 1989.

67. History, Air Force Space Command,

January–December 1988, pp. 142.

68. OSD/Acquisition, *Report of the Defense Science Board 1989 Summer Study on National Space Launch Strategy*, March 1990, pp. 1-38; History, Air Force Space Command, January–December 1990, pp. 156-159.

69. History, Space Division, October 1988–September 1989, pp. 55-57; Aeronautics and Space Report of the President. *1988 Activities* (Washington: NASA, 1989), p. 71; Speech, Maj Gen Thomas S. Moorman, Jr., Director, Space and SDI Programs, Office of the Secretary of the Air Force (Acquisition) to the National Space Club, Washington, D.C., 25 January 1989.

70. Paper, AFSPACECOM, "Point Paper on Transfer of Space Launch Operations to AFSPACECOM," 8 October 1987; History, Air Force Space Command, January–December 1990, pp. 81-84; Presentation, Maj Gen Thomas S. Moorman, Jr., Director, Space and SDI Programs, Office of the Secretary of the Air Force (Acquisition) to the National Space Club, Washington, D.C., 25 January 1989.

71. History, Air Force Space Command, January–December 1990, pp. 84-87. For discussion of the "blue suit" issue between SAC and AFSPACECOM, see Background Paper, AFSPACECOM/HO, "Transfer of DMSP Launch Operations to AFSC," 26 February 1990.

72. History, Air Force Space Command, January–December 1990, pp. 87-96; History, Space Division, October 1988–September 1989, pp. 14-25.

73. History, Air Force Space Command, January–December 1990, pp. 96-114; Secretary Rice is quoted on pp. 103-104.

74. Quoted in History, Air Force Space Command, January–December 1990, pp. 106-107.

75. History, Air Force Space Command, January–December 1990, p. xv.

76. AFSPACECOM/IM, "Air Force Space Command 1989 Posture Statement," n.d. [1989].

Chapter 7. Coming of Age

1. For general military and political studies of the Gulf War, see Dilip Hiro, *Desert Shield to Desert Storm: The Second Gulf War* (New York: Routledge, 1992) and Bruce W. Watson, ed., *Military Lessons of the Gulf War* (London: Greenhill Books, 1991). Also see "Introduction" by Defense Secretary Dick Cheney to the Interim Report to Congress on the Conduct of the Persian Gulf Conflict, July 1991, reprinted in *Defense Issues*, Vol. 6, No. 32, pp. 1-8.

2. Hiro, *Desert Shield to Desert Storm*, "Part II, The Crisis;" Watson, ed., *Military Lessons of the Gulf War*, "Part II. Diplomacy." Cheney, "Interim Report to Congress on the Conduct of the Persian Gulf Conflict;" OSAF, *Draft Report on Space Forces and National Security*, 23 September 1993; Office of the Chief of Staff of the Air Force, *Blue Ribbon Panel on the Air Force in Space in the*

21st Century, ca. February 1993, pp. 1-4 (hereafter cited as *Blue Ribbon Panel Executive Summary*).

3. *Blue Ribbon Panel Executive Summary*, pp. 1-4.

4. By 1991 defense funding in real terms had declined every year since 1986. To support the Gulf War, Congress authorized an emergency supplemental appropriation of $3.5 billion, which would not influence continued programmed future budget reductions. Dov S. Zahkeim, "Top Guns: Rating Weapons Systems in the Gulf War," *Policy Review* (Summer 1991), p. 14.

5. For a discussion of "Lessons Learned," see below, pp. 262-271.

6. Shortly after the conflict, commentators referred to Desert Storm as the first space war. Although also used by Air Force leaders such as Chief of Staff General Merrill

McPeak, the phrase was largely a journalistic label used to highlight the visible use of space systems in support of warfighting during the Gulf War. The term came to be viewed as misleading, incorrect, and somewhat threatening from the military perspective. For assessment of military usage, see George W. Bradley III, AFSPACECOM/HO, *AFSPACECOM Support to Operations Desert Shield/Desert Storm*, April 1993. See the discussion in James W. Canan, "A Watershed in Space," *Air Force Magazine* (August 1991), pp. 34-37; Sir Peter Anson BT and Dennis Cummings, "The First Space War: The Contribution of Satellites to the Gulf War," *RUSI Journal* (Winter 1991), pp. 45-53.

7. *Draft Report on Space Forces and National Security*, pp. 5-6; USSPPACECOM/HO, "Space Operations: A Decade of Support to Military/Humanitarian Contingencies/Operations (1983–1993)," June 1993.

8. Anson and Cummings, "The First Space War," p. 45.

9. Col Alan D. Campen, USAF (Ret), "Gulf War's Silent Warriors Bind U.S. Units Via Space," *Signal* (August 1991), pp. 81-84; Anson and Cummings, "The First Space War," pp. 46-48; "Satcoms success story," *Space Markets* Vol. 4, 1991, pp. 10-11; USSPACECOM, *United States Space Command Operations Desert Shield and Desert Storm Assessment*, January 1992, pp. 47-55 (hereafter cited as USSPACECOM *Assessment*); Center for Army Lessons Learned, *The Ultimate High Ground! Space Support to the Army/Lessons from Operations Desert Shield and Storm* (Fort Leavenworth, KS: U.S. Army Combined Arms Command, October 1991), pp. 15-17 (hereafter cited as *The Ultimate High Ground: Space Support to the Army*).

10. USSPACECOM *Assessment*, pp. 47-55; *The Ultimate High Ground: Space Support to the Army*, pp. 15-17. Two experimental polar orbiting MACSATs (Multiple Access Com-

munications Satellites) supported Marine communications requirements. Anson and Cummings, "The First Space War," p. 46.

11. *Aeronautics and Space Report of the President: 1989–1990 Activities* (Washington, D.C.: NASA, 1991), p. 62.

12. This description of DSCS in Desert Storm is based in large part on, AFSPACECOM/HO, History of Air Force Space Command, January–December 1990, pp. 151-155. The Defense Communications Agency was renamed the Defense Information Systems Agency (DISA) on 25 June 1991.

13. Dwayne A. Day, "A Review of Recent American Military Space Operations," *Journal of the British Interplanetary Society*, Vol. 46 (1993), pp. 468-469. Unlike DSCS II, DSCS III satellites included a specific anti-jamming feature to protect the AFSATCOM transponder it carried.

14. History, Air Force Space Command, January–December 1990, pp. 153-154.

15. *Ibid.*, p. 154; *USSPACECOM Assessment*, p. 61.

16. History, Air Force Space Command, January–December 1990, pp. 154-155; Anson and Cummings, "The First Space War," p. 46; *USSPACECOM Assessment*, pp. 25-28.

17. Anson and Cummings, "The First Space War," pp. 46-48. The Leasat positions were: L1 and L2 at 105 degrees west, L3 at 177 degrees west, and L5 at 178 degrees west. Day, "A Review of Recent American Military Space Operations," p. 469.

18. History, Air Force Space Command, January–December 1990, pp. 137-139; Anson and Cummings, "The First Space War," p. 48.

19. History, Air Force Space Command, January–December 1990, pp. 137-139; Anson and Cummings, "The First Space War," p. 48; *USSPACECOM Assessment*, pp. 25-28. Two-dimensional required three satellites in view from the ground, and three-dimensional coverage necessitated four.

20. History, Air Force Space Command,

January–December 1990, pp. 137-139; *USSPACECOM Assessment*, pp. 25-28.

21. History, Air Force Space Command, January–December 1990, pp. 140-141; *The Ultimate High Ground: Space Support to the Army*, pp. 1-2; *USSPACECOM Assessment*, pp. 28-31; Anson and Cummings, "The First Space War," p. 50; Barry Miller, "GPS Proves Its Worth in Operation Desert Storm," *Armed Forces Journal International* (April 1991), pp. 16-20.

22. History, Air Force Space Command, January–December 1990, pp. 140-141; *The Ultimate High Ground. Space Support to the Army*, pp. 1-2; Anson and Cummings, "The First Space War," p. 50; Barry Miller, "GPS Proves Its Worth in Operation Desert Storm," *Armed Forces Journal International* (April 1991), pp. 16-20. Approximately 4500 receivers actually saw service during Desert Storm, and more than 90% of these were commercial sets. Of the 5000 available, the Army used 3710 commercial and 557 military receivers. For the Navy the figures are 130 and 85, respectively; for the Air Force, 150 and 190, respectively. See *USSPACECOM Assessment*, p. 29.

23. *USSPACECOM Assessment*, pp. 28-31; Dwayne A. Day, "Transformation of National Security Space Programs in the Post-Cold War Era," Presentation to the 45th Congress of the International Astronautical Federation, 9–14 October 1994, Jerusalem, Israel, 11-12; Anson and Cummings, "The First Space War," p. 50.

24. *USSPACECOM Assessment*, p. 36.

25. History, Air Force Space Command, January–December 1990, pp. 141-143.

26. *Ibid.*, pp. 143-145; Day, "A Review of Recent American Military Space Operations," p. 472; USSPACECOM/SPJ3OS, "Point Paper on DMSP Performance in Desert Storm," 20 February 1992.

27. History, Air Force Space Command, January–December 1990, pp. 144-145;

USSPACECOM/SPJ3OS, "Point Paper on DMSP Performance in Desert Storm," 20 February 1992; *USSPACECOM Assessment*, p. 35.

28. History, Air Force Space Command, January–December 1990, pp. 144-145.

29. *The Ultimate High Ground: Space Support to the Army*, pp. 5-8; USSPACECOM/SPJ3OS, "Point Paper on DMSP Performance in Desert Storm," 20 February 1992; *USSPACECOM Assessment*, pp. 35-38.

30. Department of Defense, Final Report to Congress, *Conduct of the Persian Gulf War*, Appendices A-S, April 1992, pp. K-40, K-41.

31. *The Ultimate High Ground: Space Support to the Army*, pp. 9-13; *USSPACECOM Assessment*, pp. 35-38; Anson and Cummings, "The First Space War," pp. 52-53.

32. *The Ultimate High Ground: Space Support to the Army*, pp. 9-13; *USSPACECOM Assessment*, pp. 35-38.

33. *The Ultimate High Ground: Space Support to the Army*, pp. 9-13; *USSPACECOM Assessment*, pp. 35-38. It is estimated the Pentagon spent close to $6 million for Landsat and SPOT imagery during the conflict. "Civil Remote-Sensing Data Played Key Gulf War Role," *Space News*, 8–14 July 1991, pp. 3, 29.

34. *USSPACECOM Assessment*, p. 43.

35. Day, "A Review of Recent American Military Space Operations," pp. 465-468; John H. Cunningham, "The Role of Satellites in the Gulf War," *The Journal of Practical Applications in Space*, Vol. 2, No. 3 (Spring 1991), pp. 57-58. Prior to Desert Storm, DSP had been used to monitor Iraqi Scud launches in the Iran-Iraq War in the 1980s, and Constant Source terminals had been tested in European exercises and certified only six months prior to the Gulf conflict. Anson and Cummings, "The First Space War," p. 51.

36. Day, "A Review of Recent American Military Space Operations," pp. 465-468; John H. Cunningham, "The Role of

Satellites in the Gulf War," *The Journal of Practical Applications in Space*, Vol. 2, No. 3 (Spring 1991), pp. 57-58; Anson and Cummings, "The First Space War," p. 51.

37. Day, "A Review of Recent American Military Space Operations," pp. 465-468; John H. Cunningham, "The Role of Satellites in the Gulf War," *The Journal of Practical Applications in Space*, Vol. 2, No. 3 (Spring 1991), pp. 57-58; Anson and Cummings, "The First Space War," p. 51.

38. Cunningham, "The Role of Satellites in the Gulf War," pp. 57-58; Anson and Cummings, "The First Space War," p. 51; AFSPACECOM Draft Briefing, "Space Operations for Desert Shield/Desert Storm," January 1991.

39. Quoted in DoD, Final Report to Congress, *Conduct of the Persian Gulf War*, Appendices A-S, April 1992, p. K-30.

40. "Satcoms success story," *Space Markets* Vol. 4, 1991, p. 10; Memorandum, Deputy DSCS Program Manager, DCA, to Commander, Air Force Space Command, subj: DSCS—A Major Player in DESERT STORM, 7 August 1991.

41. Air Staff, "Space Forces," USAF Input for Defense Guidance on Space, *ca.* 1992; "Satcoms success story," *Space Markets* Vol. 4, 1991, p. 10; Memorandum, Deputy DSCS Program Manager, DCA, to Commander, Air Force Space Command, subj: DSCS—A Major Player in DESERT STORM, 7 August 1991.

42. *USSPACECOM Assessment*, p. 27.

43. Background Papers, "Background Paper on GPS Contributions to Desert Storm," and "GPS Activity in Desert Storm," in AFSPACECOM, *Desert Storm "Hot Wash,"* 12–13 July 1991; USSPACECOM/HO, *The Role of Space Forces: Quotes from Desert Shield/Desert Storm*, May 1993, p. 8.

44. Quoted in History, Air Force Space Command, January–December 1991, p. 4.

45. Presentation, Lt Gen Thomas S.

Moorman, Jr., "Military Space Systems Utility," Speech delivered to the 28th Space Congress, Cocoa Beach, Florida, 24 April 1991, p. 8; "Satcoms success story," *Space Markets* Vol. 4, 1991, p. 11; "Space Contribution to Weather Support," in AFSPACECOM, *Desert Storm "Hot Wash,"* 12–13 July 1991.

46. USSPACECOM *Assessment*, pp. 35-38; USSPACECOM/SPJ3OS, "DMSP Performance in Desert Storm," 20 February 1992.

47. *USSPACECOM Assessment*, pp. 44-45.

48. Day, "A Review of Recent American Military Space Operations," pp. 459-465; Day, "Transformation of National Security Space Programs in the Post-Cold War Era," pp. 8-9; Anson and Cummings, "The First Space War," pp. 50-51.

49. Day, "A Review of Recent American Military Space Operations," pp. 459-465; Day, "Transformation of National Security Space Programs in the Post-Cold War Era," pp. 8-9; Anson and Cummings, "The First Space War," pp. 50-51. On the other hand, General Schwartzkopf, in testimony before Congress, was not pleased with the bomb damage assessment reports he received from the U.S. intelligence community. See Watson, ed., *Military Lessons of the Gulf War*, "Part II. Diplomacy," p. 152.

50. Day, "A Review of Recent American Military Space Operations," pp. 465-468; Cunningham, "The Role of Satellites in the Gulf War," pp. 57-58; Anson and Cummings, "The First Space War," p. 51.

51. See for example, Watson, ed., *Military Lessons of the Gulf War*, Appendix C: Iraqi Scud Launches During the Gulf War. Watson refers to Gen Merrill McPeak's *Air Campaign* for his figures.

52. Presentation, Lt Gen Thomas M. Moorman, Jr., to Gen E.P. Rawlings Chapter, Air Force Association, Minneapolis, Minnesota, "Space…The Future is Now," 17 October 1991.

53. Day, "Transformation of National

Security Space Programs in the Post-Cold War Era," pp. 12-13; Paul B. Stares, *The Militarization of Space: U.S. Policy, 1945–1984* (Ithaca: Cornell University Press, 1985), especially chapters 6 and 10. For the Air Force perspective in the early 1990s, see History, Air Force Space Command, January–December 1991, pp. 121-130.

54. Horner and Widnall are quoted in Day, "Transformation of National Security Space Programs in the Post-Cold War Era," p. 13; see also History, Air Force Space Command, January–December 1991, pp. 121-130.

55. Day, "Transformation of National Security Space Programs in the Post-Cold War Era," pp. 6-8; History, Air Force Space Command, January–December 1990, pp. 159-163; History, Air Force Space Command, January–December 1991, pp. 130-132. On the need for a ballistic missile defense, see General Horner's 20 April 1994 testimony before the Senate Armed Services Committee, reprinted as Gen Charles A. Horner, "Space Systems: Pivotal to Modern Warfare," *Defense 94* (Issue 4, *ca.* Fall 1994), p. 24. Also, replying to a query from Senator John McCain, who asked General Horner and several other space leaders for the ten most important lessons emerging from Desert Storm, the General listed as number one the deployment of an operational BMD system. Memorandum, subj: Lessons Learned—Operation Desert Storm, *ca.* 1 June 1991.

56. Day, "Transformation of National Security Space Programs in the Post-Cold War Era," pp. 6-8; History, Air Force Space Command, January–December 1990, pp. 159-163; History, Air Force Space Command, January–December 1991, pp. 130-132.

57. Day, "Transformation of National Security Space Programs in the Post-Cold War Era," pp. 8-9.

58. *Ibid.*; Vice President's Space Policy Advisory Board, *A Post Cold War Assessment of U.S. Space Policy: A Task Group Report* (Advance Copy), December 1992, p. vii.

59. Presentation, Lt Gen Thomas S. Moorman, Jr., "Military Space Systems Utility," Speech delivered to the 28th Space Congress, Cocoa Beach, Florida, 24 April 1991, p. 8; "Space Contribution to Weather Support," in AFSPACECOM, *Desert Storm "Hot Wash,"* 12–13 July 1991; USSPACECOM/SPJ3OS, "Point Paper on DMSP Performance in Desert Storm," 20 February 1992.

60. Day, "Transformation of National Security Space Programs in the Post-Cold War Era," pp. 9-11.

61. *USSPACECOM Assessment*, pp. 45-46; "Need an Advanced Multi-Spectral Imagery (MSI) Capability," in AFSPACECOM, *Desert Storm "Hot Wash,"* 12–13 July 1991.

62. *USSPACECOM Assessment*, pp. 28-32; Day, "Transformation of National Security Space Programs in the Post-Cold War Era," pp. 11-12.

63. Day, "A Review of Recent American Military Space Operations," pp. 465-468; Day, "Transformation of National Security Space Programs in the Post-Cold War Era," pp. 6-8.

64. Day, "Transformation of National Security Space Programs in the Post-Cold War Era," pp. 6-8; *USSPACECOM Assessment*, pp. 65-66; Gen Charles A. Horner, "Space Systems. Pivotal to Modern Warfare," *Defense 94* (Issue 4, *ca.* Fall 1994), p. 25; Dwayne Day, "Top Cover: Origins and Evolution of the Defense Support Program, Part 3," *Spaceflight* Vol. 38 (March 1996), p. 99.

65. Department of Defense, *Annual Report to the President and the Congress*, "Space Forces," January 1994, pp. 227-228; Day, "Transformation of National Security Space Programs in the Post-Cold War Era," pp. 6-8.

66. *The Ultimate High Ground: Space Support to the Army*, p. 17; USSPACECOM *Assessment*, pp. 47-53.

67. Day, "Transformation of National Security Space Programs in the Post-Cold

War Era," pp. 3-4.

68. GAO Report, *DoD Acquisition: Case Study of the* MILSTAR *Satellite Communications System*, 31 July 1986.

69. Roger G. Guillemette, "Battlestar America: Milstar Survives A War With Congress," *Countdown* (November/December 1994), p. 22.

70. James W. Rawles, "Milstar Soars Beyond Budget and Schedule Goals," *Defense Electronics* (February 1989), pp. 66-72; Guillemette, "Battlestar America," p. 19; Day, "Transformation of National Security Space Programs in the Post-Cold War Era," pp. 3-4.

71. Guillemette, "Battlestar America," pp. 22-23; Day, "Transformation of National Security Space Programs in the Post-Cold War Era," pp. 3-4; "Satcoms success story," *Space Markets* Vol. 4, 1991, pp. 11-13; GAO Report, *DoD Acquisition: Case Study of the* MILSTAR *Satellite Communications System*,

31 July 1986; GAO Report, *Military Satellite Communications: Milstar Program Issues and Cost-Saving Opportunities*, 26 June 1992.

72. "Satcoms success story," *Space Markets* Vol. 4, 1991, p. 13. For the case against small satellites, see Col Owen E. Jensen, "Space Support to Tactical Forces," *Military Review* (November 1992), pp. 64-71.

73. *USSPACECOM Assessment*, pp. 61, 62, 64. The payloads launched between 2 August and 1 December 1990 were: three GPS, one DSP, one DMSP, and one classified.

74. "America's Future in Space," *Space Markets* (1/1991), pp. 20-25; Vice President's Space Policy Advisory Board, *The Future of the U.S. Space Launch Capability: A Task Group Report*, November 1992.

75. Department of Defense, *Annual Report to the President and the Congress, "Space Forces,"* January 1994, p. 226.

Chapter 8. An Air Force Vision for the Military Space Mission

1. Headquarters USAF, *Blue Ribbon Panel of the Air Force in Space in the 21st Century: Executive Summary* (Washington, D.C.: Headquarters USAF, 1992), p. iv, 1-2.

2. *Ibid.*, pp. 3-6.

3. *Ibid.*, pp. 7-8; Presentation, Sheila E. Widnall, Secretary of the Air Force, to the National Security Industrial Association, Washington, D.C. Chapter, 22 March 1994. See especially, Vice President's Space Policy Advisory Board, *The Future of the U.S. Space Launch Capability: A Task Group Report*, November 1992, chaired by E.C. Aldridge, Jr.; Vice President's Space Policy Advisory Board, *A Post Cold War Assessment of U.S. Space Policy: A Task Group Report*, December 1992, headed by Laurel L. Wilkening.

4. *Blue Ribbon Panel Executive Summary*, pp. 9-10.

5. *Ibid.*, pp. 14-20.

6. *Ibid.*, pp. 21-23; Also see AFSPACECOM/HO, *Establishment of the Space Warfare*

Center: A Brief History, ca. 1994.

7. History, Air Force Space Command, January 1992–December 1993, pp. 16, 19.

8. *Blue Ribbon Panel Executive Summary*, pp. 25-26.

9. *Ibid.*, p. 27.

10. Tom Cull, "GAO Frowns on USAF Bid to Lead Space as DoD Officials Pitch Idea on Hill," *Inside the Air Force*, 19 August 1994, p. 3.

11. *Ibid.*

12. Presentation, Gen Merrill A. McPeak, "Spacetalk 94," 16 September 1994, p. 1.

13. *Ibid.*, pp. 2-4.

14. *Ibid.*, pp. 4-6.

15. Office of the Secretary of the Air Force, Policy Letter, "Goals For Our Space Program," October 1994.

16. USAF Air University, *Spacecast 2020*, 22 June 1994; Speech, Gen Michael P. C. Carns, USAF (Ret), Closing Remarks to the National Security Industrial Association Spacecast 2020 Symposium, Washington,

D.C., 9–10 November 1994.

17. Speech, Gen Michael P. C. Carns, USAF (Ret), Closing Remarks to the National Security Industrial Association Spacecast 2020 Symposium, Washington, D.C., 9–10 November 1994.

18. Scientific Advisory Board, *New World Vistas: Air and Space Power for the 21st Century*, 15 vols., December 1995. See especially, "Summary Volume," and "Space Applications Volume."

19. Scientific Advisory Board, *New World Vistas: Air and Space Power for the 21st Century*, "Summary Volume," December 1995, pp. 3-4, 42-48, 57-64.

20. The report of the AFA group is condensed in, "Facing Up to Space," *Air Force Magazine* (January 1995), pp. 50-54.

21. *Ibid.*

22. Interview, George W. Bradley, AFSPACECOM/HO, with Gen Charles A. Horner, AFSPACECOM/CC, transcript, "Space in the Gulf War," 28 January 1993, Peterson AFB, CO.

Glossary

AACB	Aeronautics and Astronautics Coordinating Board
AAF	Army Air Forces
ABM	Anti-Ballistic Missile
ABMA	Army Ballistic Missile Division
ADCOM	Aerospace Defense Command
ADCSP	Advanced Defense Communications Satellite Program
AEC	Atomic Energy Commission
AFB	Air Force Base
AFBMD	Air Force Ballistic Missile Division
AFSATCOM	Air Force Satellite Communications
AFSC	Air Force Systems Command
AFSCN	Air Force Satellite Control Network
AFSTC	Air Force Space Test Center
ALARM	Alert, Locate and Report Missiles
ALDP	Advanced Launch Development Program
ALS	Advanced Launch System
AMC	Air Materiel Command
ARDC	Air Research and Development Command
ARPA	Advanced Research Projects Agency
ARS	Advanced Reconnaissance System
ASAT	Antisatellite
BMD	Ballistic Missile Division (ARDC)
BMEWS	Ballistic Missile Early Warning System
BMO	Ballistic Missile Office (AFSC)
BSTS	Boost Surveillance and Tracking System
CDC	Control Data Corporation
CELV	Complimentary Expendable Launch Vehicle
CIA	Central Intelligence Agency
CMLC	Civilian-Military Liaison Committee
COMSAT	Communications Satellite
CONAD	Continental Air Defense Command
CSOC	Consolidated Space Operations Center
DARPA	Defense Advanced Research Projects Agency
DCA	Defense Communications Agency
DCS	Defense Communications System
DCS	Deputy Chief of Staff
DDR&E	Director of Defense Research and Engineering

DEW	Distant Early Warning (radar system)
DMSP	Defense Meteorological Satellite Program
DoD	Department of Defense
DSAP	Defense Satellite Applications Program
DSCS	Defense Satellite Communications System
DSP	Defense Support Program
EHF	Extremely High Frequency
ELV	Expendable Launch Vehicle
ETR	Eastern Test Range
FEWS	Follow-on Early Warning System
FLTSATCOM	Fleet Satellite Comunications
FOB	Fractional Orbital Bombardment
FY	Fiscal Year
GALCIT	Guggenheim Aeronautical Laboratory, California Institute of Technology
GAO	General Accounting Office
GEODSS	Ground-based Electro-optical Deep Space Surveillance (System)
GOR	General Operational Requirement
GPS	Global Positioning System
HETS	Hyper-environmental Test System
ICBM	Intercontinental Ballistic Missile
IDCSP	Initial Defense Communication Satellite Program
IDSCS	Initial Defense Satellite Communications System
IGY	International Geophysical Year
INSSCC	Interim National Space Surveillance and Control Center
IOC	Initial Operational Capability
IONDS	Integrated Operational Nuclear Detonation Detection System
IRBM	Intermediate Range Ballistic Missile
IUS	Inertial Upper Stage
JATO	Jet-assisted Take-off
JCS	Joint Chiefs of Staff
JPL	Jet Propulsion Laboratory
JSC	Johnson Space Center
LES	Lincoln Experimental Satellite
LORAN	Long Range Navigation
MAC	Military Airlift Command
MAJCOM	Major Command

MER	Manned Earth Reconnaissance
MIDAS	Missile Detection Alarm System
MILSATCOM	Military Satellite Communications
MIRV	Multiple Independently Targetable Reentry Vehicle
MISS	Man-in-Space-Soonest
MIT	Massachusetts Institute of Technology
MODS	Military Orbital Development System
MOL	Manned Orbiting Laboratory
MOTIF	Maui Optical Tracking and Identification Facility
MRBM	Medium Range Ballistic Missile
MSFSG	Manned Space Flight Support Group
MSI	Multi-spectral Imagery
MX	Missile Experimental
NACA	National Advisory Committee for Aeronautics
NASA	National Aeronautics and Space Administration
NASC	National Aeronautics and Space Council
NATO	North Atlantic Treaty Organization
NOAA	National Oceanographic and Atospheric Administration
NOMSS	National Operational Meteorological Satellite System
NORAD	North American Air (or Aerospace) Defense Command
NRD	National Range Division
NRO	National Reconnaissance Office
NSA	National Security Agency
NSC	National Security Council
OAR	Office of Aerospace Research
OLS	Operational Linescan System
OOS	Orbit-to-Orbit Shuttle
ORDCIT	Ordnance, California Institute of Technology
OSD	Office of the Secretary of Defense
OSTP	Office of Science and Technology Policy
PSAC	Presidential Scientific Advisory Committee
R&D	Research and Development
RDO	Research and Development Objective
ROC	Required Operational Capability
SAC	Strategic Air Command
SALT	Strategic Arms Limitation Treaty
SAMSO	Space and Missile Systems Organization
SAMTEC	Space and Missile Test Center

SAMTO	Space and Missile Test Organization
SCF	Satellite Control Facility
SCORE	Signal Communications by Orbiting Relay Equipment
SD	Space Division (AFSC)
SDIO	Strategic Defense Initiative Organization
SHF	Super High Frequency
SIOP	Single Integrated Operations Plan
SLBM	Sea-launched Ballistic Missile
SLC	Space Launch Complex
SLGR	Small, Lightweight GPS Receiver
SLV	Standard Launch Vehicle
SMEC	Strategic Missiles Evaluation Committee
SOPC	Shuttle Operations and Planning Complex
SPADATS	Space Detection and Tracking System
SPADOC	Space Defense Operations Center
SSD	Space Systems Division (AFSC)
STC	Satellite Test Center
TAC	Tactical Air Command
TENCAP	Tactical Exploitation of National Capabilities
TIROS	Television and Infra-red Observing Satellite
TRW	Thompson, Ramo, Wooldridge, Inc.
UHF	Ultra High Frequency
UN	United Nations
USAF	United States Air Force
USAFR	United States Air Force Reserve
VHF	Very High Frequency
WDD	Western Development Division
WS	Weapon System
WTR	Western Test Range

Bibliography

Entire documents, and portions thereof, drawn from classified repositories and cited in this study have been declassified. Citations of miscellaneous government documents including correspondence and administrative documents not listed in the bibliography appear in the endnotes.

Books

The Aerospace Corporation. *The Aerospace Corporation: Its Work: 1960–1980*. Los Angeles: Times Mirror Press, 1980.

The Aerospace Corporation. *The Global Positioning System: A Record of Achievement*. Los Angeles: The Aerospace Corporation, 1994.

Ambrose, Stephen E., and Richard H. Immerman. *Ike's Spies: Eisenhower and the Espionage Establishment*. Garden City, NY: Doubleday & Company, Inc., 1981.

Arnold, H. H. *Global Mission*. New York: Harper & Brothers, Publishers, 1949.

Baar, James, and William E. Howard. *Spacecraft and Missiles of the World, 1962*. New York: Harcourt, Brace & World, Inc., 1962.

Baker, David. *The Shape of Wars to Come*. New York: Stein and Day, Publishers, 1981.

Baucom, Donald R. *The Origins of SDI, 1944–1983*. Lawrence, KS: University Press of Kansas, 1992.

Beard, Edmund. *Developing the ICBM: A Study in Bureaucratic Politics*. New York: Columbia University Press, 1976.

Bilstein, Roger E. *Orders of Magnitude: A History of the NACA and NASA, 1915–1990*. The NASA History Series. NASA SP-4406. Washington, D.C.: National Aeronautics and Space Administration, 1989.

Blaine, J. C. D. *The End of an Era in Space Exploration: From International Rivalry to International Cooperation*. Vol. 42, Science and Technology. San Diego, CA: American Astronautical Society, 1976.

Boffey, Philip M. *Claiming the Heavens: The New York Times Complete Guide to the Star Wars Debate*. New York: Times Books, 1988.

Brooks, Courtney G., James M. Grimwood, and Loyd S. Swenson. *Chariots for Apollo: A History of Manned Lunar Spacecraft*. The NASA History Series, NASA SP-4205. Washington, D.C.: NASA, 1979.

Bruce-Briggs, B. *The Shield of Faith*. New York: Simon and Schuster, 1988.

Bulkeley, Rip. *The Sputniks Crisis an Early United States Space Policy: A Critique of the Historiography of Space.* Bloomington, ID: Indiana University Press, 1991.

Burrows, William E. *Deep Black.* New York: Random House, 1986.

Cimbala, Stephen J. , ed. *Strategic Air Defense.* Wilmington, DE: Scholarly Resources, 1989.

Clarke, Arthur C. *The Promise of Space.* New York: Harper & Row, Publishers, 1968.

Coard, Edna A. *Space Travel.* Maxwell Air Force Base, AL: Air University, Air Force Junior ROTC, 1978.

Coffey, Thomas M. *Hap: Military Aviator.* New York: The Viking Press, 1982.

Collins, Martin J., and Sylvia D. Fries, eds. *A Spacefaring Nation. Perspectives on American Space History and Policy.* Washington, D.C.: Smithsonian Institution Press, 1991.

Davies, Merton E., and William R. Harris. *RAND's Role in the Evolution of Balloon and Satellite Observation Systems and Related U.S. Space Technology.* R-3692-RC. Santa Monica, CA: The RAND Corporation, September 1988.

Dethloff, Henry C. *Suddenly, Tomorrow Came...A History of the Johnson Space Center.* The NASA History Series. NASA SP-4307. Washington, D.C.: NASA, 1993.

DeVorkin, David H. *Science with a Vengeance: How the Military Created the US Space Sciences after World War II.* New York: Springer-Verlag, 1992.

Divine, Robert A. *The Sputnik Challenge.* New York: Oxford University Press, 1993.

Downs, Lt Col Eldon W., USAF, ed. *The U.S. Air Force in Space.* New York: Frederick A. Praeger, Publishers, 1966.

Dupre, Flint O. *Hap Arnold: Architect of American Air Power.* New York: The Macmillan Company, 1972.

Emme, Eugene M., ed. *The History of Rocket Technology: Essays on Research, Development, and Utility.* Detroit: Wayne State University Press, 1964.

———. *A History of Space Flight.* New York: Holt, Rinehart and Winston, Inc., 1965.

Freedman, Lawrence. *US Intelligence and the Soviet Strategic Threat.* Boulder, CO: Westview Press, 1977.

Frisbee, John L. *Makers of the United States Air Force.* USAF Warrior Studies. Washington, D.C.: Office of Air Force History, 1987.

Futrell, Robert Frank. *Ideas, Concepts, Doctrine: Basic Thinking in the United States Air Force, 1907–1960.* Vol. I. Maxwell Air Force Base, AL: Air University Press, December 1989.

———. *Ideas, Concepts, Doctrine: Basic Thinking in the United States Air Force, 1961–1984.* Vol. II. Maxwell Air Force Base, AL: Air University Press, December 1989.

Gavin, James M. *War and Peace in the Space Age.* New York: Harper & Brothers, Publishers, 1958.

Goldberg, Alfred, ed. *A History of the United States Air Force, 1907–1957.* Princeton: D. Van Nostrand Company, Inc., 1957.

Gorn, Michael H. *Harnessing the Genie: Science and Technology Forecasting for the Air Force, 1944–1986.* Air Staff Historical Study. Washington, D.C.: Office of Air Force History, 1988.

———, ed. *Prophecy Fulfilled. "Toward New Horizons" and Its Legacy.* Washington, D.C.: Air Force History and Museums Program, 1994.

———. *The Universal Man: Theodore von Kármán's Life in Aeronautics.* Smithsonian History of Aviation Series, ed. Von Hardesty. Washington, D.C.: Smithsonian Institution Press, 1992.

Gray, Colin S. *American Military Space Policy: Information Systems, Weapon Systems and Arms Control.* Cambridge, MA: Abt Books, 1982.

Green, Constance McLauglin, and Milton Lomask. *Vanguard: A History.* Washington, D.C.: Smithsonian Institution Press, 1971.

Greenstein, Fred I. *The Hidden-Hand Presidency: Eisenhower as Leader.* New York: Basic Bosks, 1982.

Hacker, Barton C., and James M. Grimwood. *On the Shoulders of Titans: A History of Project Gemini.* The NASA History Series. NASA SP-4203. Washington, D.C.: NASA, 1977.

Hallion, Richard P. *Storm over Iraq: Air Power and the Gulf War.* Washington, D.C.: Smithsonian Institution Press, 1992.

Herken, Gregg. *Counsels of War.* New York: Alfred A. Knopf, 1985.

Hiro, Dilip. *Desert Shield to Desert Storm: The Second Gulf War.* New York: Routledge, 1992.

Hulett, Louisa S. *From Cold Wars to Star Wars: Debates over Defense and Détente.* New York: University Press of America, 1988.

Hunley, J. D., ed. *The Birth of NASA: The Diary of T. Keith Glennan.* The NASA History Series. NASA SP-4105. Washington, D.C.: NASA History Office, 1993.

Jacobs, Horace, and Eunice Engelke Whitney. *Missile and Space Projects Guide 1962.* New York: Plenum Press, 1962.

Jenkins, Dennis R. *Space Shuttle: The History of Developing the National Space Transportation System.* Marceline, Missouri: Walsworth Publishing Company, 1993.

Karas, Thomas. *The New High Ground: Strategies and Weapons of Space-Age War.*

New York: Simon and Schuster, 1983.

Killian, James R., Jr. *Sputnik, Scientists, and Eisenhower: A Memoir of the First Special Assistant to the President for Science and Technology.* Cambridge: The MIT Press, 1977.

Kast, Fremont E., and James E. Rosenzweig. *Management in the Space Age.* New York: Exposition Press, 1962.

Kinsley, Michael. *Outer Space and Inner Sanctums: Government, Business, and Satellite Communication.* New York: John Wiley & Sons, 1976.

Kistiakowsky, Goerge B. *A Scientist at the White House: The Private Diary of President Eisenhower's Special Assistant for Science and Technology.* Cambridge, MA: Harvard University Press, 1976.

Lakoff, Sanford, and Herbert F. York. *A Shield in Space?* Berkeley, CA: University of California Press, 1989.

Lambright, W. Henry. *Powering Apollo. James E. Webb of NASA.* Baltimore, MD: The Johns Hopkins University Press, 1995.

Lane, Lt Col John J., Jr., USAF. *Command and Control and Communications Structures in Southeast Asia.* Maxwell AFB, AL: Air University, 1981.

Launius, Roger D. *NASA: A History of the U.S. Civil Space Program.* Malabar, FL: Krieger Publishing Company, 1994.

Levine, Arnold S. *Managing NASA in the Apollo Era.* The NASA History Series. NASA SP-4102. Washington, D.C.: National Aeronautics and Space Administration, 1982.

Levy, Lillian, ed. *Space: Its Impact on Man and Society.* New York: W.W. Norton & Co., 1965.

Logsdon, John M. *The Decision to Go to the Moon: Project Apollo and the National Interest.* Cambridge, MA: The MIT Press, 1970.

———, ed. *Exploring the Unknown: Selected Documents in the History of the U.S. Civil Space Program. Volume I: Organizing for Exploration.* Washington, D.C.: NASA History Office, 1995.

———, ed. *Exploring the Unknown: Selected Documents in the History of the U.S. Civil Space Program. Volume II: External Relationships.* Washington, D.C.: NASA History Office, 1996.

Loosbrock, John F., *et al. Space Weapons: A Handbook of Military Astronautics.* New York: Frederick A. Praeger, Publishers, 1959.

Marks, Robert W., ed. *The New Dictionary and Handbook of Aerospace.* New York: Frederick A. Praeger, Publishers, 1969.

Martin, Donald H. *Communication Satellites, 1958–1992.* El Segundo, California: Aerospace Corporation, 1991.

MacCloskey, Monro. *The United States Air Force.* New York: Frederick A. Praeger, Publishers, 1967.

McCurdy, Howard E. *Inside NASA: High Technology and Organizational Change in the U.S. Space Program.* Baltimore, MD: The Johns Hopkins University Press, 1993.

————. *The Space Station Decision: Incremental Politics and Technological Choice.* Baltimore, MD: The Johns Hopkins Press, 1990.

McDougall, Walter A. *...the Heavens and the Earth: A Political History of the Space Age.* New York: Basic Books, 1985.

Medaris, Major General John B. *Countdown for Decision.* New York: G.P. Putnam's Sons, 1960.

Miles, Howard, ed. *Artificial Satellite Observing and its Applications.* London: Faber and Faber Limited, 1974.

The Military Balance, 1974–1975. London: The International Institute of Strategic Studies, 1974.

The Military Balance, 1975–1976. London: The International Institute of Strategic Studies, 1975.

Momyer, General William W. *Air Power in Three Wars.* Washington, D.C.: Office of Air Force History, 1978.

Nebeker, Frederik. *Calculating the Weather: Meteorology in the 20th Century.* New York: Academic Press, 1995.

Needell, Allan A., ed. *The First 25 Years in Space: A Symposium.* Washington, D.C.: Smithsonian Institution Press, 1983.

Neufeld, Jacob. *The Development of Ballistic Missiles in the United States Air Force 1945-1960.* General Histories. Washington, D.C.: Office of Air Force History, 1990.

Neufeld, Michael J. *The Rocket and the Reich: Peenemuende and the Coming of the Ballistic Missile Era.* New York: The Free Press, 1995.

Newell, Homer E. *Beyond the Atmosphere: Early Years of Space Science.* The NASA History Series. NASA SP-4211. Washington, D.C.: NASA, 1980.

Newhouse, John. *Cold Dawn: The Story of SALT.* New York: Holt, Rinehart and Winston, 1973.

————. *War and Peace in the Nuclear Age.* New York: Alfred A. Knopf, 1989.

Nimmen, Jane Van, and Leonard C. Bruno, with Robert L. Rosholt. *NASA Historical Data Book, 1958-1968. Vol. I: NASA Resources.* The NASA Historical Series. NASA SP-4012. Washington, D.C.: NASA, 1976.

Ordway, Frederick I., III, and Mitchell R. Sharpe. *The Rocket Team.* New York: Thomas Y. Crowell, Publishers, 1979.

Osman, Tony. *Space History.* New York: St. Martin's Press, 1983.

Richelson, Jeffrey T. *America's Secret Eyes in Space: The U.S. Keyhole Spy Satellite Program.* New York: HarperBusiness, 1990.

Rostow, W. W. *Open Skies: Eisenhower's Proposal of July 21, 1955.* Austin, TX: University of Texas Press, 1982.

Ruffner, Kevin C., ed. *Corona: America's First Satellite Program.* Center for the Study of Intelligence. Washington, D.C.: CIA History Staff, 1995.

Schichtle, Cass. *The National Space Program: From the Fifties to the Eighties.* Washington, D.C.: National Defense University Press, 1983.

Schriever, General Bernard A., *et al. Reflections on Research and Development in the United States Air Force*, Jacob Neufeld, ed. Washington, D.C.: Center for Air Force History, 1993.

Skinner, Richard M., and William Leavitt, eds. *Speaking of Space: The Best from Space Digest.* Boston: Little, Brown and Company, 1962.

Smith, Bruce L. R. *The RAND Corporation.* Cambridge, MA: 1966.

Stares, Paul B. *The Militarization of Space: U.S. Policy, 1945–1984.* Ithica, NY: Cornell University Press, 1985.

———. *Space and National Security.* Washington, D.C.: The Brookings Institution, 1987.

Sturm, Thomas A. *The USAF Scientific Advisory Board: Its First Twenty Years, 1944–1964.* Washington, D.C.: USAF Historical Division Liaison Office, 1967.

Swenson, Loyd S. Jr., James M. Grimwood, and Charles C. Alexander. *This New Ocean: A History of Project Mercury.* The NASA Historical Series. NASA SP-4201 Washington, D.C.: National Aeronautics and Space Administration, 1966.

Tedeschi, Anthony Michael. *Live Via Satellite: The Story of COMSAT and the Technology that Changed World Communication.* Washington, D.C.: Acropolis Books, 1989.

Thomas, Shirley. *Men of Space: Profiles of the Leaders in Space Research, Development, and Exploration.* Vols. 2, 3, 4, 7. Philadelphia: Chilton Company Publishers, 1960-1965.

Trento, Joseph J. *Prescription for Disaster: From the Glory of Apollo to the Betrayal of the Shuttle.* London: Harrap Ltd, 1987.

Turnhill, Reginald. *The Observer's Spaceflight Directory.* London: Frederick Warne, 1978.

Van Dyke, Vernon. *Pride and Power: The Rationale of the Space Program.* Urbana, IL: University of Illinois Press, 1964.

von Kármán, Theodore, with Lee Edson. *The Wind and Beyond: Theodore von Kármán, Pioneer in Aviation and Pathfinder in Space.* Boston: Little, Brown and Company, 1967.

Watson, Bruce W., ed. *Military Lessons of the Gulf War.* London: Greenhill Books, 1991.

Watson, George M., Jr. *The Office of the Secretary of the Air Force, 1947–1965.* Washington, D.C.: Center for Air Force History, 1993.

Weiser, Lynne, ed. *Who's Who in Space, 1966-67.* Vol. I. Washington, D.C.: Space Publications, Inc., 1965.

Wolf, Richard I. *The United States Air Force Basic Documents on Roles and Missions.* Washington, D.C.: Office of Air Force History, 1987.

Articles

"A New Image for the Space Shuttle." *Science,* 18 January 1985, 276-277.

Aldridge, Edward C., Jr. "Defense in the Fourth Dimension." *Defense 83* (January 1983): 3-10.

———. "The Myths of Militarization of Space." *International Security* 11, no. 4 (Spring 1987): 151-164.

"America's Future in Space." *Space Markets* 1 (1991): 20-25.

Anson, Sir Peter, BT, and Dennis Cummings. "The First Space War: The Contribution of Satellites to the Gulf War." *RUSI Journal* (Winter 1991): 45-53.

Borrowman, Gerald L. "The Military Role in the Shuttle." *Spaceflight* 24 (5 May 1982): 226-229.

Brandli, Henry W. "The Use of Meteorological Satellites in Southeast Asia Operations." *Aerospace Historian* 29, no. 3 (September 1982): 172-175.

Broad, William J. "As Shuttle Orbits, a Debate Grows Over Military's Role." *The New York Times,* 6 October 1985, 1.

———. "Pentagon Leaves the Shuttle Program." *The New York Times,* 7 August 1989, 5-6.

Burrows, William E. "The Military in Space: Securing the High Ground." In *Space: Discovery and Exploration*, ed. Martin J. Collins and Sylvia K. Kraemer, 119-165. Smithsonian Institution. Hong Kong: Hugh Lauter Levin Associates, Inc., 1994.

Butterworth, Robert L. "The Case Against Centralizing Military Space." *Strategic Review* xxiv, no. 3 (Summer 1996): 41-49.

Campen, Col. Alan D, USAF (Ret). "Gulf War's Silent Warriors Bind U.S. Units Via Space." *Signal* (August 1991): 81-84.

Canan, James W. "A Watershed in Space." *Air Force Magazine* (August 1991): 34-37.

Cheney, Dick. "Introduction by Defense Secretary Dick to the Interim Report to Congress on the Conduct of the Persian Gulf Conflict." July 1991. Reprinted in *Defense Issues* 6, no. 32: 1-8.

"Civil Remote-Sensing Data Played Key Gulf War Role." *Space News*, 8–14 July 1991, 3, 29.

Cull, Tom. "GAO Frowns on USAF Bid to Lead Space as DoD Officials Pitch Idea on Hill." *Inside the Air Force*, 19 August 1994, 3.

Cunningham, John H. "The Role of Satellites in the Gulf War." *The Journal of Practical Applications in Space* 2, no. 3 (Spring 1991): 57-58.

Dash, Major Ernie R., and Major Walter D. Meyer. "The Meteorological Satellite. An Invaluable Tool for the Military Decision-Maker." *Air University Review* xxix, no. 3 (March–April 1978): 13-24.

Day, Dwayne A. "A Review of Recent American Military Space Operations." *Journal of the British Interplanetary Society* 46 (1993): 468-469.

———. "Capturing the High Ground. The U.S. Military in Space 1987–1995. Part 1." *Countdown* (January/February 1995): 17-27.

———. "Capturing the High Ground. The U.S. Military in Space 1987–1995. Part 2." *Countdown* (May/June 1995): 31-41.

———. "CORONA: America's First Spy Satellite Program." *Quest* (Summer 1995): 4-21.

———. "CORONA: America's First Spy Satellite Program, Part II." *Quest* Vol. 4, No. 3 (Fall 1995): 28-36.

———. "Lifting a Veil on History: Early Satellite Imagery and National Security." *Space Times* (July-August 1995): 7-16.

———. "Transformation of National Security Space Programs in the Post-Cold War Era." *Presentation to the 45th Congress of the International Astronautical Federation*, 9-14 October 1994, Jerusalem, Israel. Paris: American Institute Of Aeronautics and Astronautics, 1994.

————. "The Origins and Evolution of the Defense Support Program - Part 1." *Spaceflight* 37 (December 1995): 2-6.

————. "Top Cover. Origins and Evolution of the Defense Support Program Part 2." *Spaceflight* 38 (February 1996): 59-63.

————. "Top Cover. Origins and Evolution of the Defense Support Program Part 3." *Spaceflight* 38 (March 1996): 95-99.

DeMeis, Richard. "Sweetening the orbital bottom line." *Aerospace America* (August 1988): 26-30.

Dworetzky, Tom. "The Launch Gap." *Discover* (July 1988): 54-62.

Evans, Jean. "History of the School of Aviation Medicine." *The Air Power Historian* (October 1958): 245-261.

"Facing Up to Space." *Air Force Magazine* (January 1995): 50-54.

Fuller, Thomas. "DoD in Space: A Historical Perspective." *Defense 88* (November/December 1988): 26-31.

Galloway, Alec. "A Decade of US Reconnaissance Satellites." *International Defense Review* (June 1972): 249-253.

Gardner, Trevor. "How We Fell Behind in Guided Missiles." *The Air Power Historian* 5, no. 1 (January 1958): 3-13.

Garthoff, Raymond L. "Banning the Bomb in Outer Space." *International Security* 5, no. 3 (Winter 1980/81): 25-40.

Garwin, Richard L. "National Security Space Policy." *International Security* 11, no. 4 (Spring 1987): 165-173.

Getting, Ivan A. "The Global Positioning System." *IEEE Spectrum* (December 1993): 36-37.

Greenwood, John T. "The Air Force Ballistic Missile and Space Program (1954–1974)." *Aerospace Historian* 21 (Winter 1974): 190-197.

Greer, Kenneth E. "Corona." *Studies in Intelligence, Supplement,* 17 (Spring 1973): 6. Reprinted in *Corona: America's First Satellite Program,* ed. Kevin C. Ruffner, 3-39. Washington, D.C.: CIA History Staff, Center for the Study of Intelligence, 1995.

Guillemette, Roger G. "Battlestar America: Milstar Survives A War With Congress." *Countdown* (November/December 1994): 22.

————. "Vandenberg. Space Shuttle Launch and Landing Site. Part 1—Construction of Shuttle Launch Facilities." *Spaceflight* 36 (October 1994): 354-357.

————. "Vandenberg. Space Shuttle Launch and Landing Site. Part 2—Abandoned in Place." *Spaceflight* 36 (November 1994): 378-381.

Hall, R. Cargill. "Early U.S. Satellite Proposals." In *The History of Rocket Technology: Essays on Research, Development, and Utility*, ed. Eugene M.Emme, 67-93. Detroit: Wayne State University Press, 1964.

―――. "The Eisenhower Administration and the Cold War: Framing American Astronautics to Serve National Security." *Prologue* , 27, no.1 (Spring 1995): 59-72.

―――. "The Origins of U.S. Space Policy: Eisenhower, Open Skies, and Freedom of Space." *Colloquy* (December 1993): 5-6, 19-24.

Henry, Lt Gen Richard C. "Interview." *Air Force Magazine* 65, no. 6 (June 1982): 41.

Horner, Gen Charles A. "Space Systems. Pivotal to Modern Warfare." *Defense 94* no. 4 (*ca.* Fall 1994): 24-25.

Houchin, Roy F., III. "Why the Air Force Proposed the Dyna-Soar X-20 Program." *Quest* (Winter 1994): 5-12.

Jensen, Col Owen E. "Space Support to Tactical Forces." *Military Review* (November 1992): 64-71.

―――. "The Years of Decline: Air Defense from 1960 to 1980." In *Strategic Air Defense*, ed. Stephen J. Cimbala, 40-41. Wilmington, DE: Scholarly Books, 1989.

Kelly, C. Brian. "Ten Years in the Outer Realm." *Data* (June 1968): 22-24.

Kolcum, Edward H. "First USAF/McDonnell Douglas Delta 2 Launch Begins New Military Space Era." *Aviation Week & Space Technology*, 20 February 1989, 18-19.

Launius, Roger D. "Toward an Understanding of the Space Shuttle: A Historiographical Essay." *Air Power History* (Winter 1992): 3-10.

Leavitt, William, John F. Loosbrock, Richard M. Skinner, and Claude Witze. "The Space Frontier." *Air Force Magazine* 41, no. 3 (March 1958): 43-58.

Logsdon, John M., and Ray A. Williamson. "U.S. Access to Space." *Scientific American* 260, no. 3 (March 1989): 34-40.

Logsdon, John. "The Decision to Develop the Space Shuttle." *Space Policy* (May 1986): 103-118.

MacDonald, Robert A. "CORONA: Success for Space Reconnaissance, A Look into the Cold War, and Revolution for Intelligence." *PE&RS*, lxi, no. 6, 2 June 1995, 689-720.

Malina, Frank J. "Origins and First Decade of the Jet Propulsion Laboratory." In *The History of Rocket Technology. Essays on Research, Development, and Utility*, ed. Eugene M.Emme, 46-66. Detroit: Wayne State University Press, 1964.

"The Military in Space." In *Man in Space*, ed. H.J.P. Arnold, 218-227. London: Smithmark, 1993.

Mark, Hans. "The Future of NASA and the U.S. Enterprise in Space." *International Security* 11, no. 4 (Spring 1987): 174-186.

McDowell, Jonathan. "US Reconnaissance Satellite Programs. Part I: Photoreconnaissance." *Quest* (Summer 1995): 22-33.

McPeak, Gen Merrill A. "The Air Force Role in Space." *The Space Times* 32, no.4 (July-August 1993): 5-7.

Miller, Barry. "GPS Proves Its Worth in Operation Desert Storm." *Armed Forces Journal International* (April 1991): 16-20.

"Milstar Satellite Project Under Watchful Eye." *Defense Electronics* (June 1986): 64.

"NASA Shouldn't Operate Shuttle, AF Under Secretary Says." *Aerospace Daily*, 112, no. 14, 20 November 1981, 105-106.

Piotrowski, Gen John L. "C3I for Space Control" *Signal* (June 1987): 23-33.

Piotrowski, Gen John L. "The Right Space Tools." *Military Forum* (March 1989): 46-48.

Powers, Thomas. "Choosing a Strategy for World War III." *The Atlantic Monthly* (November 1982): 82-110.

Rawles, James W. "Milstar Soars Beyond Budget and Schedule Goals." *Defense Electronics* (February 1989): 66-72.

Richardson, Robert C., III, USAF (Ret). "High Frontier: 'The Only Game in Town." *Journal of Social, Political and Economic Studies.* VII, nos. 1&2 (1982): 55-66.

Sanborn, Colonel Morgan W. "National Military Space Doctrine." *Air University Review* xxvii, no. 2 (January-February 1977): 75-79.

"Satcoms success story." *Space Markets* 4 (1991): 10-11.

Scheerer, Lt Col John F., and Maj Joseph Gassmann. "Navstar GPS. Past, Present, and Future." *The Navigator* xxx, no. 3 (Winter 1983): 16-19.

Schriever, Maj Gen Bernard A. "The Battle for 'Space Superiority." *Air Force Magazine* 40, no. 4 (April 1957): 31-32, 34.

———. "Does the Military Have a Role in Space?" in *Space: Its Impact on Man and Society*, ed. Lillian Levy, 59-68. New York: W.W. Norton & Co., 1965.

Scott, William B. "Major Cultural Change on Tap in Military Space." *Aviation Week & Space Technology*, 18 September 1995, 40-63.

Siekman, Philip. "The Fantastic Weaponry." *Fortune* (June 1967): 157-159, 214, 216, 218, 223-4.

Spires, David N., and Rick W. Sturdevant. "From Advent to Milstar: The United States Air Force and the Challenges of Military Satellite Communications." In

Proceedings of the NASA Conference. Beyond the Ionosphere: The Development of Satellite Communications, ed. Andrew Butrica. Washington, D.C.: NASA, 1996.

"Statement by Lt Gen James Ferguson, Deputy Chief of Staff, Research & Technology, USAF, before the House Committee on Armed Services." *Aviation Week & Space Technology*, 5 March 1962, 75, 77, 79, 83, 87, 89, 91-92, 94, 96.

Sturdevant, Rick W. "The United States Air Force Organizes for Space: The Operational Quest." In *Organizing for the Use of Space: Historical Perspectives on a Persistent Issue*, ed. Roger D. Launius, 155-186. San Diego, CA: American Astronautical Society, 1995.

Ulsamer, Edgar. "Space Shuttle Mired in Bureaucratic Feud." *Air Force Magazine* (September 1980): 72-77.

———. "Space. High-Flying Yankee Ingenuity." *Air Force Magazine* (September 1976): 98-104.

van Keuren, David K. "Moon in Their Eyes: Moon Communication Relay (MCR) at the Naval Research Laboratory, 1951–1962." In *Proceedings of the NASA Conference. Beyond the Ionosphere: The Development of Satellite Communications*, ed. Andrew Butrica. Washington, D.C.: NASA, 1996.

Welling, William. "Questions Remain for Space Command." *Advanced Propulsion* (February/March 1986): 26-30.

White, General Thomas D. "At the Dawn of the Space Age." *The Air Power Historian* v, no. 1 (January 1958): 15-19.

Williamson, John. "Military Setbacks of Challenger Shuttle Loss." *Janes Defense Weekly*, 4 June 1986, 1109-1111.

York, Herbert F., and G. Allen Greb. "Military research and development: a postwar history." *Bulletin of Atomic Scientists* (January 1977): 13-26.

———. "Strategic Reconnaissance." *Bulletin of Atomic Scientists* (April 1977): 33-42.

Zahkeim, Dov S. "Top Guns. Rating Weapons Systems in the Gulf War." *Policy Review* (Summer 1991): 14.

Congressional Documents

U.S. Congress. House. Committee on Armed Services. *Hearings on Military Posture.* 88th Cong, 1st sess, January-February 1963. Washington: Government Printing Office, 1963.

U.S. Congress. House. Committee on Government Operations. *Government Operations in Space (Analysis of Civil-Military Roles and Relationships).* House Report No. 445. 89th Cong, 1st sess, 4 June 1965. Washington: Government Printing Office, 1965.

U.S. Congress. House. *Hearings before the Committee on Science and Astronautics.* 37th Cong, 1st sess, 16 February 1961. Washington: Government Printing Office, 1961.

U.S. Congress. House. Joint Hearings Before the Subcommittee on Space Science and Applications of the Committee on Armed Services. *Assured Access to Space During the 1990s.* 99th Cong, 1st sess, 23, 24, 25, July 1985. Washington: Government Printing Office, 1985.

U.S. Congress. House. Subcommittee on NASA Oversight of the Committee on Science and Astronautics. *The NASA-DoD Relationship.* 88th Cong, 2d sess. Washington: Government Printing Office, 1964.

U.S. Congress. Senate. Hearing before the Subcommittee of the Committee on Appropriations. *Department of Defense Appropriations for 1964.* 88th Cong, 1st sess, 24 April 1963. Washington: Government Printing Office, 1963.

U.S. Congress, Senate, Hearing Before the Subcommittee on Strategic Forces and Nuclear Disarmament of the Committee on Armed Services. *Air Force Space Launch Policy and Plans.* 100th Cong, 1st sess, 6 October 1987. Washington: Government Printing Office, 1988.

U.S. Congress. Senate. Hearings Before the NASA Authorization Subcommittee of the Committee on Aeronautical and Space Sciences. *NASA Authorization for Fiscal Year 1961, Pt 1.* 86th Cong, 2d sess, 28–30 March 1960. Washington: Government Printing Office, 1960.

U.S. Congress. Senate. Hearings Before the Subcommittee on Governmental Organization for Space Activities of the Committee on Aeronautical and Space Sciences. *Investigation of Governmental Organization for Space Activities.* 86th Cong, 1st sess, 24 March–7 May 1959. Washington: Government Printing Office, 1959.

U.S. Congress. *The National Aeronautics and Space Act of 1958.* Public Law 58-568, Section 102(c)(6).

Studies

A Post Cold War Assessment of U.S. Space Policy. A Task Group Report. (Advance Copy). Washington, D.C.: Vice President's Space Policy Advisory Board, December 1992.

A Study of Future Air Force Space Policy and Objectives. Washington, D.C.: Headquarters USAF, July 1977.

Abel, Maj Michael D. *History of the Defense Meteorological Satellite Program: Origin Through 1982.* Air Command and Staff College Report No. 87-0020. Maxwell AFB, AL: Air University Press, 1987.

Air Force Scientific Advisory Board. *The Air Force in Space. Summer Study.* Washington, D.C.: SAB, August 1980.

Aldridge, Edward C., Jr. *The Future of the U.S. Space Launch Capability. A Task Group Report.* Washington, D.C.: Vice President's Space Policy Advisory Board, November 1992.

Alford, Maj Dennis L. *History of the Navstar Global Positioning System (1963–1985).* Air Command and Staff College Report No. 86-0050. Maxwell AFB, AL: Air University Press, 1986.

Augenstein, Bruno W. *Evolution of the U.S. Military Space Program, 1945–1960: Some Key Events in Study, Planning, and Program Management.* Santa Monica, CA: Rand Corporation, September 1982.

Austerman, Wayne R. *Program 437: The Air Force's First Antisatellite System.* AFSPACECOM/HO, April 1991.

Boone, W. Fred. *The First Five Years, December 1, 1962, to January 1, 1968.* NASA Office of Defense Affairs. Historical Division, Office of Policy. Washington, D.C.: NASA, December 1970.

Bossert, Roger. Message Bob Peterson. "CORONA Program Profile." Lockheed Press Release, 24 May 1995.

Bowen, Lee. *The Threshold of Space: The Air Force in the National Space Program, 1945–1959.* Washington, D.C.: USAF Historical Division Liaison Office, September 1960.

Bradley, George W., III. *AFSPACECOM Support to Operations Desert Shield/Desert Storm.* Peterson AFB, CO: AFSPACECOM/HO, April 1993.

Carter, Launor F., Chief Scientist of the Air Force. *An Interpretive Study of the Formulation of the Air Force Space Plan.* 4 February 1963.

Cleary, Mark C. *The 6555th: Missile and Space Launches Through 1970.* Patrick AFB, FL: 45th Space Wing History Office, November 1991.

Defense Science Board. *Report of the Defense Science Board 1989 Summer Study on National Space Launch Strategy.* Washington, D.C.: OSD/Acquisition, March 1990.

Department of Defense. *Space Launch Modernization Plan: Executive Summary.* Washington, D.C.: U.S. Government Printing Office, April 1994.

Department of the Air Force. Headquarters USAF. Office of the Chief of Staff. *Blue Ribbon Panel of the Air Force in Space in the 21st Century. Executive Summary.* Washington, D.C.: 1992.

Elliott, Derek W. Elliott. "Finding and Appropriate Commitment: Space Policy Development Under Eisenhower and Kennedy, 1954–1963." Unpublished

Dissertation. Washington, D.C.: The George Washington University, 10 May 1992.

Erickson, Capt Mark A. *Highlights of United States Intercontinental Ballistic Missile Development.* Peterson AFB, CO: AFSPACECOM/HO, July 1993.

Establishment of the Space Warfare Center: A Brief History. Peterson AFB, CO: AFSPACECOM/HO, *ca.* 1994.

Frisvold, Cynthia L. *History of the Air Force Satellite Control Facility.* Space Systems Management 500. 5 May 1987.

Geiger, Clarence J. *History of the X-20A Dyna-Soar.* Vol. 1. AFSC Historical Publications Series 63-50-I. Wright-Patterson AFB, OH: Air Force Systems Division, Aeronautical Systems Division, October 1963.

General Proposal for Organization for Command and Control of Military Operations in Space, [n.d.], Box 15, White House Office. Office of the Special Assistant for Science and Technology. Records (James R. Killian and George B. Kistiakowsky, 1957–1961). "Space [July–December 1959] (7)." Eisenhower Library.

The Great Frontier. Miliary Space Doctrine. The Final Report from the United States Air Force Academy Military Space Doctrine Symposium. USAF Academy, CO: USAFA, 1–3 April 1981.

Hall, R. Cargill. "The Air Force in Space." Unpublished draft chapter, May 1984.

Hungerford, Maj John B., Jr. *Organization for Military Space. A Historical Perspective.* Air Command and Staff College Report No. 82-1235. Maxwell AFB, AL: Air University Press, 1982.

Kidd, Col John, and 1Lt Holly Caldwell. *Defense Support Program: Support to a Changing World.* Huntsville, AL: AIAA Space Programs and Technologies Conference, 24–27 March 1992.

Killebrew, Maj Timothy D. *Military Man in Space: A History of Air Force Efforts to Find a Manned Space Mission.* Air Command and Staff College Report No. 87-1425. Maxwell AFB, AL: Air University Press, 1987.

Lamping, Lt Col Neal E., USAF, and Lt Col Richard P. MacLeod, USAF. *Space—A National Security Dilemma: Key Years of Decision.* Washington: National Defense University, June 1979.

Lee, Maj Robert E. *History of the Defense Satellite Communications System (1964–1986).* Air Command and Staff College Report No. 87-1545. Maxwell AFB, AL: Air University Press, 1987.

Mangold, Maj Sanford D. *The Space Shuttle—A Historical View from the Air Force Perspective.* Air Command and Staff College Report No. 83-1540. Maxwell AFB, AL: Air University Press, 1983.

McPeak, Gen Merrill A., US Air Force Chief of Staff. *Presentation to the Commission on Roles and Missions of the Armed Forces.* Washington, D.C.: Headquarters USAF, 14 September 1994.

Narducci, Henry M. *Strategic Air Command and the Space Mission, 1977–1984.* Historical Monograph No. 209. Offutt AFB, NE: Strategic Air Command/HO, 10 October 1985.

National Space Council. *Final Report to the President on the U.S. Space Program.* Washington, D.C.: National Space Council, January 1993.

Peebles, Curtis. *High Frontier: The United States Air Force and the Military Space Program.* Washington, D.C.: Air Force History and Museums Program, 1997.

Perry, Robert L. *Origins of the USAF Space Program, 1945–1956. V, History of DCAS 1961.* AFSC Historical Publications Series 62-24-10. Los Angeles, CA: Air Force Systems Command, Space Systems Division, 1961.

Piper, Robert F. *History of the Titan III, 1961-1963.* Vol.1, AFSC Historical Publications Series 64-22-1. Washington, D.C.: Air Force Systems Command, Space Systems Division, June 1964.

Proposal for Man-in-Space (1957–1958). AFSC Historical Publications Series 66-11-1. 1966.

Scientific Advisory Board. *New World Vistas: Air and Space Power for the 21st Century.* 15 vols. Washington, D.C.: SAB, December 1995.

Space Missions Organizational Planning Study (SMOPS). Washington, D.C.: Headquarters USAF, February 1979.

Space Policy and Requirements Study (SPARS). Washington, D.C.: Headquarters USAF, May 1981.

Space Surveillance Architecture Study. Washington, D.C.: Headquarters USAF, 10 June 1983.

Spacecast 2020. Maxwell AFB, AL: Air University Press, 22 June 1994.

Sprague, Maj Barkley G.,USAF. *Evolution of the Missile Defense Alarm System (MIDAS) 1955–1982.* Air Command and Staff College Report No. 85-2580. Maxwell AFB, AL: Air University Press, 1985.

The Role of Space Forces. Quotes from Desert Shield/Desert Storm. Peterson AFB, CO: USSPACECOM/HO, May 1993.

Torgerson, Maj Thomas A., USAF. *Global Power through Tactical Flexibility. Rapid Deployment Space Units.* Research Report No. AU-ARI-93-6. Maxwell AFB, AL: Air University Press, June 1994.

The Ultimate High Ground! Space Support to the Army/Lessons from Operations Desert Shield and Storm. Center for Army Lessons Learned. Fort Leavenworth, KS: U.S. Army Combined Arms Command, October 1991.

United States Space Command Operations Desert Shield and Desert Storm Assessment. Peterson AFB, CO: USSPACECOM, January 1992.

USAF Space Plan(draft). Washington, D.C.: Headquarters USAF, September 1961.

USAF Space Plan(draft working paper). Washington, D.C.: Headquarters USAF, revised July 1962.

von Kármán, Theodore. *Toward New Horizons: Science, the Key to Air Supremacy. Commemorative Edition, 1950–1992.* Wright-Patterson AFB, OH: Air Force Systems Command, 1992.

Wilkening, Laurel L. *A Post Cold War Assessment of U.S. Space Policy: A Task Group Report.* Washington, D.C.: Vice President's Space Policy Advisory Board, December 1992.

Internal Reports

AFBMD staff. *Relationship between Ballistic Missile Programs and Space.* January 1959.

AFBMD. *Impact of the OSD 'Space Operations' Memorandum on the AFBMD Space Program.* October 1959.

AFSC/SSD Historical Division. *Chronology of Early Air Force Man-In-Space Activity.* AFSC Historical Publications Series 65-21-1. January 1965

AFSPACECOM. *Desert Storm 'Hot Wash.'* 12–13 July 1991.

AFSPACECOM/Commander's Group. *History of Space Policy.* 1 June 1992.

AFXPD-PA-LRP. *Air Force Space Program FY 1963: Panel One, Ferguson Task Force.* 22 December 1961.

Air Force/NASA Space Program Management Panel 3 Report (Ferguson Task Force) USAF Space Program FY 63-64. n.d. [December 1961–January 1962].

Air Staff, *Space Forces. USAF Input for Defense Guidance on Space. ca.* 1992.

HQ USAF/XO. *Aerospace Forum Report.* 22 May 1986.

HQ USSPPACECOM/HO. "Space Operations: A Decade of Support to Military/Humanitarian Contingencies/Operations (1983-1993)." June 1993.

Joint DoD/NASA Study of Space Transportation Systems. *Summary Report.* 16 June 1969.

OSAF. *Draft Report on Space Forces and National Security.* 23 September 1993.

Project RAND. *Preliminary Design of an Experimental World-Circling Spaceship.* Santa Monica, CA: Douglas Aircraft Co, 2 May 1946.

Schriever, Bernard A. Committee on Space Policy. *Report on National Space Policy: Enhancing National Security and U.S. Leadership Through Space.* January 1981.

USAF Scientific Advisory Board. *Report of the Scientific Advisory Board Ad Hoc Committee on Space Technology.* 6 December 1957.

Published Government Reports

General Accounting Office

General Accounting Office. *DoD Acquisition: Case Study of the MILSTAR Satellite Communications System.* GAO/NSIAD-86-45S-15. 31 July 1986.

General Accounting Office. *Military Space Programs: An Unclassified Overview of Defense Satellite Programs and Launch Activities.* GAO/NSIAD-90-154FS. June 1990.

General Accounting Office. *Issues Concerning the Future of the Space Transportation System.* GAO/MASAD-83-6, 28 December 1982.

General Accounting Office. *Report to the Chairman, Committee on Government Operations, House of Representatives. Implications of Joint NASA/DoD Participation in Space Shuttle Operations.* GAO/NSIAD-84-13, 7 November 1983.

General Accounting Office. *Report to the Chairman, Legislation and National Security Subcommittee, Committee on Government Operations, House of Representatives. Military Satellite Communications: Opportunity to Save Billions of Dollars.* GAO/NSIAD-93-216. July 1993.

General Accounting Office. *Report to the Chairman, Subcommittee on Defense, Committee on Appropriations, House of Representatives. Military Satellite Communications: DoD Needs to Review Requirements and Strengthen Leasing Practices.* GAO/NSIAD-94-48. February 1994.

General Accounting Office. *Testimony Before the Subcommittee on Defense, Committee on Appropriations, House of Representatives. Military Satellite Communications: Potential for Greater Use of Commercial Satellite Capabilities.* GAO/T-NSIAD-92-39. 22 May 1992.

General Accounting Office. *Briefing Report to the Ranking Minority Member, Subcommittee on Military Construction, Committee on Appropriations, United States Senate. Space Shuttle: Issues Associated With the Vandenberg Launch Site.* GAO/NSIAD-87-32BR. October 1986.

Department of Defense

Department of Defense. *Annual Report of the Secretary of Defense and the Reports of the Secretary of the Army, Secretary of the Navy, Secretary of the Air Force, July 1, 1958 to June 30, 1959.* Washington, D.C.: U.S. Government Printing Office, 1960.

Department of Defense. *Annual Report to the President and the Congress, Space Forces.* January 1994.

Department of Defense. *Annual Reports for Fiscal Years 1961–1968,* including the *Reports of the Secretary of Defense, Secretary of the Army, Secretary of the Navy, and Secretary of the Air Force.* Washington: U.S. Government Printing Office, 1962–1969.

Department of Defense. *Conduct of the Persian Gulf War: Final Report to Congress.* Appendices A-S. April 1992.

Department of Defense. *Report of Defense Secretary Harold Brown to the Congress on the FY 1981 Budget, FY 1982 Authorization Request and FY 1981-1985 Defense Programs, January 29, 1980.* Washington D.C.: U.S. Government Printing Office, 1980.

Department of Defense. *Report of Secretary of Defense James R. Schlesinger to the Congress on the FY 1976 and FY 1977T (transition) Budgets February 5, 1975.* Washington, D.C.: U.S. Government Printing Office, 1975.

Department of Defense. Under Secretary of Defense for Acquisition. *Report of the Defense Science Board 1989 Summer Study on National Space Launch Strategy.* March 1990.

Executive Branch

Aeronautics and Space Report of the President, Transmitted to the Congress January 1970. Washington, D.C.: U.S. Government Printing Office, 1970.

Aeronautics and Space Report of the President, Transmitted to the Congress January 1971. Washington, D.C.: U.S. Government Printing Office, 1971.

Aeronautics and Space Report of the President: 1972 Activities–1981 Activities. Washington, D.C.: U.S. Government Printing Office, 1973–1982.

Aeronautics and Space Report of the President: 1982 Activities–1993 Activities. Washington, D.C.: NASA, 1983–1994.

Message from the President of the United States: United States Aeronautics and Space Activities, 1961. Washington, D.C.: U.S. Government Printing Office, 1962.

Report to the Congress from the President of the United States. United States Aeronautics and Space Activities. Washington, D.C.: National Aeronautics and Space Council, 1962–1968.

Miscellaneous

Smith, Marcia S. Science Policy Research Division. Congressional Research Service Report for Congress. *Space Activities of the United States, C.I.S., and Other Launching Countries/Organizations: 1957–1992*. Report No. 93-379 SPR. 31 March 1993.

Interviews

Allen, General Lew, Jr., USAF (Ret). USAF Oral History Interview by Dr. James C. Hasdorff. Albert F. Simpson Historical Research Center. No. K239.0512-1694. Maxwell AFB, AL: Air University, 8–10 January 1986.

Beamer, Col Samuel, USAF (Ret), ADCOM/XO, USAF (Ret). Interview by the author. Tape recording. Colorado Springs, CO, 10 July 1991.

Estes, General Howell M., Jr., USAF (Ret). USAF Oral History Interview by Lt Col Robert G. Zimmerman and Lt Col Lyn R. Officer. Albert F. Simpson Historical Research Center. No. K239.0512-686. Maxwell AFB, AL: Air University, 27–30 August 1973.

Ferguson, General James, USAF (Ret). USAF Oral History Interview by Maj Lyn R. Officer and James C. Hasdorff. Albert F. Simpson Historical Research Center. No. K239.0512-672. Maxwell AFB, AL: Air University, 8-9 May 1973.

Glasser, Lt Gen Otto J., USAF (Ret). USAF Oral History Interview by Lt Col John J. Allen. Albert F. Simpson Historical Research Center. No. K239.0512-1566. Maxwell AFB, AL: Air University, 5–6 January 1984.

Hartinger, General James V., USAF (Ret). USAF Oral History Interview by Capt Barry J. Anderson. Albert F. Simpson Historical Research Center. No. ⅜239.0512-1673. Maxwell AFB, AL: Air University, 5–6 September 1985.

Hill, General James E., USAF (Ret), USAF Oral History Interview by James C. Hasdorff. Albert F. Simpson Historical Research Center. No. K239.0512-1324. Maxwell AFB, AL: Air University, 3–4 May 1982.

Horner, General Charles, AFSPACECOM/CC. Interview by G. W. Bradley. Peterson AFB, CO. 28 January 1993.

McLucas, John L., Secretary of the Air Force. Interview by the author. Tape recording. Washington, D. C., 30 March 1995.

Moorman, Maj Gen Thomas S., Jr., USAF. Interview by Robert Kipp and Thomas Fuller. Transcript. Peterson AFB, CO, 27 July 1988.

Schriever, General Bernard A., USAF (Ret). USAF Oral History interview with Maj Lyn R. Officer and Dr. James C. Hasdorff. No. K239.0512-676. Albert F. Simpson Historical Research Center. Maxwell AFB, AL: Air University, 20 June 1973.

Van Inwegan, Brig Gen Earl S., USAF (Ret). Interview by Rick W. Sturdevant. Transcript. 1 November 1995.

Zuckert, Eugene M., Secretary of the Air Force. Oral History Interview for the Kennedy Memorial Library by Lawrence McQuade. Albert F. Simpson Historical Research Center. File No. 168.7050-1. Maxwell AFB, AL: Air University, May–June 1964.

Chronologies

Air Force Satellite Control Facility. *Historical Brief and Chronology, 1954-Present.* Sunnyvale, CA: AFSCF/HO, n.d.

Air Force Space Command. *Aerospace Defense: A Chronology of Key Events, 1945–90.* Peterson AFB, CO: HQ AFSPACECOM/HO, 1 October 1991.

Air Force Space Command. *Space Detection and Tracking. A Chronology, 1957–1983.* Peterson AFB, CO: HQ AFSPACECOM/HO, March 1990.

Space Systems Division (AFSC). *Chronology of Early Air Force Man-in-Space Activity, 1955-1960.* AFSC Historical Publications Series 65-21-1. Los Angeles, CA: Space Systems Division, January 1965.

Space Systems Division (AFSC). *Space and Missile Systems Organization: A Chronology, 1954-1979.* Los Angeles, CA: Space Systems Division, n.d.

Space Systems Division (AFSC). *Space Division: A Chronology, 1980-1984.* Los Angeles, CA: Space Systems Division, n.d.

Handbooks/Readings Books

Readings in Astronautics and Space Operations. Air Force ROTC. Maxwell AFB, AL: Air University Press, 1971.

Selected Readings on Space. Warfare Systems School. Maxwell AFB, AL: Air University Press, January 1968.

Space Handbook. A Warfighter's Guide to Space. Vol I. AU-18. Maxwell AFB, AL: Air University Press, December 1993.

Space Handbook. AU-18. Maxwell AFB, AL: Air University Press, August 1977.

Space Handbook. AU-18. Maxwell AFB, AL: Air University Press, July 1972.

Official Histories

Air Force Space Command

Space Command. History of Space Command, ADCOM, ADC, 1982.

Space Command. History of Space Command, ADCOM, January–December 1983.

Space Command. History of Space Command, ADCOM, January–December 1984.

Air Force Space Command. History of USSPACECOM, ADCOM, Air Force Space Command, January–December 1985.

Air Force Space Command. History of Air Force Space Command, January–December 1986.

Air Force Space Command. History of Air Force Space Command, January–December 1987.

Air Force Space Command. History of Air Force Space Command, January–December 1988.

Air Force Space Command. History of Air Force Space Command, January–December 1989.

Air Force Space Command. History of Air Force Space Command, January–December 1990.

Air Force Space Command. History of Air Force Space Command, January–December 1991.

Air Force Space Command. History of Air Force Space Command, January 1992–December 1993.

Air Research and Development Command

Air Research and Development Command. History of Headquarters, Air Research and Development Command, Vol I, 1 July–31 December 1959.

Air Research and Development Command. History of Headquarters, Air Research and Development Command. Vol I, 1 January–31 March 1961.

Space and Missile Systems Organization

Space and Missile Systems Organization. History of Space and Missile Systems Organization, 1 July 1972–30 June 1973.

Space and Missile Systems Organization. History of Space and Missile Systems Organization, 1 July 1973–30 June 1975.

Space Division

Space Division. History of Space Division, October 1981–September 1982.

Space Division. History of Space Division, October 1982–September 1983.

Space Division. History of Space Division, October 1983–September 1984.

Space Division. History of Space Division, October 1984–September 1985.

Space Division. History of Space Division, October 1985–September 1986.

Space Division. History of Space Division, October 1986–September 1987.

Space Division. History of Space Division, October 1987–September 1988.

Space Division. History of Space Division, October 1988–September 1989.

Space Division. History of Space Division, October 1989–September 1990.

Space Division. History of Space Division, October 1990–September 1991.

United States Space Command

US Space Command. United States Space Command, Command History, September 1985–December 1986.

US Space Command. United States Space Command, Command History, January 1987–December 1988.

Other

Aerospace Defense Command. History of ADCOM/ADC, 1 January–31 December 1979.

Berger, Carl. The Air Force in Space, Fiscal Year 1961. Vol. SHO-S-66/142 Washington, D.C.: USAF Historical Division Liaison Office, April 1966.

Berger, Carl. The Air Force in Space, Fiscal Year 1962. Vol. SHO-S-66/198 Washington, D.C.: USAF Historical Division Liaison Office, June 1966.

Cantwell, Gerald T. The Air Force in Space, Fiscal Year 1964. Vol. SHO-S-67/52 Washington, D.C.: USAF Historical Division Liaison Office, June 1967.

Cantwell, Gerald T. The Air Force in Space, Fiscal Year 1965, Vol. SHO-S-68/186 Washington, D.C.: USAF Historical Division Liaison Office, April 1968.

Cantwell, Gerald T. The Air Force in Space, Fiscal Year 1967, Part I. Washington, D.C.: USAF Historical Division Liaison Office, May 1970.

Neufeld, Jacob. The Air Force in Space, 1970–1974. Washington, D.C.: Office of Air Force History, August 1976.

Rosenberg, Max. The Air Force in Space, 1959–1960. Vol. SHO-S-62/112 Washington, D.C.: USAF Historical Division Liaison Office, June 1962.

Index